AF572481

PHASE DIAGRAMS

Materials Science and Technology

VOLUME V

Crystal Chemistry, Stoichiometry, Spinodal Decomposition, Properties of Inorganic Phases

REFRACTORY MATERIALS

A SERIES OF MONOGRAPHS

John L. Margrave, *Editor*
DEPARTMENT OF CHEMISTRY
RICE UNIVERSITY, HOUSTON, TEXAS

VOLUME 1. L. R. McCreight, H. W. Rauch, Sr., and W. H. Sutton
Ceramic and Graphite Fibers and Whiskers
A Survey of the Technology

VOLUME 2. Edmund K. Storms
The Refractory Carbides

VOLUME 3. H. W. Rauch, Sr., W. H. Sutton, and L. R. McCreight
Ceramic Fibers and Fibrous Composite Materials

VOLUME 4. Larry Kaufman and Harold Bernstein
Computer Calculation of Phase Diagrams
With Special Reference to Refractory Metals

VOLUME 5. Allen M. Alper, Editor
High Temperature Oxides
Part I: Magnesia, Lime, and Chrome Refractories
Part II: Oxides of Rare Earths, Titanium, Zirconium, Hafnium, Niobium, and Tantalum
Part III: Magnesia, Alumina, Beryllia Ceramics: Fabrication, Characterization, and Properties
Part IV: Refractory Glasses, Glass–Ceramics, and Ceramics

VOLUME 6. Allen M. Alper, Editor
Phase Diagrams: Materials Science and Technology
Volume I: Theory, Principles, and Techniques of Phase Diagrams
Volume II: The Use of Phase Diagrams in Metal, Refractory, Ceramic, and Cement Technology
Volume III: The Use of Phase Diagrams in Electronic Materials and Glass Technology
Volume IV: The Use of Phase Diagrams in Technical Materials
Volume V: Crystal Chemistry, Stoichiometry, Spinodal Decomposition, Properties of Inorganic Phases

VOLUME 7. Louis E. Toth
Transition Metal Carbides and Nitrides

PHASE DIAGRAMS

Materials Science and Technology

Edited by ALLEN M. ALPER

Director of Research and Engineering
Chemical and Metallurgical Division
GTE Sylvania, Incorporated
Towanda, Pennsylvania

VOLUME V

Crystal Chemistry, Stoichiometry, Spinodal Decomposition, Properties of Inorganic Phases

1978

ACADEMIC PRESS New York San Francisco London
A Subsidiary of Harcourt Brace Jovanovich, Publishers

ACADEMIC PRESS, INC.
111 Fifth Avenue, New York, New York 10003

United Kingdom Edition published by
ACADEMIC PRESS, INC. (LONDON) LTD.
24/28 Oval Road, London NW1 7DX

Library of Congress Cataloging in Publication Data

Main entry under title:

Phase diagrams.

(Refractory materials, v. 6)
Includes bibliographical references
CONTENTS: v. 1. Theory, principles, and techniques of phase diagrams.--v. 2. The use of phase diagrams in metal, refractory, ceramic, and cement technology. [etc.]
1. Phase diagrams. I. Alper, Allen M., Date ed.
QD503.P48 541'.363 76-15326
ISBN 0-12-053205-0 (v. 5)

PRINTED IN THE UNITED STATES OF AMERICA

TO MY UNCLE

Irving Frohlich

for the profound influence he had in inspiring my career in science and technology by sharing with me the innovative work he has done in the field of plastics

Contents

List of Contributors

Numbers in parentheses indicate the pages on which the authors' contributions begin.

S. T. BULJAN*(287) Ceramics Department, GTE Sylvania Incorporated, Chemical and Metallurgical Division, Towanda, Pennsylvania 18848

LARRY E. DRAFALL† (185), Materials Research Laboratory and Department of Materials Science and Engineering, The Pennsylvania State University, University Park, Pennsyivania 16802

H. HERMAN (127), Department of Materials Science, State University of New York, Stony Brook, New York 11794

K. H. JACK (241), Wolfson Research Group for High-Strength Materials, Crystallography Laboratory, The University, Newcastle upon Tyne, England

C. M. F. JANTZEN‡ (127), Department of Materials Science, State University of New York, Stony Brook, New York 11794

R. N. KLEINER (287),Ceramics Department, GTE Sylvania Incorporated, Chemical and Metallurgical Division, Towanda, Pennsylvania 18848

R. E. NEWNHAM (1), Materials Research Laboratory, The Pennsylvania State University, University Park, Pennsylvania 16802

* Present address: GTE Laboratories, 400 Sylvan Road, Waltham, Massachusetts 02154.

† Present address: Lambda/Airtron, 200 East Hanover Avenue, Morris Plains, New Jersey 07950

‡ Present address: University of Aberdeen, Department of Chemistry, Old Aberdeen, Scotland AB9 2UE

Present address: Coors Porcelain Company, 17750 32nd Avenue, Golden, Colorado 80401.

DELLA M. ROY (185), Materials Research Laboratory and Department of Materials Science and Engineering, The Pennsylvania State University, University Park, Pennsylvania 16802

RUSTUM ROY (185), Materials Research Laboratory and Department of Materials Science and Engineering, The Pennsylvania State University, University Park, Pennsylvania 16802

O. TOFT SØRENSEN (75), Metallurgy Department, Risø National Laboratory, Denmark

Foreword

Perhaps no area of science is regarded as basic in so many disciplines as that concerned with phase transitions, phase diagrams, and the phase rule. Geologists, ceramists, physicists, metallurgists, materials scientists, chemical engineers, and chemists all make wide use of phase separations and phase diagrams in developing and interpreting their fields. New techniques, new theories, computer methods, and an infinity of new materials have created many problems and opportunities which were not at all obvious to early researchers. Paradoxically, formal courses and modern, authoritative books have not been available to meet their needs.

Since it is the aim of this series to provide a set of modern reference volumes for various aspects of materials technology, and especially for refractory materials, it was logical for Dr. Allen Alper to undertake this new coverage of "Phase Diagrams: Materials Science and Technology" by bringing together research ideas and innovative approaches from diverse fields as presented by active contributors to the research literature. It is my feeling that this extensive and intensive treatment of phase diagrams and related phenomena will call attention to the many techniques and ideas which are available for use in the many materials-oriented disciplines.

John L. Margrave

Preface

This volume is a continuation of the use of phase diagrams in the understanding and development of inorganic materials. In order to create materials with properties that are required for specific applications, it is necessary to understand how to form the desired phases by controlling composition, temperature, atmosphere, etc. Also, phase diagrams are useful in giving us insight in understanding how the created phases will change under different environments such as high temperatures, cycling temperatures, corrosive environments, and atmospheric changes (reducing, oxidizing, inert).

This volume contains some excellent articles by R. E. Newnham, Della and Rustum Roy, and Larry E. Drafall on the relationship of phase diagrams to crystal chemistry that should be helpful to all material scientists and engineers. The field of spinodal decomposition has been extremely active in the last few years. The contribution by C. M. Jantzen and H. Herman analyzes spinodal decomposition in metallic, halide, oxide, glasses, and geologic systems. This should be of importance to most scientists and engineers who are investigating metals and ceramics.

The paper by O. Toft Sørensen on nonstoichiometric phases should be of great value to material scientists and engineers who are studying oxide systems.

The use of phase diagrams in ceramic systems that relate to applications where energy saving is critical is discussed by K. H. Jack, T. Buljan, and R. Kleiner. Recent developments in sialons are discussed by K. H. Jack. These materials have very high potential as parts in turbine engines. The cordierite and spodumene systems discussed by R. Kleiner and T. Buljan have excellent potential as heat-exchanger materials.

The editor wishes to thank GTE Sylvania for its assistance.

Contents of Other Volumes

Volume I: Theory, Principles, and Techniques of Phase Diagrams

Volume II: The Use of Phase Diagrams in Metal, Refractory, Ceramic, and Cement Technology

Volume III: The Use of Phase Diagrams in Electronic Materials and Glass Technology

Volume IV: The Use of Phase Diagrams in Technical Materials

I

Phase Diagrams and Crystal Chemistry

R. E. NEWNHAM

MATERIALS RESEARCH LABORATORY
THE PENNSYLVANIA STATE UNIVERSITY
UNIVERSITY PARK, PENNSYLVANIA

ISBN 0-12-053205-0

I. INTRODUCTION

In relating phase diagrams to crystal chemistry, we seek an atomistic understanding of the geometry of the diagram and of the thermodynamic parameters on which the diagram is based. Among the questions to be considered are the following:

Can the number of intermediate phases in a composition diagram be predicted?
Which structure types will occur?
Can melting points and boiling points be predicted?
What types of phase transformations occur with temperature and pressure?
When are crystallochemical factors important in kinetics?
What determines solid solution limits?
How are entropy and other thermodynamic quantities related to structure?
When do liquid crystals, glasses, and other noncrystalline states form?

Such questions can be approached at several levels, ranging from the sublime to the empirical. We shall adopt a crystallographic viewpoint, attempting to relate thermochemical observations to atomic structure. The aim is to develop physical insight and to recognize trends, not to explain every observation. Crystal chemistry is a sloppy science which should not be taken too seriously. Solids are such complicated collections of electrons and nuclei that it is presumptuous to attempt explanations in terms of simple-minded notions such as ionic radii and atomic polarizabilities. This is especially true for phase diagrams where the cohesive energies of competing phases are often nearly identical.

But the simplicity of the crystallochemical approach is a strength as well as a weakness. A useful theory is not only accurate but easy to use and of general applicability as well. Arguments based on crystal chemistry can be quickly applied to a large number of hypothetical situations. New experiments and new materials can be predicted in this way. Of course some of the predictions will be wrong, but if an appreciable number are right, then the concepts are worthwhile. Simplicity and utility go hand in hand with accuracy and beauty in nature's grand design.

A. Miscibility and Compound Formation

The principal relationship between phase diagrams and crystal chemistry is this: *miscibility occurs when atoms have similar size and valence, and*

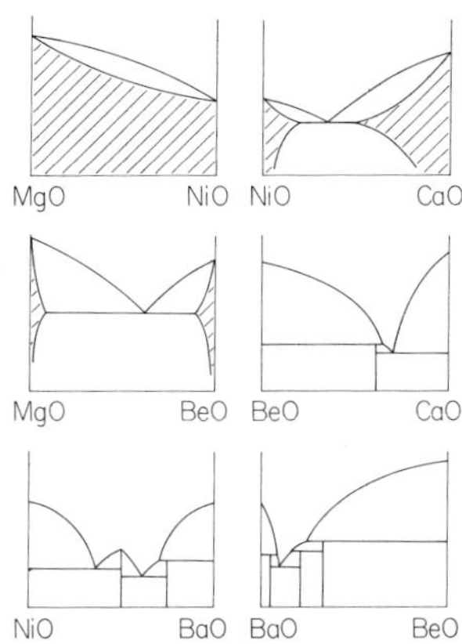

Fig. 1. Six binary phase diagrams illustrating the importance of ionic size. Complete solid solution occurs in the MgO–NiO system where the cations are similar in size. This gives way to extensive compound formation when one cation is small and the other large. Diagrams are from Levin *et al.* (1964).

compounds form when they do not. The importance of ionic size can be illustrated with the six oxide binary diagrams in Fig. 1. Solid solutions form when the ions are similar in size; hence the oxides of Ni^{2+} (0.70 Å) and Mg^{2+} (0.72 Å) are completely soluble. The Ca^{2+} (1.00 Å) ion is 0.3 Å larger than Ni^{2+}, and NiO–CaO are only partially soluble. A deep eutectic and only very limited solid solution occur in the BeO–MgO binary. Be^{2+} (0.27 Å) is 0.45 Å smaller than Mg^{2+}. Solid solubility is negligible in the remaining three diagrams, as compound formation develops. CaO–BeO, with a size difference of 0.73 Å, shows one intermediate compound, $Be_3Ca_2O_5$. Even more intermediate phases are stable in the BaO–NiO and BaO–BeO binaries. Ba^{2+} (1.36 Å) is 0.66 Å larger than Ni^{2+}, and 1.09 Å larger than Be^{2+}. There are two intermediate phases in the BaO–NiO system and three for BaO–BeO. Thus the tendency toward compound formation increases with size mismatch, as the extent of solid solution decreases. In this preliminary discussion of solid solution we are referring to substitutional solid solution where one atom replaces another in a crystal structure. Interstitial solid solutions behave very differently.

The influence of valence on oxide phase diagrams is less obvious, but the number of intermediate phases appears to increase with the difference in valence. Consider phase equilibria in oxide systems where the cations are similar in size but differ in valence. Al^{3+} (0.53 Å), Mg^{2+} (0.72), and Ti^{4+} (0.61) are generally found in octahedral coordination. Spinel ($MgAl_2O_4$) is the only intermediate phase between MgO and Al_2O_3 where the valence difference is one. The Al_2O_3–TiO_2 system also has one intermediate compound and a difference in valence of one. In the MgO–TiO_2 binary there are three compounds, showing an increased tendency toward compound formation with valence difference. Large differences lead to a large number of intermediate phases and deep eutectics. The Li_2O–MoO_3 system used as a flux in growing crystals is an important example with at least four intermediate phases (Hoermann, 1928), despite the fact that Li^{+} and Mo^{6+} are about the same size.

Solid solutions between ions with different valence are uncommon because of the importance of electric neutrality. Only a few very stable structures tolerate defect concentrations of more than a few percent. Among the more notable exceptions to this rule are the extensive (though incomplete) solid solutions in the $MgAl_2O_4$–Al_2O_3 and CaO–ZrO_2 binaries. Substitution of a few percent calcia in zirconia stabilizes the cubic fluorite structure, avoiding the disruptive phase transition near 1000°C found in pure zirconia and making "stabilized" cubic zirconia a superior refractory to pure ZrO_2. The spinel–alumina solid solution is stable because cation vacancies are tolerated. One of the metastable polymorphs of alumina, γ-Al_2O_3, has a structure resembling spinel, but with cation vacancies. Thus the solid solution extending from $MgAl_2O_4$ toward Al_2O_3 can be written as $Mg_{1-x}Al_{2+(2x/3)}\square_{x/3}O_4$, emphasizing the cation vacancies. For the flame-fusion spinel crystals used in costume jewelry, x is about 0.5.

B. Dietzel's Correlation

Using field strength as a parameter, Dietzel (1942) made an attempt to correlate ionic size and valence with compound formation in inorganic materials. In the theory of ionic crystals, Coulombic fields are of the form (charge)/(distance)^2, a quantity sometimes referred to as field strength. In applying this parameter to inorganic salts the field strength parameter can be represented by Z/d^2, where Z is the cation valence and d is the interatomic distance, the sum of the cation and anion ionic radii. The basic idea is that each cation attempts to shield itself from other cations, thereby reducing the Coulomb energy. Shielding is accomplished by surrounding the cation with anions, and field strength parameter is a measure of this effect. Using this concept, correlations can be established with the extent of immiscibility in ionic melts and with the number of compounds in binary and ternary systems.

The number of compounds in a binary system is directly proportional to the field strength difference of the two cations. When $\Delta(Z/d^2)$ is less than 10%, extensive or complete solid solution takes place. As $\Delta(Z/d^2)$ increases, a simple eutectic is achieved, and still further increases result in the formation of subsolidus or incongruently melting compounds. Intermediate compounds with two eutectics occur for still larger differences in field strength. When the difference is very large, binary systems with many intermediate compounds occur. Examples of this behavior are shown in Fig. 2. These

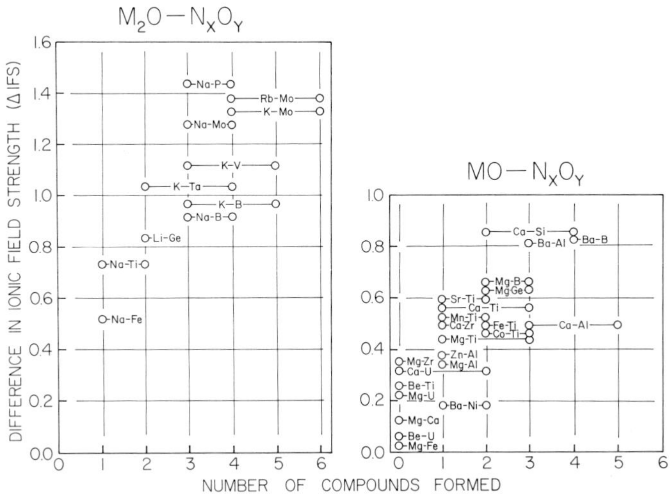

Fig. 2. Correlation between the number of compounds formed and the difference in ionic field strength of the cations for oxide systems. Large differences in field strength lead to extensive compound formation. In the CaO–Al_2O_3 oxide system, for instance, there are five intermediate compounds, three of which are congruently melting. Data compiled by Berkes and Roy (1970).

trends were first outlined by Dietzel (1942) who showed that the number of intermediate compounds is proportional to the field strength difference. Vorres (1965) has extended the study to a large number of oxide and halide binaries with the same conclusion. Using data from 160 oxide systems, Berkes and Roy (1970) have correlated several characteristics of binary phase diagrams with the electrostatic field strength differences. Following Dietzel's definition, field strength (fs) was defined as cation valence divided by the square of the cation–anion distance. The number of compounds in the binary system increases as a function of $\Delta(Z/d^2)$, the difference in field strength of the end-member cations. As might be expected, the extent of solid solution is a maximum when $\Delta(Z/d^2) = 0$, and decreases rapidly as $\Delta(Z/d^2)$ increases. Binary systems with $\Delta(Z/d^2) > 0.4$ exhibit no solid solution. For the oxide systems analyzed, liquid immiscibility was most common when $0.5 \leq \Delta(Z/d^2) \leq 1.0$.

Similar principles appear to govern ternary systems, although few correlations have been examined in detail. Among silicate ternaries, the field strength difference between the other two cations (excepting Si) determines the number of compounds. No compounds form when $\Delta(Z/d^2)$ is below 0.05–0.07, while up to three or four compounds appear when $\Delta(Z/d^2)$ lies between 0.7 and 0.8. Such predictions are less reliable for ions with large polarizibilities.

C. Model Structures

Goldschmidt (1926) showed that crystal structures are determined by sizes and polarizabilities of the constituent ions, and introduced the concept of model structures. Model structures have similar radius ratios and similar crystal structures, but differ in valence, and therefore in bond strengths. Zinc orthosilicate (Zn_2SiO_4) is a strengthened model structure of Li_2BeF_4 with doubled valences. Other model pairs include BeF_2–SiO_2, LiF–MgO, MgF_2–TiO_2, CaF_2–ThO_2, $KMgF_3$–$SrTiO_3$, $RbBF_4$–$BaSO_4$, and CdI_2–$ZrSe_2$. The weakened structures (halides) generally have lower hardnesses, lower melting points, and lower refractive indices, together with increased chemical reactivity and solubility.

As might be expected, phase diagrams involving model structures are often similar. Compare the KF–MgF_2 and SrO–TiO_2 systems shown in Fig. 3. Melting points are much higher in the oxide system because of the larger valences (McCarthy *et al.*, 1969). Both systems contain intermediate compounds of composition ABX_3 and A_2BX_4. $SrTiO_3$ and $KMgF_3$ have the perovskite structure, while the other two compounds have a layer structure. The perovskites melt congruently and the layer structure incongruently in both systems (DeVries and Roy, 1953). Eutectic compositions are also similar.

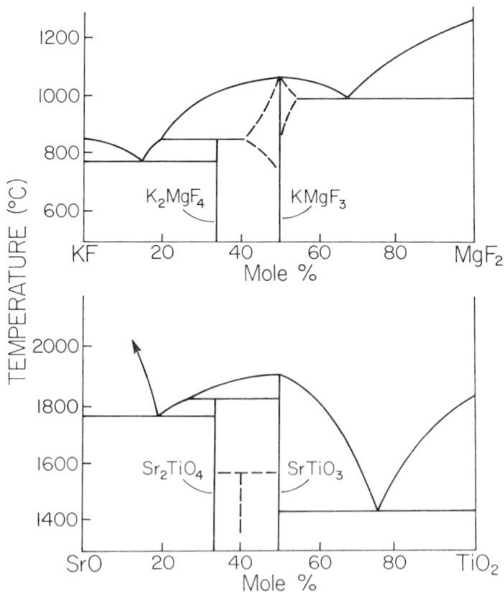

Fig. 3. KF–MgF_2 and SrO–TiO_2 diagrams as model systems. Ionic radii are similar but valences are doubled in the oxide system, which therefore has higher melting points. Note the correspondence in intermediate compounds and eutectic compositions. Diagrams from DeVries and Roy (1953) and from McCarthy *et al.* (1969).

II. SOLID SOLUTIONS

The most common type of solid solution is the substitutional solid solution in which one atom substitutes for another in a crystal structure. Some of the crystallographic restrictions limiting the substitution are considered in this section. The restrictions are somewhat different for interstitial solid solutions and other defect solid solutions which differ from simple substitution.

There is no such thing as a perfect solid solution, one with complete randomness. Consider an alloy of composition RX with a close-packed structure. In an ideal solid solution, each atom position has equal probability of being occupied by R or X, and each atom is surrounded by six R and six X, on the average. If the atoms differ sufficiently in scattering power, the numbers and species of near neighbors can be experimentally determined by x-ray diffuse scattering measurements.

Studies of a number of intermetallic systems have shown that the departures from randomization are substantial. In the Cu–Au, Ag–Au, and Au–Ni binaries, short-range order exists in which *unlike* atoms have a higher probability of being neighbors than *like* atoms. Another type of deviation occurs in the Al–Ag and Al–Zn systems, one in which like atoms tend to be neighbors and unlike atoms begin to segregate. This is called *clustering*. Solid solutions can therefore be thought of as a range of configurations, tending toward clustering and phase segregation on one side, and extending toward short-range order and eventually long-range order (compound formation) on the other. All real solutions exhibit either clustering or short-range ordering to some degree, though many are close to being random, especially at high temperatures.

Since the bonding forces are strongest for near neighbors, the internal energy can be crudely considered as resulting from energies associated with neighboring pairs (Slater, 1939). Let W_{RX}, W_{RR}, and W_{XX} be the energies for the neighboring pairs RX, RR, and XX. For a perfect solid solution of composition RX, there will be twice as many RX pairs as RR or XX pairs. The internal energy is then proportional to $2W_{RX} + W_{RR} + W_{XX}$. For an RX system with segregated R and X phases, the total internal energy is proportional to $2W_{RR} + 2W_{XX}$ and for one with long-range order it is $4W_{RX}$. To include short-range order and clustering these results can be generalized to an energy of

$$U = 4SW_{RX} + 2(W_{RR} + W_{XX})(1 - S)$$

where S is an ordering parameter ranging from 0 (complete segregation) to 1 (long-range order). $S = \frac{1}{2}$ is an ideal solid solution in which R and X are distributed at random. Clustering and short-range order correspond to

$S < \frac{1}{2}$ and $S > \frac{1}{2}$, respectively. If $2W_{RX} < W_{RR} + W_{XX}$, the energy is minimized for $S > \frac{1}{2}$, a situation favoring order because of strong attractive forces between R and X atoms. Clustering occurs if $2W_{RX} > W_{RR} + W_{XX}$.

This discussion presupposes that the internal energy can be written as a sum of pair energies, that the number of nearest neighbors is the same in all phases, and that $T = P = 0$, so that the Gibbs free energy is equal to the internal energy.

A. Substitutional Solid Solutions

Atoms sometimes substitute for one another in crystals, forming a solid solution—a homogeneous crystal of variable composition. Forsterite (Mg_2SiO_4) and fayalite (Fe_2SiO_4) form a complete solid-solution series. Both end members and all intermediate compositions possess the olivine structure. Oxygen ions make up a close-packed array with Si^{4+} occupying tetrahedral interstices, and Mg^{2+} and Fe^{2+} in octahedral interstices. Magnesium and iron are distributed nearly at random over the octahedral positions.

Solid solubility depends on a number of factors: the structure type, the radii and charges of the ions, and the temperature. Some structures are much more stable than others, and tolerate extensive atomic substitution. Many examples of mixed crystals occur in the spinel, perovskite, and rock salt families. On the other hand, quartz and diamond crystals are noted for their purity because of their intolerance to substitution. Regarding radii, it has been found that ions of the same valence substitute freely when the radii differ by less than 15%. Iron and magnesium occur together in minerals because the radii correspond closely: Fe^{2+} (0.77 Å), Mg^{2+} (0.72 Å), and Fe^{3+} (0.65 Å). Valence is important also. As a rule, little or no substitution occurs when the ions differ by more than one in valence. Coupled substitutions tend to increase solubility limits by maintaining charge neutrality. The plagioclase feldspars ($Ca_{1-x}Na_xAl_{2-x}Si_{2+x}O_8$) are a good example in which calcium and aluminum are replaced by sodium and silicon. Solubility limits increase with temperature because of the entropy of mixing. The large entropy arising from atomic disorder tends to stabilize mixed crystals at high temperatures.

Unit cell dimensions vary smoothly with composition in a solid-solution series. For a cubic crystal the lattice parameter can be represented by

$$(a_{ss})^n = (a_1)^n c_1 + (a_2)^n c_2$$

where a_{ss}, a_1, and a_2 are the lattice parameters of the solid solution and the two end members 1 and 2. Mole fractions c_1 and c_2 are the respective concentrations and n is an arbitrary power describing the variation.

Vegard suggested that for many substances $n = 1$, while theoreticians have predicted n to be considerably larger, in the range 3–8. For additive volumes, $n = 3$ is a relation known as Retger's law. Accurate experimental values are needed to determine n because solid solutions seldom form if a_1 and a_2 differ by more than 15%. This is why Vegard's law fits most data fairly well, though in many cases it is not exactly obeyed. Measurements on the KCl–KBr series (Slagle and McKinstry, 1966) support Retger's law showing that the volume of the anion, rather than its radius determines the lattice constant.

In substitutional solid solutions, guest atoms do not always have exactly the same crystallographic coordinates as host atoms. Ruby, $Al_{2-x}Cr_xO_3$, is a solid solution used extensively in laser and maser devices. In dilute ruby, Cr does not occupy the Al site, but takes up a position displaced by 0.1 Å along c (Moss and Newnham, 1964). Trivalent Cr is larger than Al^{3+}, and the displacement leads to more reasonable interatomic distances. This type of off-center substitution is likely to occur when the site has variable parameters such as the z coordinate of aluminum in corundum. Size difference between host and solute atom is also important, and possibly bonding differences too.

Unusual solid solutions with important biological implications occur in the apatite family. The chemical formula is $Ca_5(PO_4)_3X$, where X = F, Cl, OH. Fluorapatite is a common mineral and chlorapatite exhibits unusual dielectric properties. Hydroxyapatite is the chief constituent of teeth and bones, though the beneficial effect of fluoridation is well known.

The three X anions lie along the 6.88 Å c axis but with significantly different positions. The z coordinates for Cl, F, O, and H are 0.444, 0.250, 0.196, and 0.061, respectively, giving very different structures as shown in Fig. 4. The anions are bonded to calcium ions which form triangles about the c axis. Chlorine, being a large anion, takes a position nearly midway between the calcium groups. Fluorine lies directly in the triangles and hydroxyls are slightly displaced from this position. In fluoridated hydroxyapatite, NMR experiments indicate that the hydroxyl groups form hydrogen bonds to fluorine with important biological consequences.

In the dissolution of tooth enamel by acids, the X-ion column provides the easiest diffusion path, with hydroxyl ions exhibiting especially high mobilities. The formation of H bonds to the strongly bound fluorine ions greatly inhibits diffusion, controlling dissolution, and preventing caries (Young *et al.*, 1969).

Another uncommon substitution occurs in the hydrogarnet–grossularite series which is a product of cement hydration. The chemical formula of the hydrogarnets can be written as $3CaO \cdot Al_2O_3 \cdot xSiO_2 \cdot (6 - 2x)H_2O$ or $Ca_3Al_2Si_xH_{12-4x}O_{12}$, with $0 \leq x \leq 3$. Calcium aluminum hydroxide,

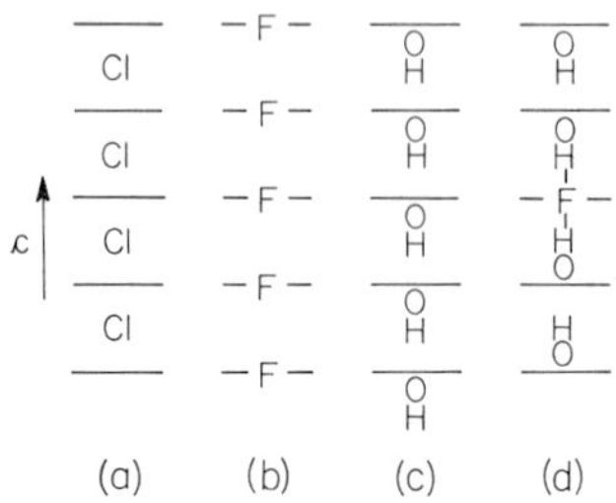

Fig. 4. Unusual solid solutions form between members of the apatite family: $Ca_5(PO_4)_3X$, X=Cl, F, OH. The univalent anions are located in channels along the *c* crystallographic axis. Horizontal lines indicate the heights of Ca^{2+} ions surrounding the channels. In chlorapatite (a) the large Cl^- ions occupy sites between the calcium rings, while the smaller F^- ions in fluorapatite (b) lie in the plane of the surrounding cation. In hydroxyapatite (c) the asymmetric OH^- group takes an off-center position with protons pointing up or down along *c*. NMR results on solid solutions suggest hydrogen-bond formation between anions in fluoridated hydroxyapatite (d). The bonds anchor the hydroxyl groups and thereby inhibit tooth decay (Young *et al.*, 1969).

$Ca_3Al_2(OH)_{12}$ is transformed to grossularite, $Ca_3Al_2(SiO_4)_3$ by substituting Si^{4+} for $4H^+$ ions. Calcium, aluminum, and oxygen positions remain virtually unchanged throughout. In the aluminate, the H^+ ions are found at the vertices of a tetrahedron inscribed within a second tetrahedron of oxygens. In converting it to grossularite, a Si^{4+} ion replaces the tetrahedron formed by four H^+ ions. Recent x-ray studies indicate several discontinuities in the solid-solution series (Marchese *et al.*, 1972).

B. Miscibility Limits

Forty years ago Hume-Rothery showed that solid solubility is very restricted when atomic radii differ by more than 15%. The 15% rule has come to be recognized as a necessary but not sufficient condition rule since extensive miscibility does not always occur between atoms of the same size; other factors such as valency and electronegativity are important as well. Darken and Gurry (1953) took electronegativity and size into account by plotting metallic radius against electronegativity for various elements. Predictions regarding miscibility were made by drawing an ellipse about the solvent element with diameters ± 0.4 units of electronegativity difference and $\pm 15\%$ size difference for solute and solvent. Elements falling within the ellipse generally form extensive solid solution, while those outside do not. The Darken–Gurry plot for iron shown in Fig. 5 is typical. Of the 20 elements within the ellipse, 19 are more than 5% miscible in iron, while 28 out of 36 outside the ellipse are less than 5% miscible.

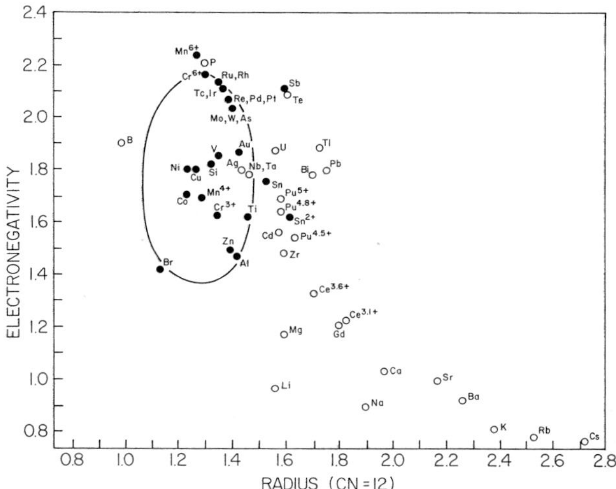

Fig. 5. A Darken–Gurry plot for metallic elements alloyed with iron. Solid circles denote elements with a solubility of more than 5 at. % in either of the polymorphic modifications of iron. Open circles represent elements which do not form extensive solid solutions with iron. Iron itself is located at the center of the ellipse, showing that solubility is greatest between atoms of similar size and electronegativity. Data collected by Waber and co-workers (1963).

Waber and co-workers (1963) made Darken–Gurry plots for a large number of metals and compared the results with experiment, using 5 at.% as the dividing line between extensive and restricted solid solution. Of the systems predicted to have extensive solubility, 62% were correct, and of those outside the limiting ellipses 85% showed less than 5% solid solution. The overall percentage of correct predictions for 850 alloy systems was 77%. There is some indication that size is more important than electronegativity, since more than 90% of the elements falling outside the 15% size limits are insoluble. On the other hand only 50% of those within the size limits show extensive miscibility.

The 15% rule holds for nonmetals as well as metals. Consider the $M^{2+}O$ binaries in Fig. 1. Complete solid solution occurs in the NiO–MgO system where the bond lengths differ by only 1%, but not in the NiO–CaO system where the Ca—O bonds are 15% longer than the Ni—O bonds. Miscibility is negligible in the remaining systems where size differences are even larger. It is interesting to speculate on the origin of the 15% rule. Lindemann observed that many solids melt when the thermal vibration amplitude is about 15% of the interatomic distance, and it is also a fact that most solids expand by about 10% before melting. It therefore appears that most crystals become unstable when the bond lengths are changed by 10–15%. To understand *why*, we examine the potential energy function.

For ionic crystals, the Born model leads to a lattice energy $-Ar^{-1} + Br^{-n}$ where A is the Madelung coefficient, r the interatomic distance, B the repulsive coefficient, and n is about 10. At equilibrium, $r = r_0$ and the Coulomb energy $-Ar^{-1}$ is about ten times larger than the repulsive energy Br^{-n}. However when the interatomic distance is decreased, the repulsive energy increases rapidly because n is large. Decreasing r to $0.9r_0$ makes the repulsive energy as large as the attractive energy, destabilizing the crystal. Hence variations in interatomic distance of 10 or 15%, whether caused by temperature or composition changes, can lead to dissociation.

C. Defect Solid Solution

Steel, an alloy of iron and carbon, is a billion dollar example of the importance of interstitial sites. Three phases play a role in developing the hardness and ductility of steel: body-centered cubic (bcc) α-Fe, face-centered cubic (fcc) γ-Fe, and iron carbide Fe_3C, called ferrite, austenite, and cementite, respectively. Steels contain less than 2 wt % carbon, the amount of C which austenite accepts in solid solution (Fig. 6). Carbon enters the largest interstitial sites of austenite, the octahedral holes in the cubic close-packed structure. The metallic radii of C and Fe are 0.75 and 1.24 Å, respectively, giving a radius ratio 0.6. This exceeds the radius ratio of the octahedral site to the close-packed sphere, $\sqrt{2} - 1 = 0.414$. Thus carbon is slightly large for the fcc interstitial site so that only a few percent can be filled before the phase becomes unstable. The interstices are even smaller in α-Fe. This is somewhat surprising since the bcc structure is more open than the fcc structure, but the largest interstice (a deformed tetrahedral site) is about 25% smaller than in the fcc structure. The interstitial to sphere radius ratio is only 0.29, much too small for C. Thus α-Fe tolerates only a very small amount of carbon in solid solution (Fig. 6).

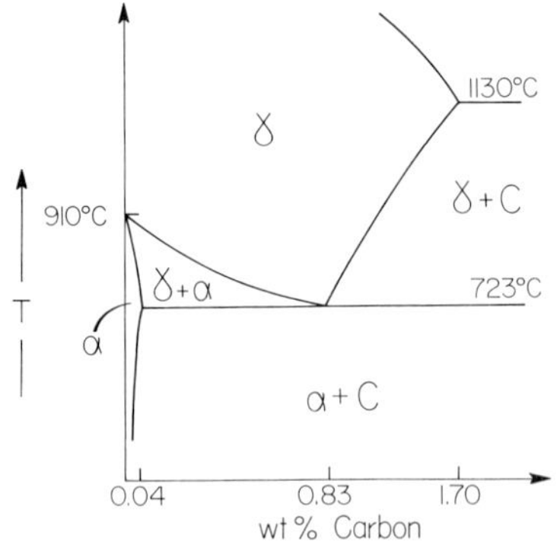

Fig. 6. Eutectoid region of the iron–carbon system used in steel making. α and γ refer to the bcc and fcc forms of iron, while C is the cementite phase of composition Fe_3C. γ-Fe accepts a much larger amount of carbon than α-Fe because of the larger interstitial sites in the fcc structure (Van Vlack 1959).

Steels are often prepared near the eutectoid at 0.8 wt % carbon. As the steel cools below 723°C, austenite converts to ferrite and cementite, the two forming simultaneously in an intimate mixture, giving a lamellar microstructure known as pearlite. Cementite is hard but brittle, α-iron is softer and more ductile, giving the composite hardness and ductility. Quenching the samples quickly from the austenite range gives a metastable phase called martensite which retains the carbon in solid solution. Martensite has a deformed bcc structure of tetragonal symmetry and is very strong.

Massive nonstoichiometry with defect concentrations of 10% or more occurs in several ways. NiTe–$NiTe_2$ and other transition-metal chalcogenide systems show extensive solid solubility because of the compatibility of the end-member structures. $NiTe_2$ has the CdI_2 structure, and NiTe is isostructural with NiAs. Both structures are hexagonal with similar lattice parameters. An intermediate composition midway between NiTe and $NiTe_2$ could be described as $NiTe_2$ with 25% anion vacancies, or as NiTe with 50% anion interstitials.

A homogeneous array of defects is found in high-temperature titanium monoxide, with the rock salt structure. The stability range extends from TiO to $TiO_{1.3}$. X-ray diffraction and density measurements reveal that even in "stoichiometric" TiO more than 15% of the atomic sites are vacant. Less than half of the titaniums are coordinated to six oxygens. This type of behavior is in stark contrast to that of other rock salt type oxides. The defect concentration in CoO is only 3×10^{-3} at 1400°C, while that of MgO is below the limits of detectability, less than 10^{-10} at 1700°C. Ti^{2+} behaves differently from Co^{2+} and Mg^{2+} because of the overlapping d-orbitals. The d electrons are delocalized in conduction bands giving added stability to the crystal and providing a source or sink for the electrons involved in nonstoichiometric behavior. The d-orbitals in CoO are more contracted because of increased nuclear charge, and do not overlap with neighboring metal ions. The absence of nonstoichiometry in MgO stems from the inaccessibility of higher oxidation states and the high energy required to force Mg^{2+} into interstitial sites.

Although the point defect description is valid at concentrations normally found in semiconductors, defect interactions and clustering become evident at concentrations of 10^{-4} or higher. Defect conglomerates can lead to coherent intergrowths and nonstoichiometric phases (Greenwood, 1968). Wüstite, $Fe_{1-x}O$, has a defect rock salt structure, but with clusters rather than isolated cation vacancies. Tetrahedral sites begin to fill as octahedral sites empty, forming Fe^{3+} clusters as oxidation proceeds. Figure 7 shows a Koch cluster of four tetrahedrally coordinated iron atoms and thirteen octahedral vacancies. The oxygen sublattice is continuous throughout the host structure and the defect cluster. Koch clusters intergrow with the

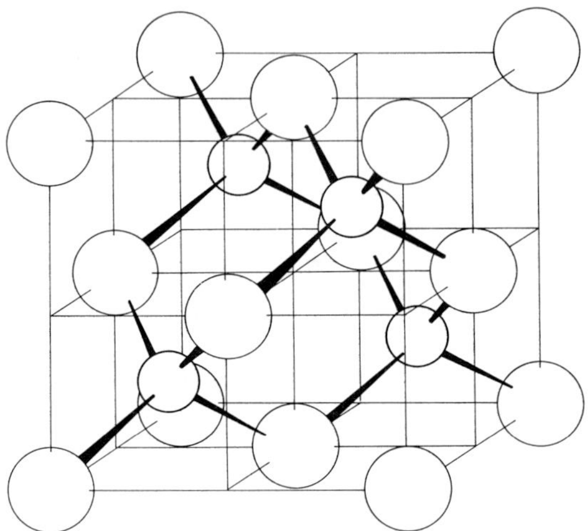

Fig. 7. The Koch defect cluster in nonstoichiometric $Fe_{1-x}O$ with the rock salt structure. Open circles represent oxygen, solid circles tetrahedrally coordinated Fe^{3+}, and crosses empty octahedral sites. Defect clusters such as this promote massive nonstoichiometry (Greenwood, 1968).

rock salt structure but are not electrostatically neutral and must be compensated by additional Fe^{3+} ions in the immediate vicinity. The clusters tend to produce long-range order, generating superlattice structures.

Defect clusters occur in other nonstoichiometric compounds as well. In VO_{2+x} oxygens are displaced from their normal sites to give an interstitial complex. A 2:1:2 cluster is typical with two displaced oxygen atoms and one additional oxygen occupying two kinds of low-symmetry interstitial positions, associated with two oxygen vacancies. Other types of defect clusters occur in CaF_2–YF_3 and MH_2–MH_3 mixed crystals. The vanadium carbides form defect rock salt structures similar to wüstite, V_6C_5 and V_3C_7 contain clusters of vacant sites ordered in spirals along the body diagonal directions.

Excellent examples of coherent intergrowth occur in Magneli phases with compositions Ti_nO_{2n-1} (Anderson, 1971). These are shear structures with rutilelike regions joined by lamellae of edge-sharing octahedra. For large *n*, the shear planes are widely spaced so that the driving force for ordering is small, and the compounds order only sluggishly. Random fluctuations in shear plane spacing occur under these circumstances giving rise to nonstoichiometry. The manner in which shear phases develop is illustrated in Fig. 8.

Structural coherence is the key to the development of nonstoichiometry. Whether the defects are isolated, clustered, or in domains such as shear

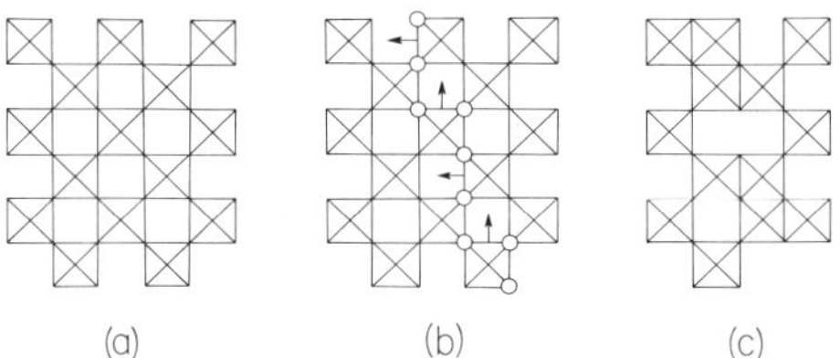

Fig. 8. Formation of a shear phase from a transition-metal oxide with corner-shared octahedra (a). The introduction of oxygen vacancies, indicated by circles in (b), is followed by a rearrangement of octahedra shown by arrows. The shear phase (c), contains shared octahedral edges as well as shared corners (Anderson, 1971).

planes, there must be coherence between the matrix and the defect region. This requires a dimensional match at the boundary together with a correspondence of atomic positions making diffusion easy. Generally one sublattice runs continuously through the composite, such as the fcc oxygen lattice in defect wüstite.

D. Trapped Gases

Gas storage in crystals is another interesting use of interstitial sites. Gaseous hydrogen has many applications but is not easy to store in a safe and economic fashion. It can be held as a compressed gas, or as a liquid at temperatures below 20°K, but both methods are expensive and dangerous. Recently a new technique has been developed in which the hydrogen is stored as a hydride, to be subsequently released as hydrogen gas and reabsorbed at room temperature and pressures of a few atmospheres.

Intermetallic compounds such as $LaNi_5$ are capable of incorporating large amounts of hydrogen, converting to the hydride $LaNi_5H_6$. The hydride and $LaNi_5$ are in equilibrium with each other, at a given temperature and pressure, and the hydrogen content can be varied within wide limits at an equilibrium pressure that is nearly constant. When $LaNi_5$ is placed in contact with H_2 at a pressure slightly greater than the equilibrium pressure and the temperature is lowered to compensate for the heat liberated during the reaction, H_2 is adsorbed by $LaNi_5$ until it is entirely converted to hydride. If gaseous H_2 is then allowed to escape from the vessel, the pressure decreases rapidly to the equilibrium pressure and remains there while the hydrogen drains away from the hydride, and $LaNi_5H_6$ returns to $LaNi_5$. Only when the hydride disappears does the H_2 gas pressure drop below the equilibrium value.

In addition to safety and expense, there are several other advantages to storing hydrogen in hydrides. At an external H_2 pressure of only 4 atm, the density of hydrogen in $LaNi_5H_6$ is equivalent to 1000 atm. Moreover, selective absorption by the intermetallic compounds rejects other gases, resulting in purification of the hydrogen gas.

Some crystals have interstitial cavities large enough to accept molecules. One of the interesting features about trapped molecules is that in some ways they behave like a gas and in other ways like a solid. Crystals such as cordierite ($Mg_2Al_4Si_5O_{18}$) contain about one cavity in a volume of 100 $Å^3$. When all the cavities are occupied by gas molecules (often H_2O or CO_2 in mineral specimens) the density of molecules is equivalent to 200 atm pressure, and yet the molecules are never in contact with one another. The degree of interaction between molecule and cage ranges from tight bonding through hindered rotation to free rotation. The infrared spectrum of water in cordierite (Farrell and Newnham, 1967) shows all the sharp overtone and combination bands of water vapor with one important difference—the spectra depend on the polarization vector, showing that the molecules are oriented in the cages. Trapped gases constitute an unusual state of matter—a dense noninteracting gas with preferred orientation. Trapped molecules are also interesting geologically. In tight cages like those of cordierite, the molecules have little chance of escaping. Like insects trapped in amber, they were present when the mineral formed, and are therefore representative of the fluids and gases of the past.

E. Precipitation in Solids

Precipitation reactions have been exploited by metallurgists to optimize the physical properties of steel, but the use of this phenomenon in ceramic systems has been slower to develop. Many materials with similar crystal structures form extensive solid solutions at high temperatures, but decompose on cooling to form two phases. Star sapphires and moonstones are glamorous examples of precipitation phenomena, in this case exsolution from solid solution.

There are two paths by which supersaturated solid solutions undergo decomposition through composition fluctuations. The fluctuations may be large in degree and small in volume, or small in degree and large in volume. The first type requires nucleation because of the large surface energy between precipitate and matrix. Dislocations and other structural imperfections generally promote heterogeneous nucleation so that control of nucleation sites is often the key to controlling precipitation.

The spinel crystals used as jewel bearings are an example of precipitation strengthening. The crystals are first ground and polished, and then hardened by heat treatment resulting in a considerable savings in diamond abrasive.

At high temperatures, spinel ($MgO \cdot Al_2O_3$) accepts a large excess of Al_2O_3 in solid solution. Two types of precipitation occur at lower temperatures where the solubility limit decreases. A metastable monoclinic phase, approximately $MgAl_{26}O_{40}$ in composition, forms initially on annealing near 1000°C. The structure conforms closely to spinel, forming lamellae within the spinel crystals. Further heat treatment converts the metastable phase to the stable spinel and corundum phases, thus strengthening the solid (Fine, 1972).

In the second type of decomposition, only a gradual change in composition occurs on traversing the fluctuation so that no surface energy term is involved. Nucleation is not required under these conditions, leading to spinodal decomposition.

Spinodal decomposition has been observed both in glasses and crystalline ceramics. Vycor silica glass and the $CoFe_2O_4$ system are interesting examples. Magnetic cobalt–iron ferrite precipitates have extremely high coercive fields, comparable to the commercially important Al–Ni–Co metallic magnets. In spinodal decomposition, the supersaturated solid solution contains periodic composition fluctuations with the fluctuation spacing being determined by a balance between diffusion length and energy gradient. Strain energy is important in spinodal decomposition since it adds to the free energy. Periodic composition fluctuations generally occur along low modulus directions.

Intimate microstructures are also found near eutectic points. Composite materials with useful properties can be created by judicious choice of the two phases and the solidification conditions. The microstructure of a eutectic is sensitive to cooling rate and crucible shape. During the crystallization process certain crystallographic directions of the two phases tend to align, giving needlelike or platelike patterns of the two phases. Such morphology can be used to produce strong materials by growing stiff fibers embedded in a ductile matrix. Other uses for eutectics with tailored microstructures include permanent magnets, polarization filters, and superconductor composites.

An important type of nucleated precipitation called cellular growth occurs in the eutectoid decomposition of wüstite ($Fe_{0.9}O$) into metallic iron and magnetite. α-Fe and Fe_3O_4 form alternate lamellae with a spacing of 0.1 μm when annealed at 490°C. Aging at slightly higher temperatures near the eutectoid of 570°C produces coarser lamellae easily observed with an optical microscope. The nucleation rate is determined by that of α-Fe since Fe_3O_4 nucleates easily because of its structural similarity to wüstite. Cellular decomposition does not occur if the grain size is too small; under these conditions α-Fe precipitates along grain boundaries.

III. PREDICTION OF PHASES

There are a number of simple but effective techniques for predicting crystal structures. Empirical correlations based on atomic size have been moderately successful because of the importance of radius ratios to near-neighbor coordinations.

The existence or nonexistence of various rare-earth boride structures is correlated with ionic radii in Table I. Phases with the UB_{12}, AlB_2, and YB_{66} structures are stable for small rare-earth ions while the CaB_6 structure is stable only for the larger rare earths. The ThB_4 structure is found in nearly all rare-earth boron binaries.

Additional inferences can be drawn from the correlations in Table I. For example, when the metal radius lies near the limit for a particular structure type, the compound generally shows a tendency to decompose. GdB_2, the largest diboride phase, is stable only above 1200°C, and the largest tetraboride phase (LaB_4) has the lowest melting point of any rare-earth boride.

Another useful way of correlating radii with structure type is the *structure-field map*. A map for fluorides and oxides of composition ABX_4 is shown in

TABLE I

CORRELATION OF RARE-EARTH BORIDE PHASES WITH IONIC RADII[a]

Ion	Radius	UB_{12}	AlB_2	YB_{66}	ThB_4	CaB_6
Eu^{2+}	1.15 Å					×
La^{3+}	1.13				×	×
Ce^{3+}	1.08				×	×
Pr^{3+}	1.06				×	×
Nd^{3+}	1.05			×	×	×
Pm^{3+}	1.04			+	+	+
Sm^{3+}	1.03			×	×	×
Gd^{3+}	1.00		×	×	×	×
Yb^{3+}	0.99	×	×	×	×	×
Tb^{3+}	0.98	×	×	×	×	
Y^{3+}	0.97	×	×	×	×	
Dy^{3+}	0.96	×	×	×	×	
Ho^{3+}	0.95	×	×	×	×	
Er^{3+}	0.93	×	×	×	×	
Tm^{3+}	0.92	×	×	×	×	
Yb^{3+}	0.91	×	×	×	×	
Lu^{3+}	0.90	×	×	×	×	
Sc^{3+}	0.81	×	×	×	×	

[a] Stable phases are designated by ×, predicted phases by + (Spear, 1975).

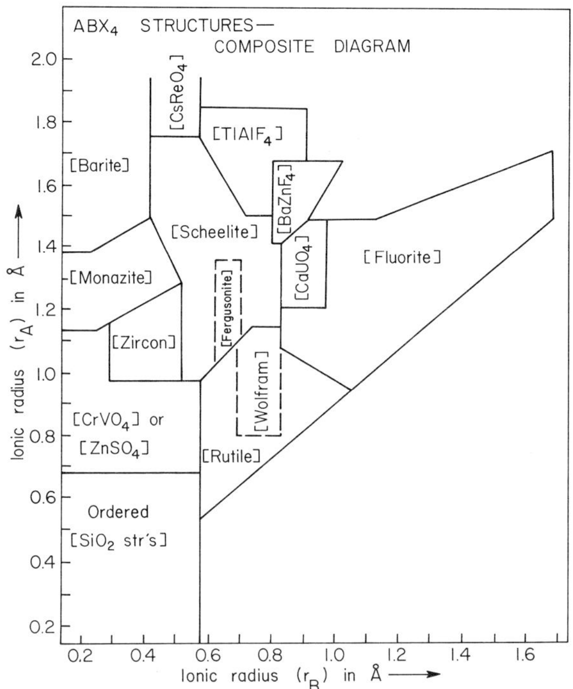

Fig. 9. Structure-field maps can be used to predict structures with remarkable reliability. The map shown above is for oxides and fluorides of composition ABX_4 and was constructed empirically (Muller and Roy, 1974).

Fig. 9. Here A is the cation of larger radius, B the smaller, and X is fluorine or oxygen. The barite structure is favored when A is very large and B very small. When both are small, a silicalike structure is very likely. Other radii stabilize different ABX_4 structures. Dashed lines indicate structural variations depending on valence. There is considerable size overlap between the rutile and wolframite structures, for instance. The rutile structure is favored for $A^{3+}B^{5+}O_4$ compounds where the A and B ions are disordered over the octahedral sites. However, the wolframite structure replaces rutile for $A^{2+}B^{6+}O_4$ oxides since the charge difference is apparently too great to allow disorder among the A and B ions. Structure-field maps are useful in predicting unknown structures and phase transformations. A number of examples have been discussed by Muller and Roy (1974).

It is difficult to predict the relative stability of different crystal structures from first principles because the cohesive energies for different structures are often nearly identical. The Born model has been applied to alkali halides having the NaCl and CsCl crystal structures, but even with extensive refinement the correct structure is not always predicted (Tosi, 1964). Relatively minor contributions to the cohesive energy are often sufficient to alter the delicate balance of energies for different structures. For the alkali halides, the Madelung energy favors slightly the CsCl structure, which has the larger Madelung constant, but the repulsive interactions of nearest neighbors favor the NaCl structure. Van der Waals interactions again favor CsCl. Despite the uncertainty, the Born concept of an ionic solid has been useful in setting up predictive rules. The importance of radius ratio, for instance, finds its justification in electrostatic energy.

In the following sections we consider some of the simpler approaches to structure prediction. Valence bonds and the $8 - n$ rule help in understanding the structures of nonmetallic elements. Among metals, the electron-to-atom ratio appears to be an effective predictor of certain structure types. Pauling's rules provide a qualitative understanding of minerals and other inorganic structures, and packing efficiency is important in organic and inorganic structures alike.

A. Valence Bond Theory

Valence bond theory provides the most straightforward explanation of the thermodynamic stability and crystal structures of the elements. The heat of atomization—the amount of heat required to vaporize one mole—is a good measure of bond strength. The values quoted here are expressed in kilocalories per mole, and refer to solid elements at 300°K or at the melting point, whichever is lower.

The heats required to atomize rare gas solids are small: He, 0.5; Ar, 1.8; Kr, 2.6; and Xe, 3.6 kcal/mole. The atoms have closed electron shells so that only van der Waals forces act between atoms. Close-packed structures are favored by the nondirectional van der Waals forces. Solid helium is hexagonal close packed, and the other inert-gas solids are cubic close packed.

Halogens have seven electrons per atom and bond together to form diatomic molecules. The energy of the electron pair bond can be estimated from the heats of atomization: F, 20; Cl, 32; Br, 28; and I, 26 kcal/mole. The bond energy is an order of magnitude larger than that of van der Waals solids. The halogens form molecular solids consisting of diatomic molecules. Melting points are low since the forces between molecules are weak.

Column VI elements have even higher heats of atomization: O, 60; S, 66; Se, 49; Te, 46; and Po, 35 kcal/mole. The bonding energies are roughly twice those of the halogens since two pairs of electrons are involved. The crystal structures of sulfur, selenium, and tellurium consist of rings or chains in which each atom is bonded to two others. Each bond is a single electron-pair bond. Solid oxygen contains O_2 molecules with double electron-pair bonds.

Heats for group V elements are as follows: N, 114; P, 80; As, 69; Sb, 62; and Bi, 50 kcal/mole. These elements lack three electrons for a filled shell. Phosphorus, arsenic, antimony, and bismuth crystallize in puckered layers with each atom forming three single bonds, each involving an electron pair. As expected, the heat of atomization is about three times that of column seven elements. Solid nitrogen contains N_2 molecules with three electron pairs concentrated between nitrogen atoms. Multiple bonds are common in first row elements, but not elsewhere.

Carbon, silicon, and the other elements of group IV lack four electrons for a filled octet. The large heats of atomization (C, 171; Si, 108; Ge, 90; Sn, 72; and Pb, 47 kcal/mole) reflect the increase in the number of bonding electrons. Several of the elements crystallize in the diamond structure in which tetrahedrally coordinated atoms form four single bonds. The tendency for first row elements to form multiple bonds is again reflected in graphite, a common polymorph of carbon.

Before discussing the structures of metals, we summarize the bonding in nonmetallic elements. Most nonmetals obey the *8 − n rule*: elements in column n of the periodic system form $8 - n$ covalent bonds. Many of the crystal structures can be explained by this rule. Each atom forms $8 - 4 = 4$ covalent bonds in diamond, $8 - 5 = 3$ bonds in bismuth, $8 - 6 = 2$ in sulfur, $8 - 7 = 1$ in bromine, and $8 - 8 = 0$ in argon. The multiple bonds formed by first row elements are exceptions to the $8 - n$ rule since fewer (but stronger) double and triple bonds are formed.

Heats of atomization among the nonmetals are proportional to the number of bonding electrons. The heats increase steadily from column VIII elements in which there are no bonding electrons to column IV elements with four. There is also marked dependence on the row of the periodic table as well as the column. Within a given column, the heats of atomization generally decrease with increasing atomic number. The value for carbon, for instance, is nearly four times that of lead. This behavior can be ascribed to the influence of the inner closed electron shells. Inner electrons contribute little to covalent bonding while increasing the interatomic distances because of overlap repulsion. Note that the trend is reversed in rare gas elements: xenon has a greater heat of atomization than argon. Inner electrons enhance the dipole interactions responsible for van der Waals attraction.

The energies required to atomize metals are comparable to those of nonmetals, showing that the bonding energies are similar. For alkali metals in column I, the heats of atomization (Li, 38; Na, 26; K, 22; Rb, 20; and Cs, 19 kcal/mole) span the same range as the halogens in column VII. In both cases there is one electron available for bonding.

Similar correlations exist between columns II and VI, and between III and V. Heats for the alkaline earth elements (Be, 78; Mg, 36; Ca, 42; Sr, 39;

and Ba, 43 kcal/mole) are comparable to those of the sulfur family, and about twice as large as corresponding alkali metals. Group IIB elements are slightly lower: Zn, 31; Cd, 27; and Hg, 15 kcal/mole, indicating fewer bonding electrons. Atomization energies for group IIIA elements (B, 135; Al, 78; Ga, 69; In, 58, and Tl, 43 kcal/mole) are nearly the same as those for the nitrogen family. There are three electrons per atom in both groups, even though some are metals and others are not. The elements of group IIIB have somewhat larger heats of atomization (Sc, 88; Y, 98; and La, 102).

The similarity in energies points out the similarity between covalent and metallic bonding. Metallic bonding can be visualized as resonating electron-pair bonds.

B. Alloy Chemistry

The understanding of the phase behavior of metals, particularly transition metals, is complicated by the large number of factors to be considered. It appears, however, that the primary factor fixing the thermodynamic properties of metallic solutions is the *electronic configuration* of the components. Secondary factors such as size, electronegativity, and solubility parameters are dependent on electronic structure.

One of the puzzling features of alloy structures is the appearance of the same structure for dissimilar systems and dissimilar compositions. Consider, for example, the β phase (bcc structure) found for CuZn, Cu_3Al, and Cu_5Sn. All three have a 3:2 electron-to-atom ratio, and are examples of "electron" compounds. Though Hume-Rothery (1936) originally advanced these ideas empirically, they have since been explained in terms of band theory. Systems in which Brillouin zones are just filled without overlap into higher zones across forbidden energy gaps are especially stable.

The Engel (1949) correlation between electronic configuration and crystal structures is useful in predicting the phase diagrams of intermetallic systems (Brewer, 1967). It is found that bcc, hcp, and fcc occur for metals with 1, 2, and 3 s and p electrons per atom, respectively. The d electrons are important in bonding but the structure type correlates best with the total number of s and p electrons. Closer examination of stability ranges for intermetallic compounds and solid solutions of known valence shows that bcc metals are stable up to 1.5 electrons per atom. Hexagonal close-packed (hcp) structures lie between 1.7 and 2.1 and the stability range for fcc structures is 2.5–3.0.

Solubility limits are determined by the electron–atom limits for each structure type (Brewer, 1967). As an example, consider the solid solution limits of rhenium dissolved in bcc molybdenum. The ground state of Mo is $4d^5 5s^1$. In rhenium the configuration is d^5sp and so a composition of 50 at. % Re and 50% Mo corresponds to an s + p electron concentration of 1.5, the expected solubility limit for the bcc structure. The observed solubility limit of rhenium in molybdenum is 43%. The predictions are fairly good for many intermetallic solid solutions.

As stated previously, d electrons do not determine structure in the Engel theory. One reason why s and p electrons are more effective than d electrons is because of their larger radii. The fcc and hcp structures are nearly identical for nearest neighbors but differ significantly in more distant neighbors. The s and p electrons interact over longer ranges than d electrons. Body-centered cubic structures predominate at least at high temperatures for the first six groups of the periodic system. This is because the unfilled d shells act as electron sinks, keeping the s and p electron densities below 1.7, the lower limit for hexagonal close-packed structures.

In the Engel approach, the ground states of some atoms are regarded as suitable for bonding, while in others it is not. Sodium with the outer electron configuration $3s^1$ is well suited to bonding with an unpaired electron, but in magnesium electrons are paired in the $3s^2$ ground state. The crystal structure of Mg is regarded as being derived from the excited state $3s^1 3p^1$ with two electrons per atom available for bonding. Sodium is bcc, and magnesium is hcp in accordance with the Engel correlation. Aluminum has the fcc structure with three bonding electrons when the atom is in excited state $3s^1 3p^2$. For transition metals, d electrons contribute to bonding but do not determine structure. In iron, for instance the bcc structure is associated with excited state $3d^7 4s^1$.

Although the Engel–Brewer correlations have led to valuable predictions of phases in multicomponent systems, some of the objections to the correlation have never been satisfactorily answered. The fundamental postulate that bcc, hcp, and fcc structures correspond to 1, 2, and 3 s and p electrons is made only by ignoring some of the polymorphic phases of certain metals. Lithium and sodium, for instance, have close-packed structures at low temperatures, as well as the high-temperatures bcc structure. This is difficult to reconcile with the Engel correlation because of the relatively simple electronic structure of the alkali metals. Hume-Rothery (1967) has pointed out that many of Brewer's phase diagram predictions could have been made without using the Engel correlation.

Several additional rules have been developed which provide guidelines regarding the occurrence of various structures (Sinha, 1972). The tendency of two elements to form intermetallic compounds increases with electronegativity difference. Many of the close-packed phases involve an A_xB_y compound in which A belongs to a group left of column VIIB (Mn, Tc, Re), and B to the right extending as far as Bi and Sb in column VA. Manganese and rhenium sometimes act as an A component, and other times as B. In general the B component is more electronegative than A.

These transition-metal compounds crystallize in a family of close-packed structures related to the β–W structure. Nb_3Sn and many of the technologically important superconductors belong to this family, as does the sigma phase which causes embrittlement in alloy steels. The three Laves phases—typified by $MgZn_2$, $MgCu_2$, and $MgNi_2$—possess similar structures. Such structures are characterized by a high packing density containing only tetrahedral interstices. The octahedral interstices found in normal close packing are absent. Greater packing densities are possible when spheres of two sizes are present. Only a small number of coordinations are in this family of intermetallic compounds; the four Kasper polyhedra with coordination numbers 12, 14, 15, and 16 are especially common.

An important consequence of close packing is the likelihood of a sharp peak in the electronic density of states near the Fermi level, giving rise to superconductivity and band ferromagnetism.

As with inorganic materials, size factors are often important in determining the stability of close-packed intermetallic compounds. Using metallic radii derived from interatomic distances in metallic elements, it is found that the β–W compounds have radius ratios $0.87 \leq r_A/r_B \leq 1.11$, close to the ideal value of 0.99. The spread is somewhat larger for Laves phases, $1.05 \leq r_A/r_B \leq 1.68$, bracketing the ideal value 1.225.

C. Lattice Energy

In ionic crystals the binding energy arises chiefly from the Coulomb attraction between cations and anions. Rock salt contains sodium atoms ionized to Na^+ with the stable $1s^2 2s^2 2p^6$ configuration, while the electron thus released completes the $1s^2 2s^2 2p^6 3s^2 3p^6$ configuration of a Cl^- anion. Interatomic spacings are determined by the size of the ion cores which are compressed slightly in the crystal. The compression results in a small antibonding contribution amounting to about 10% of the total energy. Neglecting this repulsive term, the bonding energy is inversely proportional to the interatomic distance since Coulomb forces are dominant. The Na–Cl distance in NaCl is 2.81 Å and the binding energy per ion pair is 7.7 eV. Cation–anion distances are smaller in KF (2.66 Å) and the bonding energy larger (8.2 eV). Binding energies are about four times larger in crystals comprised of divalent ions. The lattice spacing in CaS is 2.84 Å, about the same as NaCl, but the bonding energy is considerably larger, 31.0 eV per ion pair. Salts such as CuCl, ZnS, AlN, and TiC follow a similar pattern in bonding energy, even though they are not generally regarded as ionic (Brown, 1972).

Differences in Coulomb energy for various crystal structures are often slight. Bonding energies for the CsCl structure are about 1% greater than for the NaCl arrangement. Nevertheless rock salt structures are much more common. The antibonding energy due to compression of the ion cores is larger for the CsCl structure where there are eight nearest neighbors rather than six as in NaCl. Overlap energy increases with the number of near neighbors.

The cohesive energy of an ionic crystal can be calculated by summing the Coulomb energy for all ion pairs. There is also an important contribution from the repulsive potential caused by electronic overlap between neighboring ions. Calculation of cohesive energy is not difficult for rock salt structures, but becomes rather involved for more complex structures where Madelung coefficients are unknown. The empirical Kapustinskii relation provides an estimate of the lattice energy.

$$U_0 = \sum_i \frac{287.2 N_i Z_a Z_c}{r_c + r_a}\left(1 - \frac{0.345}{r_c + r_a}\right)$$

N_i is the number of ions in the ith constituent molecule, Z_a and Z_c are the valence of anion and cation, r_a and r_c are ionic radii. As an example, Houlihan and Roy (1974) have estimated the lattice energy of the Magneli phase Ti_4O_7. The constituent molecules are $2TiO_2 + Ti_2O_3$. For TiO_2, $N_i = 3$, $Z_c = 4$, $Z_a = -2$, and $r_c + r_a = 1.99$ Å; and for Ti_2O_3, $N_i = 5$, $Z_c = 3$, $Z_a = -2$, and $r_c + r_a = 2.05$ Å, giving $U_0 = -9200$ kcal/mole.

This estimate may be compared with the results of detailed calculations by Anderson and Burch (1971). Depending on the exponent of the repulsive potential, they obtained lattice energies ranging from -8700 to -9400 kcal/mole for Ti_4O_7.

D. Ramberg's Rules

Thus far only compounds with one type of anion have been considered. Ramberg (1954) developed concepts for treating more than one type of anion by using ionic radii to predict ion assemblages. As an illustration, consider an assemblage of two cations and two anions. If a melt contains equimolar portions of K^+, Li^+, Cl^-, and F^-, which phases will form on solidification, KCl + LiF or KF + LiCl? Calculating internal energies gives $-162 - 239 = -398$ kcal/mole for the first combination and $-189 - 191 = -380$ kcal/mole for the second. Thus KCl + LiF is the stable combination because it has the lowest internal energy. In terms of Dietzel's concepts, the stable assemblage has the greatest difference in field strengths. The LiF + KCl combination is more stable because of the more efficient packing of the *two smallest* and the *two largest* ions. This leads to the first of Ramberg's rules.

Rule 1. The stable assemblage always pairs the two smallest ions and the two largest ions: $LiCl + NaF \rightarrow LiF + NaCl$.

Ramberg's other rules are based on similar reasoning.

Rule 2. The more stable assemblage contains pairs of ions with equal charges, or in the more general case, the highly charged cation is paired with the highly charged anion. This rule maximizes the product of charges in the numerator of the coulomb equation. As examples, consider some reactions where ionic radii are similar throughout, so that the effect of charge can be separated from size: $MgF_2 + Li_2O \rightarrow MgO + 2LiF$.

Rule 3. Small cations combine with highly charged anions, and large cations with lower charged anions. Again this rule tends to minimize Coulomb energy. Consider two reactions involving cations of different size and similar charge, and anions of similar size but different valence: $2LiF + Na_2O \rightarrow Li_2O + 2NaF$.

Rule 4. The inverse of the previous situation: If the anions have the same charge and the cations the same size, the smaller anion is paired with the highly charged cation: $Li_2O + MgS \rightarrow MgO + Li_2S$.

E. Pauling's Rules

Ionic crystals are made up of a collection of charges. Coulomb energy contributes to the internal energy and stabilizes certain ionic arrangements. In general, the structures with the smallest neutral units are most stable.

Pauling's rules (1929) provide an evaluation of structural stability. To a first approximation, the structures of most inorganic crystals can be pictured as an array of large anions with cations occupying various interstices. The first of Pauling's rules states that a coordination polyhedron of anions is formed around each cation. The distance between cation and anion is equal to the sum of their ionic radii and the type of polyhedron is determined by the radius ratio. If the radius ratio of cation to anion is less than 0.225, triangular coordination is favored. Ratios between 0.225 and 0.414 favor tetrahedral coordination; 0.414–0.732 octahedral; 0.732–1.00 cubic; ratios of 1.00 or greater favor close packing of cation and anion. Numerical values of the critical radius ratios are determined geometrically by the "rattle" criterion. A given polyhedron becomes unstable when the cation and anion are no longer in contact. To illustrate the rule, consider magnesium oxide; the radius of Mg^{2+} is 0.7 and O^{2-} is 1.4 Å. The radius ratio is 0.5, for which the predicted cation coordination is octahedral.

Coordination numbers are not always determined unambiguously by the first rule. The radius ratio for aluminum and oxygen falls near the critical value of 0.414, hence Al^{3+} occurs in both octahedral and tetrahedral sites in oxides. Shannon and co-workers (1975) have examined the conditions which determine site preference for such ions. For compounds of composition $M_aAl_bO_c$, Al prefers tetrahedral coordination when the ratio a/b is greater than one. Also, the greater the M–O bond strength, the greater the tendency for octahedral coordination. The coordination of Te^{6+}, V^{5+}, As^{5+}, Ge^{4+}, Ti^{4+}, Fe^{3+}, Ga^{3+}, B^{3+}, Be^{2+}, and Zn^{2+} in many oxides are consistent with these ideas. Exceptions occur for highly stable structures such as spinel and perovskite.

Pauling's second rule is sometimes called the "electrostatic valence" rule. Each cation–anion bond is assigned a bond strength equal to the cation valence divided by the coordination number of the cation. The bond strength of Mg^{2+} in octahedral coordination, for example, is $+\frac{2}{6}$. The second rule states that the sum of the bond strengths for the bonds to a given anion is equal to the magnitude of its valence. MgO has the rock salt structure with O^{2-} bonded to six Mg^{2+}. There are six bonds of strength $+\frac{2}{6}$ to each oxygen so the sum of the bond strengths is $6(\frac{2}{6}) = 2$, the magnitude of the oxygen valence.

Pauling's rules for ionic structures tend to minimize electrostatic energy, thereby promoting stability. Consider the first rule which says the radius ratio determines the coordination number. Figure 10 illustrates two extreme situations in which the cation is very big compared to the four coordinating anions, and then very small. The electrostatic energy depends on the number of neighbors n and the interatomic distances d. The contribution of nearest neighbor cation–anions to the Coulomb energy tends to stabilize the structure, and is usually the largest term in the summation. Referring to Fig. 10a where the cation is large, the cation–anion distance is as small as possible. This helps minimize the Coulomb energy. But the energy also depends on the number of near neighbors. If the cation is sufficiently large, more anions can be fitted around the cation, without increasing the cation–anion distance. Hence large cations have large coordination numbers.

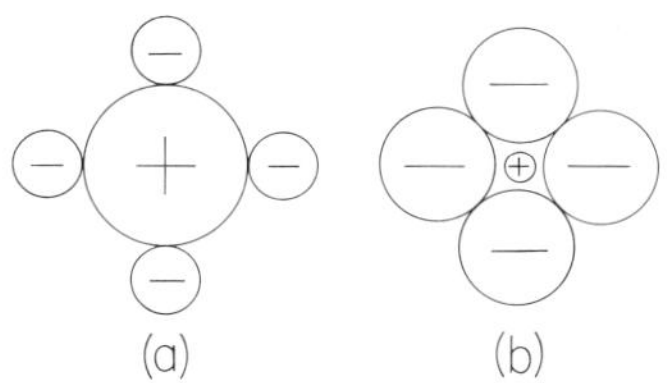

Fig. 10. Diagrams illustrating the physical basis of Pauling's first rule. The total attractive energy between a cation and its anion neighbors is directly proportional to the number of anions, and inversely proportional to the distance between cation and anion. Configuration (a) is unstable because there are too few anion neighbors; (b) is unstable because there are too many anions, making the cation–anion distance unnecessarily large.

At the other extreme (Fig. 10b), small cations have small coordination numbers for two reasons. When the cation is too small to fill the space between anions, the cation and anions are not in contact, thus destabilizing the structure. Additional destabilization results from the fact that anions are in contact, thus increasing the electrostatic repulsion energy between anions.

The stable situation lies between these extremes, with cations and anions in contact and the coordination number maximized. This is what Pauling's first rule accomplishes.

The second rule also rests upon Coulomb's law. Electrostatic energy is minimized when charges add to zero in the smallest volumes. If obeyed, Pauling's second rule leads to charge neutralization around every anion.

Consider a solid of composition A^+B^-. Two possible structures are illustrated Fig. 11. Assume that R is much larger than d. The structure in Fig. 11a satisfies Pauling's second rule and also has the smaller electrostatic energy. For the diatomic molecules in Fig. 11a the second rule gives $(1)(\frac{1}{1}) = 1$ and the electrostatic energy for N cations and N anions is $-Ne^2/d$ when R is very large. For the other structure (Fig. 11b) Pauling's second rule is not satisfied at either type of anion, giving 2 for the anion coordinated to two cations and 0 for the other. The Coulomb energy is $\frac{1}{2}N[-(2e^2/d) + (e^2/d)]$ which is greater than the value for the structure shown in Fig. 11a. Hence the structure with diatomic molecules is the more stable of the two structures, according to Pauling's rules and according to energy calculation.

The first and second are the most useful of Pauling's five rules. The third states that shared faces (and to a lesser extent, shared edges) decrease the stability of a structure. Electrostatic repulsion is reduced by eliminating short cation–cation distances. The fourth rule can be justified by a similar argument. According to the fourth rule, cations of high valence and small coordination tend not to share anions.

Pauling's third rule is not always obeyed. The highly charged Al^{3+} ions in corundum share octahedral faces, and yet the structure is favored electro-

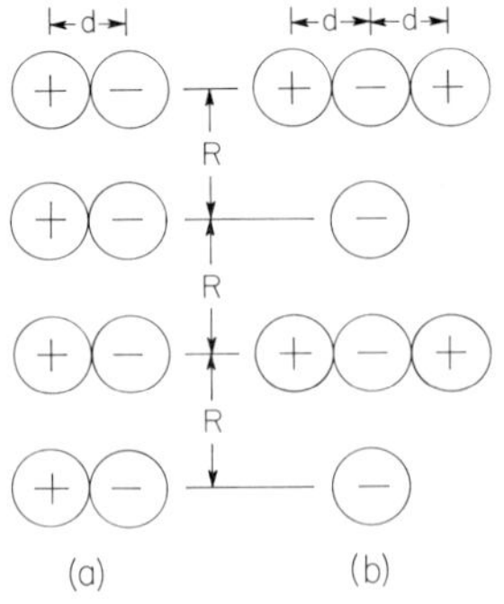

Fig. 11. Two A^+X^- structures demonstrating the relationship between Pauling's rules and electrostatic energy. Model (a) has the lower Coulomb energy and satisfies the second rule.

statically over the Sc_2S_3 structure which consists of edge-shared octahedra (Ludwiczek and Zemann, 1973).

The fifth (or parsimony) rule states that the number of different polyhedra is small. Like the other rules, the fifth helps ensure that charge will be neutralized in the smallest possible volume. The parsimony rule also implies that ternary compounds will be less common than binary compounds. Examination of 41 ceramic ternary systems shows that this is true in every case (Levin *et al.*, 1964). The $CaO–FeO–SiO_2$ system is typical with six binary compounds and one ternary compound. The 41 ternary systems contain 198 binary compounds and 36 ternary compounds, a ratio of 5.5:1.

1. Application of Pauling's Rules: Topaz

Topaz is a handsome gem mineral with chemical composition $Al_2SiO_4F_2$. Using the first and second rules it is possible to predict the coordination of all four ions in the structure.

First the radius ratios are calculated to predict probable coordination numbers for the cations. Since the radii of F^- and O^{2-} are about equal, the ratios are independent of the nature of the anion. Ratios calculated from the radii give the expected result that Si^{4+} is four coordinated, while Al^{3+} is a borderline case with both four or six coordination possible. A decision can be made applying the electrostatic valence rule.

Assume that the structure is a simple one with as few different polyhedra as possible (parsimony rule). Then every oxygen ion will have the same coordination of Si and Al, and each F will have identical surroundings also. In the following equations we make use of the notation n_{Si}^{O}, the number of Si bonded to each oxygen. The other symbols are defined in an analogous fashion. From the chemical formula it can be seen that

$$4n_{Si}^{O} = n_{O}^{Si}, \qquad 2n_{Al}^{O} = n_{O}^{Al}, \qquad n_{Al}^{F} = n_{F}^{Al}, \qquad \text{and} \qquad 2n_{Si}^{F} = n_{F}^{Si} \tag{1}$$

The electrostatic valence rule for O and F gives

$$n_{Si}^{O}(\tfrac{4}{4}) + n_{Al}^{O}(\tfrac{3}{4} \text{ or } \tfrac{3}{6}) = 2 \tag{2}$$

$$n_{Si}^{F}(\tfrac{4}{4}) + n_{Al}^{F}(\tfrac{3}{4} \text{ or } \tfrac{3}{6}) = 1 \tag{3}$$

Equations (2) and (3) can only be satisfied if the coordination number of Al is six. The values of n must of course be positive integers so that n_{Si}^{O}, for example, has three possible values: 0, 1, 2. If $n_{Si}^{O} > 2$, Eq. (2) cannot be satisfied. It can be shown that only $n_{Si}^{O} = 1$ is possible. If $n_{Si}^{O} = 0$, then every Si is surrounded by four fluorine and since the Si:F ratio in topaz is 1:2, there must be two silicons bonded to each F, $n_{F}^{Si} = 4 = 2n_{Si}^{F}$. Equation (3) then becomes $2(\frac{4}{4}) + n_{Al}^{F}(\frac{3}{6}) = 1$ which is impossible so that $n_{Si}^{O} \neq 0$.

If $n_{Si}^{O} = 2$, Eq. (2) shows that $n_{Al}^{O} = 0$, so that every aluminum is completely surrounded by fluorine. From the chemical formula $n_{Al}^{F} = n_{F}^{Al} = 6$ and Eq. (3) again leads to an impossibility: $n_{Si}^{F}(\frac{4}{4}) + 3 = 1$.

The only remaining possibility is $n_{Si}^{O} = 1$. Equation (2) then becomes $1 + n_{Al}^{O}(\frac{3}{6}) = 2$ or $n_{Al}^{O} = 2$. Since the Si:O ratio is 1:4, all four anions around Si are oxygen, making $n_{F}^{Si} = 0 = n_{Si}^{F}$. Equation (3) then gives $0 + n_{Al}^{F}(\frac{3}{6}) = 1$ or $n_{Al}^{F} = 2$.

Combining these results with Eq. (1) we find that each Si is coordinated to four O, each Al to four O and two F, every O to one Si and two Al, and F to two Al. Thus all coordinations are correctly predicted.

F. Close Packing

In metals and ionic crystals where the bonding forces are largely nondirectional, there are many examples of close-packed structures (Table II). The densest possible structure is often the most stable, but our understanding of close packing is very primitive. There has been little progress in describing the close packing of nonspherical groups, or even of spheres of several different sizes. Many of the structures referred to as close packed are not really very densely packed. Forsterite, for instance, with its close-packed oxygen positions is not packed as efficiently as pyrope garnet, a structure which is not based on close packing. Space filling fractions for a number of oxides listed in Table III show that close-packed and non-close-packed structures are comparable in packing efficiency and that few exceed the packing density of 74% expected for identical close-packed spheres.

Most of the crystallochemical ideas concern the closest packing of spheres of equal size. In this type of packing, each sphere contacts twelve others, six within the close-packed layer, three above and three below. The A positions in Fig. 12a represent one layer of a close-packed structure. Each A atom touches six other A atoms arranged in a hexagon. Atoms in the adjacent layer occupy the B or C positions, but not both. Three B atoms contact each A atom in the layer below.

An infinite number of stacking sequences are possible in close-packed structures but only three are at all common: hexagonal close packed (ABABAB · · ·), cubic close packed (ABCABC · · ·), and "double" hexagonal close packed (ABACABAC · · ·). Hexagonal close packed, typified by magnesium, has a two-layer repeat, while cubic close-packed structures such as copper have three, and double hcp four. There are many other possible sequences with longer repeat patterns but examples are rare.

Examining the inorganic compounds in Table II, it can be seen that anions are usually the close-packed ions. This is not surprising since O^{2-},

TABLE II

STRUCTURES BASED ON CLOSE PACKING[a]

Structure	Close-packed atoms	Stacking sequence	Octahedral atoms	f_o	Tetrahedral atoms	f_t
Mg (HCP)	Mg	AB	—	0	—	0
K_2GeF_6	K_2F_6	AB	—	$\frac{1}{8}$	—	0
UCl_6	Cl_6	AB	U	$\frac{1}{6}$	—	0
$Cs_3Tl_2Cl_9$	Cs_3Cl_9	AB	Tl_2	$\frac{1}{6}$	—	0
$CsNiCl_3$	Cs_3Cl_3	AB	Ni	$\frac{1}{4}$	—	0
PdF_3	F_3	AB	Pd	$\frac{1}{4}$	—	0
BiI_3	I_3	AB	Bi	$\frac{1}{3}$	—	0
CdI_2(C6)	I_2	AB	Cd	$\frac{1}{2}$	—	0
TiO_2 (rutile)	O_2	AB	Ti	$\frac{1}{2}$	—	0
α-Al_2O_3	O_3	AB	Al_2	$\frac{2}{3}$	—	0
NiAs	As	AB	Ni	1	—	0
$AlBr_3$	Br_3	AB	—	0	Al	$\frac{1}{6}$
Al_2Se_3	Se_3	AB	—	0	Al_2	$\frac{1}{3}$
Al_2ZnS_4	S_4	AB	—	0	Al_2Zn	$\frac{3}{8}$
ZnS(2H)	S	AB	—	0	Zn	$\frac{1}{2}$
Mg_2SiO_4	O_4	AB	Mg_2	$\frac{1}{2}$	Si	$\frac{1}{8}$
Cu(fcc)	Cu	ABC	—	0	—	0
K_2PtCl_6	K_2Cl_6	ABC	Pt	$\frac{1}{8}$	—	0
$Cs_3As_2Cl_9$	Cs_3Cl_9	ABC	As_2	$\frac{1}{6}$	—	0
$SrTiO_3$	SrO_3	ABC	Ti	$\frac{1}{4}$	—	0
ReO_3	O_3	ABC	Re	$\frac{1}{4}$	—	0
$CrCl_3$	Cl_3	ABC	Cr	$\frac{1}{3}$	—	0
TiO_2 (anatase)	O_2	ABC	Ti	$\frac{1}{2}$	—	0
$CdCl_2$(C19)	Cl_2	ABC	Cd	$\frac{1}{2}$	—	0
$Cu_2(OH)_3Cl$	$(OH)_3Cl$	ABC	Cu_2	$\frac{1}{2}$	—	0
NaCl	Cl	ABC	Na	1	—	0
SnI_4	I_4	ABC	—	0	Sn	$\frac{1}{8}$
HgI_2	I_2	ABC	—	0	Hg	$\frac{1}{4}$
α-Ga_2S_3	S_3	ABC	—	0	Ga_2	$\frac{1}{3}$
Ag_2HgI_4	I_4	ABC	—	0	Ag_2Hg	$\frac{3}{8}$
ZnS(3C)	S	ABC	—	0	Zn	$\frac{1}{2}$
Bi_2O_3	Bi_2	ABC	—	0	O_3	$\frac{3}{4}$
CaF_2	Ca	ABC	—	0	F_2	1
Co_9S_8	S_8	ABC	Co	$\frac{1}{8}$	Co_8	$\frac{1}{2}$
$MgAl_2O_4$	O	ABC	Al_2	$\frac{1}{2}$	Mg	$\frac{1}{8}$
$BiLi_3$	Bi	ABC	Li	1	Li_2	1
K_2MnF_6	K_2F_6	ABAC	Mn	$\frac{1}{8}$	—	0
CdI_2(C27)	I_2	ABAC	Cd	$\frac{1}{2}$	—	0
TiO_2 (brookite)	O_2	ABAC	Ti	$\frac{1}{2}$	—	0
SiC(4H)	Si	ABAC	—	0	C	$\frac{1}{2}$
$Al_2SiO_4F_2$	O_4F_2	ABAC	Al_2	$\frac{1}{3}$	Si	$\frac{1}{12}$
$K_3W_2Cl_9$	K_3Cl_9	ABCACB	W	$\frac{1}{6}$	—	0
$BaTiO_3$	BaO_3	ABCACB	Ti	$\frac{1}{4}$	—	0

[a] f_o and f_t denote the fractional filling of octahedral and tetrahedral interstices [after Wells (1958)].

TABLE III

PACKING PERCENTAGES FOR SOME COMMON OXIDES CALCULATED FROM PAULING–AHRENS RADII[a]

Corundum, Al_2O_3	81%	Diopside, $CaMgSi_2O_6$	67%
Pyrope, $Mg_3Al_2Si_3O_{12}$	76	Enstatite, $MgSiO_3$	66
Rutile, TiO_2	71	Forsterite, Mg_2SiO_4	65
Spinel, $MgAl_2O_4$	69	Beryl, $Be_3Al_2Si_6O_{18}$	59
Zircon, $ZrSiO_4$	69	Quartz, SiO_2	55

[a] The packing efficiencies of pyrope garnet and several other non-close-packed structures are comparable to corundum, forsterite, and spinel where the oxygens are close packed.

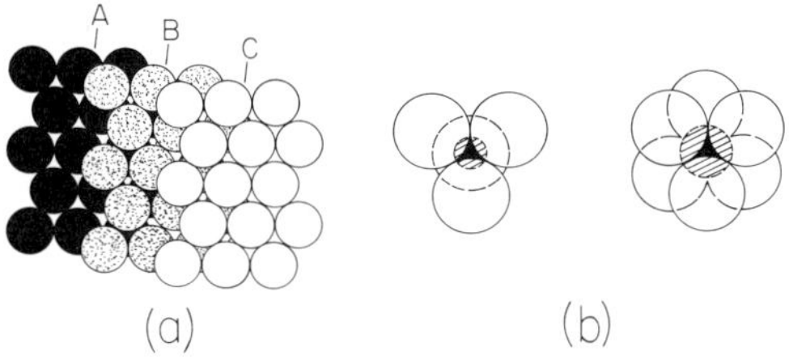

Fig. 12. (a) Cubic close packing of identical spheres in which each sphere contacts twelve others. Tetrahedral and octahedral interstices (b) are common in all structures based on close packing of identical spheres. For each close-packed sphere, there are two tetrahedral and one octahedral interstitial sites.

F^-, Cl^-, and the other anions are large ions, larger than most of the cations. Three common types of cations are found in close-packed oxides:

(1) large cations like K^+ enter in close packing and are 12 coordinated,
(2) Mg^{2+} and other octahedrally coordinated ions are bonded to six anions, and
(3) small cations such as Si^{4+} in tetrahedral sites bonded to four oxygens.

Tetrahedral and octahedral interstices in close-packed structures are illustrated in Fig. 12b.

Not all interstitial sites in close-packed structures are tetrahedral or octahedral. The minerals kotoite and magnetoplumbite provide examples of three- and fivefold coordinations. Kotoite, $Mg_3B_2O_6$, contains interlinked MgO_6 octahedra and BO_3 triangles. Oxygens are hexagonally close packed with each oxygen bonded to three magnesiums and one boron. Magnetoplumbite ($PbFe_{12}O_{19}$) is an important member of the hexagonal ferrite family used as permanent magnets. Oxygen ions together with Pb^{2+} ions form a close-packed array with a ten-layer stacking sequence, ABACB-

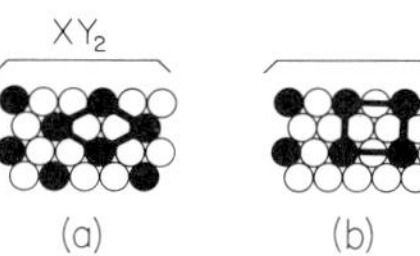

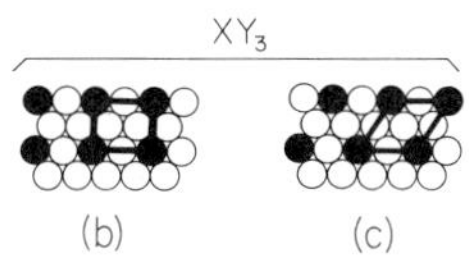

Fig. 13. Two elements of comparable size can crystallize in ordered close-packed arrays. The superstructures illustrated above are found in intermetallic compounds.

CBCAB. The ferric ions in magnetoplumbite occupy three different sites. Along with octahedral and tetrahedral sites, there are trigonal bipyramid positions in which Fe^{3+} is coordinated to five oxygens, three in triangular coordination, and two others above and below, slightly further away.

Examining the close-packed structures in Table II, it is interesting to speculate on the absence of certain structures. Why, for instance, are there no inorganic compounds based on hexagonal close packing with more than half the tetrahedral sites filled? The probable explanation involves the distance between tetrahedral sites. In hcp structures, the distance between neighboring tetrahedral sites is only about $0.4D$, where D is the distance between close-packed atoms. In cubic close-packed (ccp) structures the tetrahedral holes are separated by a considerably larger value, $0.7D$. Thus if neighboring tetrahedral sites are occupied by ions of like charge, the Coulomb energy will be higher for the hcp structure, explaining why fluorite is more stable than its hexagonal analog. The argument can be extended to any fraction of filling greater than $\frac{1}{2}$ and to other close-packed structures containing an hcp component, for instance the ABAC sequence.

Close packing of two different elements occurs when the atoms are comparable in size. Layers of composition XY_2 and XY_3 are known. The XY_2 layers are found in WAl_5 where close-packed layers of composition WAl_2 alternate with Al_3 layers. It is impossible to superpose XY_2 layers without bringing X atoms into contact, hence examples are not common among ionic compounds. Two XY_3 patterns (Fig. 13) are found in metals and inorganic substances. Examples include Al_3Ti, Ni_3Sn, and $BaTiO_3$.

G. Molecular Packing in Organic Crystals

Three types of bonding are important in organic crystals. Within each molecule, *covalent* bonds of single, double, or triple bond character dominate. Bond lengths and angles vary little from compound to compound. Bond angles for the sp^3 hybrid are always near 109.5°, and near 120° for sp^2; forces from neighboring molecules seldom cause distortions larger than a few degrees. Covalent bond lengths between carbons, nitrogens, and oxygens are about 1.1 Å for triple bonds, 1.3 Å for double bonds, and 1.5 Å for single bonds. All are considerably shorter than the nonbonded distances between atoms calculated from van der Waals radii, which range from 1.2 Å for H

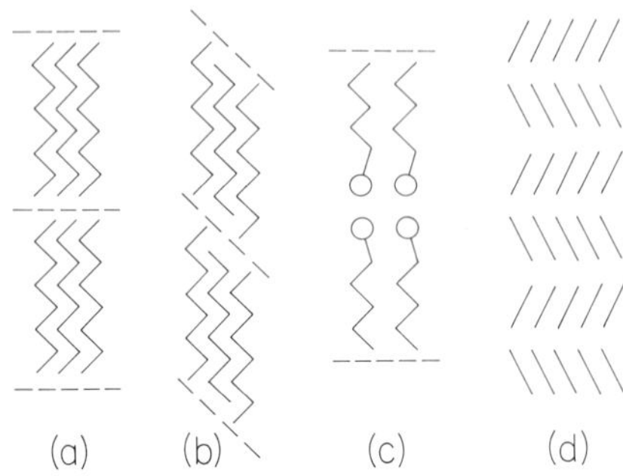

Fig. 14. Structures of long-chain hydrocarbons are governed by packing considerations. Polymorphism, (a) and (b), sometimes occurs when the chains are tilted differently with respect to the sheets. Polar molecules (c) tend to pack head-to-head, while herring bone patterns (d) are observed in aromatic hydrocarbons.

to 2.2 for I, with C, N, and O in the range 1.4–1.8 Å. Distances between molecules generally exceed 2.8 Å.

Hydrogen bonds—in which a hydrogen covalently bonded to one electronegative atom is attracted to a second anion—also contribute to intermolecular bonding. Hydrogen bonds are usually a little shorter than van der Waals distances, O–H---O bonds, for instance, range from 2.4 to 3.1 Å.

Straight chain hydrocarbons pack together like tiny gas cylinders, often in hexagonal arrangements with the molecules parallel to one another (Fig. 14) forming layers with easy cleavage. Some of the saturated straight chain hydrocarbons have a number of polymorphic forms in which the chains form different angles with the cleavage plane. When the chains terminate with functional groups such as –COOH or $-NH_2$, they form bimolecular sheets with the molecules packing head to head, usually joined by hydrogen bonds. Cleavage then takes place between the tails. Since the hydrocarbon chains are zigzag in nature, it is not surprising to find odd-numbered chains packing differently from even-numbered molecules. Odd-numbered molecules from $C_{19}H_{40}$ to $C_{29}H_{60}$ are orthorhombic, while even numbers from $C_{18}H_{38}$ to $C_{36}H_{74}$ are triclinic or monoclinic. Similar effects are responsible for the raggedness in the melting point curve (see Fig. 29).

Aromatic ring molecules are generally planar and pack in staggered herringbone arrays (Fig. 14d) to achieve maximum density. Anthracene and naphthalene crystallize this way.

Kitaigorodskii (1961) has applied close-packing concepts to organic crystals with considerable success, showing that they are even more useful than in inorganic or metallic systems. Some of these concepts were slow in developing because hydrogen atom positions were not determined in most of the early structure analyses. In most organic crystals, individual molecules

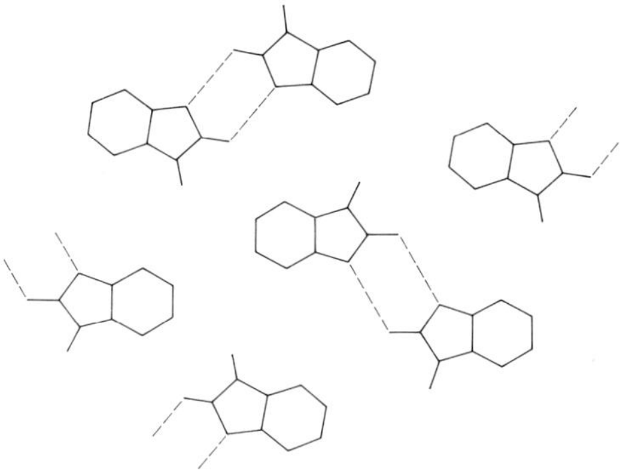

Fig. 15. Close packing in Isatin, showing six molecules in contact with the center molecule. Density of packing is an important criterion in organic structures.

are in contact with 10–14 other molecules, 12 being very common coordination number. Moreover, it is possible to identify close-packed planes (Fig. 15) in which each molecule is surrounded by six others. Even though the molecules are of complicated shapes, high packing densities are achieved by positioning the molecules in interlocking patterns, thereby minimizing the empty space.

To compute the packing density it is necessary to first calculate the molecular volume. Bonding and nonbonding distances in organic molecules are well known and vary little from molecule to molecule. The nonbonding distances can be used to calculate a set of intermolecular radii, which together with the covalent bond lengths, can be used to calculate molecular volumes V_m. Intermolecular radii for C, H, N, and O are 1.80, 1.17, 1.57, and 1.36 Å, respectively. V_m is computed by summing the volumes associated with each its atoms. The atom components are approximated by spheres which are truncated in the directions of valence bonds.

To illustrate, consider the molecule AB in Fig. 16: $V_m = V_A + V_B$ and $V_A = \frac{4}{3}\pi R_A{}^3 - (\pi/3)h_{AB}^2\,(3R_A - h_{AB})$, where h is the height of the truncated segment of sphere A, calculated from $h_{AB} = R_A - (R_A{}^2 + d_{AB}^2 - R_B{}^2)/2d_{AB}$. V_B is calculated in an analogous fashion. In aromatics with C—C = 1.40 Å and C—H = 1.08 Å, V_C is about 8 Å^3, and V_{C-H} about 14 Å^3. From these numbers the molecular volume for anthracene $C_{14}H_{10}$ is calculated as 4(8) + 10(14) = 172 Å^3.

The packing efficiency k is given by $k = ZV_m/V$ where Z is the number of molecules per unit cell and V is the unit cell volume. Anthracene is monoclinic with $a = 8.561$, $b = 6.036$, $c = 11.163$ Å, $\beta = 124°42'$, $Z = 2$, giving $k = (2)(172)/(474) = 0.73$, comparable to packing densities for close-packed spheres.

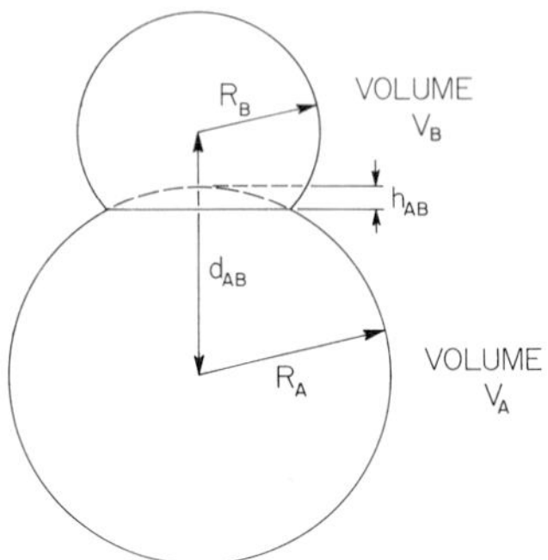

Fig. 16. A diatomic molecule AB represented as truncated spheres in calculating molecular volume (Kitaigorodskii, 1961).

Packing coefficients for aromatic compounds range between 0.6 and 0.8. Compounds with low packing efficiencies such as dioctynaphthalene often form glasses and crystallize only with difficulty. Certain irregular molecular shapes can only be packed in low density arrays, providing insufficient energy to render the crystalline state stable. This is true for compounds with large aromatic rings and long side chains.

The primary determinant is shape: more regular shapes approximating spheres or ellipsoids yield higher packing efficiencies. Thus anthracene

has a higher k value than phenanthrene

Not only does the packing efficiency depend on molecular shape and symmetry but the converse is also true: the symmetry of a molecule in a crystal depends on achieving a high packing density. As a consequence, the symmetry of a molecule is usually lowered in the solid state. A molecule only occupies a position of symmetry higher than 1 in the crystal if the retention of this symmetry involves no substantial loss in packing density.

In triclinic and monoclinic crystals, point symmetries 1, 2, m, 1, and $2/m$ are possible, so that molecules may retain one of these symmetries when incorporated in a triclinic or monoclinic unit cell. If the molecule occupies a general position it has six degrees of freedom; for $\bar{1}$, three rotational degrees; m has one rotational and two translational; 2, one of each; and $2/m$, just one rotational degree of freedom.

Ease of close packing increases with the number of degrees of freedom, except that the loss of the three translational degrees of freedom imposes no restrictions on the contacts between molecules because lattice parameters provide the necessary translational freedom. 1 and $\bar{1}$ are the preferred symmetries for close packing. Molecules tend to retain inversion symmetry in the solid state but other symmetry elements are often lost.

As examples, *p*-dibromobenzene has *mmm* symmetry as a molecule and $\bar{1}$ in the solid state. Chrysene and oxalic acid are reduced from $2/m$ to $\bar{1}$, hexachlorocyclohexane from *m* to 1, and tartaric acid from 2 to 1. Only very regular-shaped molecules such as adamantane are able to retain high symmetry ($\bar{4}3m$) when solidified.

There are many molecular crystals in which polymorphs with identical molecular layers differ only in the stacking of close-packed layers. A similar phenomenon occurs in the crystal structures of racemic and optically active pairs (Pedone and Benedetti, 1972), making it possible to derive the one structure from the other. The structure of L-alanine can be derived from DL-alanine by reversing the sense and direction of a column of D molecules. In such structures it is usually possible to recognize similar planes of identical molecules, all of the same handedness.

IV. POLYMORPHISM

Isomorphs are different compounds with the same structure; *polymorphs* are different structures of the same compound. NaCl, KCl, and NaBr are isomorphs having the rock salt structure; calcite, vaterite, and aragonite are polymorphs of $CaCO_3$. Polymorphs differ in physical properties as well as structure. Melting point, density, hardness, morphology, refractive indices, and electrical and magnetic properties all depend on crystal structure.

Polymorphism is a common phenomenon since most compounds possess more than one crystal structure. The factors governing the stability of polymorphs are complex, although some contributions to the energy are apparent from the crystal structure.

As an example, consider the kaolin clay minerals which are of great importance in china and other whiteware ceramics. The polymorphs found in kaolin minerals can be explained in terms of interlayer bonding (Newnham, 1961). The basic structural unit of the kaolin minerals is the compound aluminosilicate sheet illustrated in Fig. 17. The chemical formula can be written as $O_3Si_2O_2(OH)Al_2(OH)_3$ to emphasize the planar groupings of the structure. Each Si atom is tetrahedrally surrounded by three oxygens in the basal plane and a fourth in the layer above. The latter is shared with two

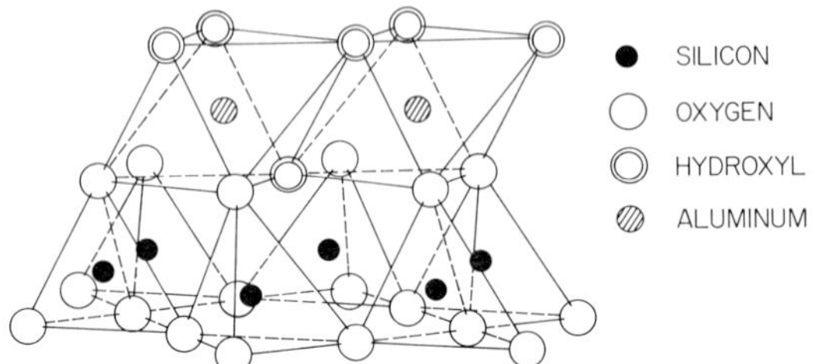

Fig. 17. Aluminosilicate sheet found in kaolinite and other clay minerals. Silicon is bonded tetrahedrally to four oxygens and aluminum octahedrally to two oxygens and four hydroxyls. Hydrogen bonds from the hydroxyl groups link adjacent kaolin layers in several different stacking sequences.

aluminum ions, each of which is octahedrally coordinated to two oxygens and four hydroxyls, three in the uppermost layer and one coplanar with the apex oxygens of the silica sheet. Silica tetrahedra share corners with three other tetrahedra, while each Al octahedron shares three edges in common with other octahedra. Successive kaolin layers are held together by long hydrogen bonds extending from hydroxyls to oxygen in the next kaolin layer.

Three polymorphic forms of kaolin-layer silicates are found in nature: kaolinite, nacrite, and dickite. In addition to the crystalline varieties, there are several minerals showing a more or less irregular sequence of kaolin layers. The key feature of the interlayer coordination is the manner in which oxygens and hydroxyls approach one another in pairs to form long hydrogen bonds. In forming such bonds there are 36 different ways of positioning the second kaolin layer over the first. These sequences generate six one-layer unit cells and 108 two-layer cells, of which only three are observed. The stability of kaolinite, nacrite, and dickite can be explained by considering the interlayer forces in more detail.

Highly charged Si^{4+} and Al^{3+} ions tend to avoid one another as much as possible because of electrostatic repulsion. The Coulomb energy of two kaolin layers is lowered by minimizing the cation–cation superposition in adjacent layers. Rejecting the stacking arrangements showing more than the minimum amount of overlap leaves two one-layer cells and 12 two-layer cells. The two single-layer cells are those of right- and left-handed kaolinite, explaining the stability of kaolinite over other polytypes.

Dickite and nacrite are two of the 12 two-layer polymorphs with minimum Coulomb energy. Their stability arises from the fact that the kaolin layers are not perfectly flat, but slightly puckered because of the rotation of tetrahedra and the shortening of shared octahedral edges. Matching the ridges and grooves in neighboring layers gives shorter, stronger O---H–O bonds, stabilizing the sequences observed in dickite and nacrite.

Polytypes are a special class of polymorphs. When identical layers are stacked in different ways, the sequences are referred to as polytypes. Several origins for polytypism have been suggested: impurities, nonstoichiometry, screw dislocatons, and higher-order phase transformations. Silicon carbide, zinc sulfide, cadmium iodide, and lead iodide are found in many polytypes, some with exceedingly large repeat distances. The step heights measured for the spiral surface structure of some polytypes appear to be related to the repeat distance, but other measurements suggest a thermodynamic origin.

Schneer (1955) has proposed a higher-order transformation theory based on the fact that the internal energies of various polytypes are almost identical. The theory resembles the Bragg–Williams treatment of order–disorder in binary alloys. The transformation from hexagonal close packing to cubic close packing is presumed to proceed in infinitesimal steps over a finite temperature range, with long-period polytypes representing intermediate steps in the transition. Assuming that Boltzmann statistics govern the relative number of hexagonal and cubic states, Schneer concluded that the stable polytypes possess the maximum number of interaction contacts between unlike states. Thus the 4H polytype (ABAC) is a likely intermediary between 3C (ABC) and 2H (AB) but (ABCBCAB) is not. He found this to be true for the 15 known polytypes of SiC.

A. Phase Transformation

Buerger (1971) has described the crystallographic aspects of phase transitions and related them to the thermodynamic description. There are two different kinds of transformations: *Reconstructive transformations* in which the two structures are different markedly, and *displacive transformations* where the atomic positions differ by only small displacements (Fig. 18). The energy change associated with the transition can be correlated with the

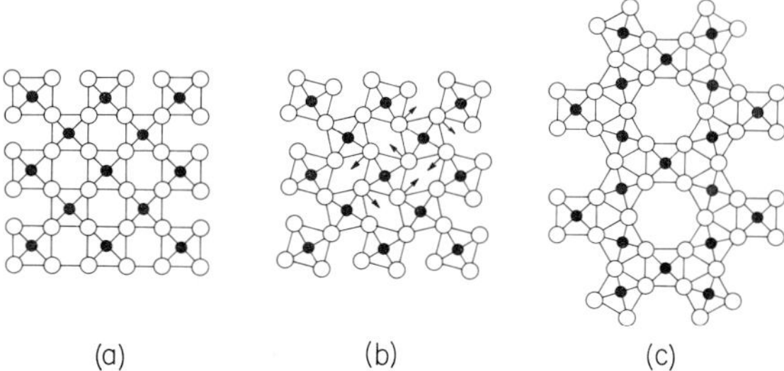

Fig. 18. Square planar networks illustrating reconstructive and displacive phase transitions. Only small displacements are required to convert (a) to (b), but the reconstructive transformation from (a) to (c) requires that near-neighbor bonds to be broken (Buerger, 1971).

change in bonding energy, so that there is a correlation between the two types of structural transformations and first- and second-order thermodynamic transformations. First-order transitions are characterized by a latent heat, a discontinuous change in volume, and substantial changes in structure. The more subtle second-order transitions show only discontinuities in the temperature derivatives, specific heat, and thermal expansion.

For thermal transformations at low pressures, the change in internal energy $\Delta U \simeq T\,\Delta S$. The high-temperature phase has the larger entropy and the larger internal energy. The change in internal energy is due principally to the change in bonding energy, with the nearest-neighbor bonds making the grestest contribution. Bonds between more distant neighbors are weaker, contributing less to the internal energy.

Most transformations are accompanied by a change in coordination number, either primary or secondary. When the primary coordination changes there is usually a large change in bonding energy resulting in a sizeable latent heat. On heating, cesium chloride changes from the CsCl to NaCl structures with a decrease in coordination number from 8 to 6. The latent heat is 1.8 kcal/mole. Primary coordination does not change at the α–β quartz transition. Each silicon is bonded to four oxygens in both the high- and low-temperature polymorphs. The slight changes in secondary coordination are accompanied by a latent heat of only 0.15 kcal/mole. Transformations of primary coordination generally involve larger latent heats.

Transformations proceed by a variety of structural mechanisms. Iron transforms from fcc to bcc by differential dilation affected by expansion in the (001) plane and contraction in the perpendicular direction. There are no energy barriers between the polymorphs, so the changeover is rapid. On heating, calcium is converted from fcc to hcp by shearing close-packed planes, a process which is somewhat slower since bonds must be broken. At higher temperatures, hcp calcium transforms to bcc by a combination of differential dilation and shear.

Certain distortions and displacive transitions can be traced to structural collapse around a small cation. The perovskite and feldspar families provide excellent examples (Megaw, 1965). The feldspar structure consists of a three-dimensional aluminosilicate framework with Na^{+}, K^{+}, Ca^{2+}, or Ba^{2+} in large interstices. There are ten oxygens surrounding the large cations, a comfortable coordination for the large K^{+} and Ba^{2+} ions, but too large a number for the intermediate-sized Ca^{2+} and Na^{+} ions. As a result the Na–Ca plagioclase feldspars distort to give these ions a smaller coordination number, thereby lowering the symmetry from monoclinic to triclinic.

In the perovskite family (RXO_3), large R ions form a cubic close-packed configuration with oxygen and the smaller X cations occupy octahedral interstices. The R ion is twelve coordinated in the ideal perovskite structure but when the ion is too small, distortions take place. Deformations of this type are found in $CaTiO_3$ (the mineral perovskite) and $NaNbO_3$ as the structure collapses about the Na^+ and Ca^{2+} ions in the low-temperature phase. Transformation temperatures can be raised or lowered by substituting ions of different size.

In some cases, the origin of a phase transition can be traced to the electronic configuration of certain atoms. Degeneracies occur in ground state of several transition-metal and rare-earth ions. The degeneracy can be lifted and the energy lowered by displacive phase transitions in which the local crystal field is altered. A Jahn–Teller phase transition such as this occurs in $DyVO_4$ at low temperatures. The transition is driven by a coupling between low-lying electronic energy levels of the rare-earth ions and lattice phonons. The rare-earth vanadates crystallize in the tetragonal zircon structure with the trivalent dysprosium ion bonded to eight oxygens, and V^{5+} in tetrahedral coordination with four oxygens. The $4f^9$ electron configuration of Dy^{3+} has a $^6H_{15/2}$ ground state, which in $DyVO_4$ leads to two nearly degenerate doublets. A phase transition takes place at 14°K, producing an orthorhombic deformation which lifts the degeneracy. One of the elastic stiffness coefficients approaches zero at the transition, showing that the transformation from tetragonal to orthorhombic symmetry is near second order.

For reconstructive transformations there is no special relation between the symmetries of the high- and low-temperature polymorphs. More often than not, however, an up-temperature transition results in increased symmetry because the large number of equivalent positions contributes to entropy. In the case of displacive transitions, the symmetry of the low-temperature phase is usually a subgroup of the high-temperature form because the distortion suppresses certain symmetry elements. The α–β transitions in quartz, cristobalite, and tridymite (Table IV) are examples, as are most ferroelectrics, although the low-temperature transitions in $BaTiO_3$ and Rochelle salt are exceptions.

Transformation speed depends on the energy barrier between the two polymorphic states. The breaking of bonds in a reconstructive transformation produces sluggishness. Solvents sometimes accelerate the process dramatically. Strong Si—O bonds must be broken in the quartz–tridymite transition—a very slow process which brick manufacturers speed up by adding a small amount of lime solvent.

Aluminum–silicon ordering occurs in the tetrahedral sites of many minerals and provides a measure of the thermal history of the specimen. Most ordered arrangements obey the aluminum-avoidance rule which states that AlO_4 tetrahedra never share corners. The Al–O–Al linkage is unstable

TABLE IV

SYMMETRY AND DENSITY OF THE SILICA POLYMORPHS[a]

Polymorph	Stability	Density	Space group	Si site symmetry and equivalent directions
High cristobalite	1470–1723°C	2.22	$Fd3m$	$\bar{4}3m$ (24)
High tridymite	867–1470	2.17	$P6_3/mmc$	$3m$ (6)
High quartz	573–867	2.51	$P6_22$	222 (4)
Low quartz	<573	2.65	$P3_12$	2 (2)
Low cristobalite	metastable $<200°$	2.33	$P4_12\,2$	2 (2)
Coesite	30 kbar, low T	2.89	$B2/b$	1 (1)

[a] High temperature forms have lower density and higher symmetry while pressure stabilizes the denser polymorphs.

because Al is slightly too large for tetrahedral coordination and also because of the deficiency of valence electrons at the oxygen link. In calcium and barium feldspars, the Al–Si ordering is especially simple: each Al tetrahedron is connected to four Si tetrahedra, and vice versa. Microcline ($KAlSi_3O_8$) and albite ($NaAlSi_3O_8$) have more complicated arrangements in which the Al tetrahedra are again surrounded by four Si tetrahedra, but two-thirds of the Si tetrahedra have one Al and three Si neighbors while the remaining one-third of the Si tetrahedra have two Al and two Si tetrahedra as neighbors. The difference between the two types of distribution is demonstrated by the two-dimensional networks in Fig. 19, where solid circles represent tetrahedra of high Al content and open circles those which are primarily silicon (Newnham and Megaw, 1960). Though the analogy between the drawing and the three-dimensional feldspar structures has obvious limitations, it serves to illustrate nearest-neighbor configurations. It also brings out the cause of exsolution in the $NaAlSi_3O_8$–$CaAl_2Si_2O_8$ series, the plagioclase feldspars common in many rock types. At high temperatures

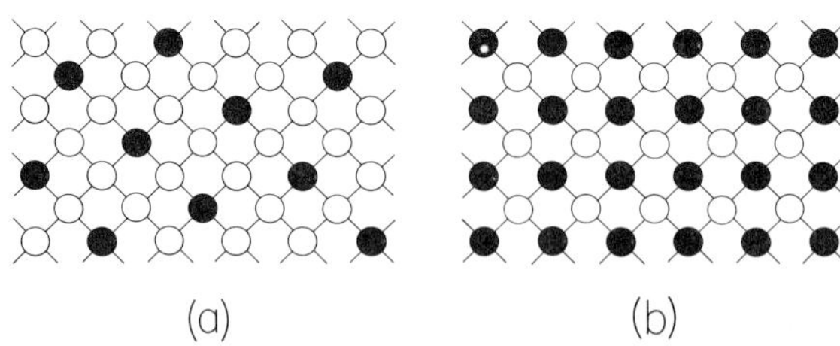

Fig. 19. Two-dimensional networks representing the aluminum–silicon distributions in feldspars. Si and Al tetrahedra are represented by open solid circles, respectively. The ordering pattern of alkali feldspars (a) is different from that of the alkaline earth feldspars (b), making low temperature solid solution impossible (Newnham and Megaw, 1960).

a complete solid solution forms, but at low temperatures exsolution produces Na- and Ca-rich lamellae as two phases exsolve. The Al–Si pattern in Fig. 19a cannot be derived from that in Fig. 19b by a simple substitution of Si for Al; there must simultaneously occur a redistribution in which Al replaces Si in other sites. The two patterns cannot form solid solutions because of their crystallographic incompatibility.

Order–disorder phenomena are more sluggish in inorganic solids than in metals because it is usually next nearest neighbors which exchange places, rather than nearest neighbors. Intermediary atoms impede diffusion, making ordering more difficult.

B. Entropy and Structure

The Gibbs free energy determines phase stability, $G = U - TS + PV$. At high temperatures the entropy term becomes important since polymorphs with large entropy S become more stable as T increases. Unfortunately, the relation between entropy and structure type is not a simple one, though a few generalities apply.

Open structures become *less* stable at high pressure because of the large molar volumes and *more* stable at high temperature because of their large entropy. In P–T diagrams the phase boundary separating polymorphs usually has a positive slope, as in the $CaCO_3$ diagram in Fig. 20.

The dependence of entropy on volume is given by the thermodynamic relation $(\partial S/\partial V)_T = -(\partial V/\partial T)_P/(\partial V/\partial P)_T$ which is equivalent to the volume thermal expansion divided by the compressibility. Since V usually increases with increasing T, and decreases with increasing P, the entropy S usually *increases* with volume. Thus open structures with larger volumes and lower densities, have larger entropies than dense structures. In the example in Fig. 20, the density of calcite, the high-temperature phase, is 2.71, somewhat less than that of aragonite, 2.93 (Jamieson, 1957).

In relating entropy to structure it is important to realize that there are two contributions to the entropy, thermal and configurational. *Thermal entropy*

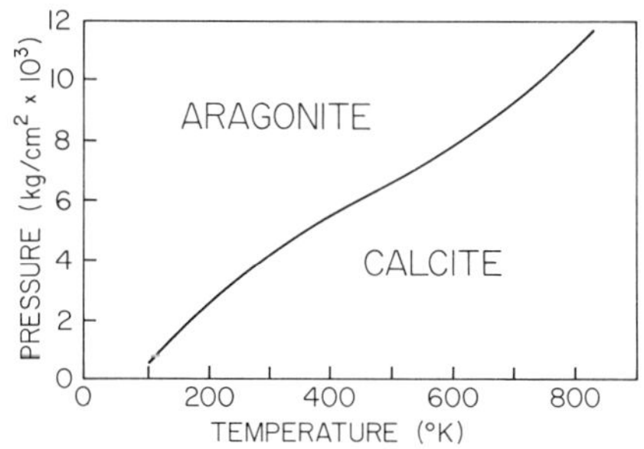

Fig. 20. Pressure–temperature diagram for calcium carbonate showing the stability regions of calcite and aragonite. Calcite is the more stable polymorph at low pressures and high temperatures. Its structure is less dense and more symmetric than that of aragonite (Jamieson, 1957).

is proportional to the number of different ways in which the vibrational energy of the crystal can be distributed over vibrational modes, whereas *configurational entropy* is determined by the number of different ways in which atoms can be distributed over the available sites. More equivalent sites in a structures leads to greater mode degeneracy, and therefore *high-symmetry* crystals tend to be stabilized at high temperatures. This is illustrated by the aragonite (orthorhombic) to calcite (rhombohedral) transition, and by the several polymorphs of SiO_2 (Table IV). In summary, high-temperature forms are usually open, high-symmetry structures in which the atoms have plenty of vibration room and many different ways to disorder.

The relation between configurational entropy and structure can be illustrated with ice. The structure of ice resembles that of tridymite with oxygens forming hydrogen bonds to four nearby oxygens arranged tetrahedrally. Hydrogen positions were uncertain until neutron-diffraction experiments substantiated Pauling's half-hydrogen mode (Fig. 21). This is a partially disordered structure in which hydrogens have equal probability of occupying two positions along O—O bonds. The O—O distance is 2.76 Å and the preferred hydrogen sites lie at 0.96 Å from the oxygens. The disorder is not complete because one and only one H position on each bond is filled. It is highly improbable to have both sites empty or both sites occupied, so that oxygen has two close hydrogen neighbors and two distant hydrogen neighbors.

The configurational entropy for ice can be estimated using the relation $S = k \ln W$. Here k is Boltzmann's constant and W is the number of equivalent ways of arranging the molecules. N molecules of H_2O contain N oxygen atoms and 2N hydrogens. The oxygen positions are completely specified so that $W_0 = 1$. There are many ways of arranging the hydrogens because of disorder. Each of the 2N hydrogens has two possible positions, O–H– – –O or O– – –H–O, giving $(2)^{2N}$ arrangements. Many of these arrangements are

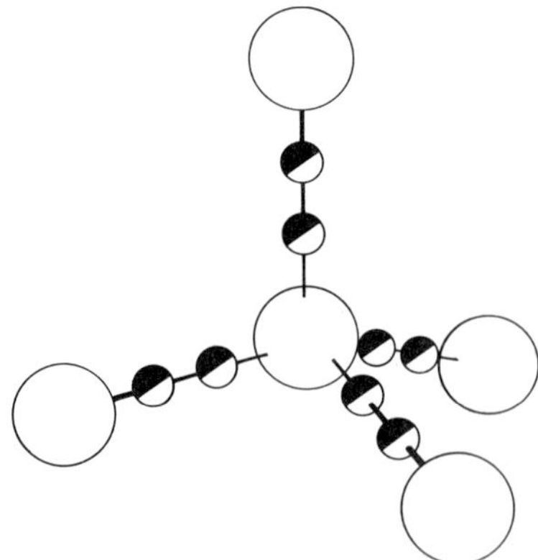

Fig. 21. A portion of the ice I structure, showing oxygen positions (large circles) and half-hydrogen sites. Configurational entropy can be calculated from the Boltzmann relation by enumerating the possible hydrogen configurations.

impossible, however, because they do not allocate two hydrogens to each oxygen. Consider the central oxygen in Fig. 21, and count the arrangements for the four neighboring hydrogens. There is one arrangement in which all four H are close to the central O ion, four arrangements with three close H, six with two, four with one, and one with none, a total of sixteen different arrangements. Only the six arrangements with two neighbors are possible, and since the same situation arises at each of the N oxygens, the number of hydrogen arrangements is $W_H = (\frac{6}{16})^N (2)^{2N} = (\frac{3}{2})^N$. Experimental measurements have verified this result to within 1%.

C. Morphotropic Phase Boundaries

Lead zirconate–titanate (PZT) and other ferroelectric perovskites are widely used in sonar and other transducer applications. Poling is accomplished by cooling a ferroelectric through the Curie temperature under an applied electric field, thereby reorienting the spontaneous polarization vectors of individual grains by domain wall motion.

Several of the most useful perovskite systems show morphotrophic phase boundaries near which the ceramic can be readily poled (Fig. 22), giving large

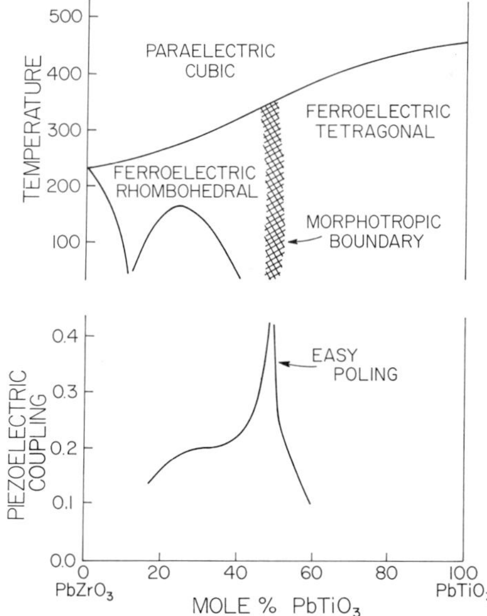

Fig. 22. Ceramics based on the $PbZrO_3$–$PbTiO_3$ binary system are widely used as transducers, capacitors, and in electrooptic applications. Poling is especially easy near the morphotropic phase boundary between rhombohedral and tetragonal ferroelectric phases (Jaffe *et al.*, 1971).

electromechanical coupling factors. On a binary phase diagram, a morphotropic phase boundary appears as a nearly vertical line (or pair of lines) separating two ferroelectric phases (Jaffe *et al.*, 1971). The chemical composition of the boundary remains nearly constant over a wide temperature range up to the Curie temperature of the ferroelectric. Ceramics poled near such boundaries show enhanced electromechanical coupling factors because of the proximity in free energy of an alternate ferroelectric structure.

The best known example is the morphotropic boundary in the $PbZrO_3$–$PbTiO_3$ system where large coupling factors of 0.7 have been observed. At 46 mole % Ti the structure changes from rhombohedral ferroelectric to tetragonal ferroelectric. Poling is easily accomplished at a ferroelectric–ferroelectric phase boundary because of the large number of directions in which individual grains can polarize. Near the morphotropic boundary in the PZT system (Fig. 22) there are fourteen possible polarization directions: six $\langle 001 \rangle$ axes for the tetragonal ferroelectric and eight $\langle 111 \rangle$ axes for the rhombohedral phase. Under these circumstances mechanical and electric boundary conditions across grain boundaries can be satisfied very precisely, minimizing strain and polarization mismatch. Morphotropic boundaries also appear in the $PbTiO_3$–$PbSnO_3$, $PbTiO_3$–$PbHfO_3$, and $PbTiO_3$–$BiFeO_3$ systems.

It is interesting to inquire into the origin of morphotropic phase boundaries since temperature-independent boundaries are not common in ferroelectric systems. The most important diagrams for ferroelectrics are those based on $PbTiO_3$ and $BaTiO_3$. Perovskites can be grouped into those with active octahedral ions (octa-active), those with active cubooctahedral ions (cubo-active), those with active ions in both sites, and those without active ions. By "active" ions, we mean those which promote ferroelectric distortions: Pb^{2+} and Bi^{3+} in the cubo-octahedral site, Ti^{4+} and Nb^{5+} in the octahedral site.

Morphotropic phase boundaries occur between $PbTiO_3$ in which both sites are active and several of the cubo-active perovskites. They do not occur in the systems based on $BaTiO_3$, an octa-active perovskite with three ferroelectric phases: tetragonal, orthorhombic, and rhombohedral. In diagrams involving barium titanate, the orthorhombic phase invariably occurs between the rhombohedral and tetragonal phases. The $BaTiO_3$–$BaHfO_3$ system is a good example (Payne and Tennery, 1965). Such boundaries are not vertical, though they sometimes show some interesting "pinch effects" in which the cubic, tetragonal, orthorhombic, and rhombohedral phases appear to join together at a point (Fig. 23). A survey of more than 50 systems shows that cubic–orthorhombic and tetragonal–rhombohedral transitions never occur with temperature. Only cubic–tetragonal, cubic–rhombohedral, tetragonal–orthorhombic, and orthorhombic–rhombohedral transformations are ob-

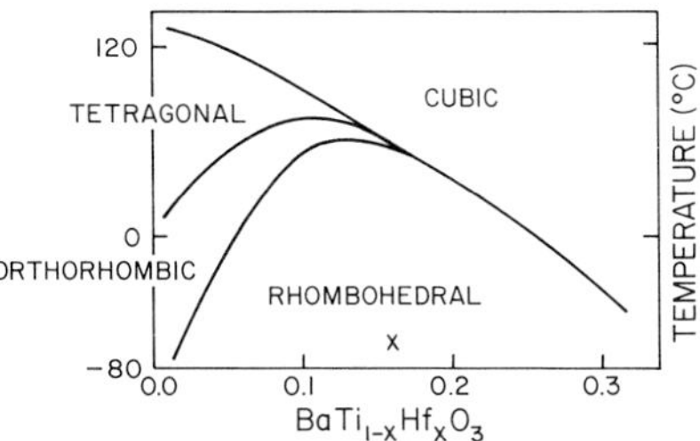

Fig. 23. A portion of the $BaTiO_3$–$BaHfO_3$ phase diagram showing the cubic paraelectric phase together with the tetragonal, orthorhombic, and rhombohedral ferroelectric phases. Note that the orthorhombic phase always lies between the tetragonal and rhombohedral phases. Such behavior is common in $BaTiO_3$ based systems (Payne and Tennery, 1965).

served between the simple ferroelectric phases. It appears that the symmetry hierarchy is always obeyed.

The structural reason for the morphotropic boundary in *PZT* and similar systems, may be related to the instability of the orthorhombic phase. Structure analyses of ferroelectric $PbTiO_3$, $PbZr_{0.6}Ti_{0.4}O_3$, and $BiFeO_3$ show that the lone-pair ions Pb^{2+} and Bi^{3+} undergo large polar displacements, causing the coordination to change from cubo-octahedral to pyramidal. In tetragonal $PbTiO_3$, each Pb^{2+} ion has four oxygen neighbors at 2.53 Å, four at 2.80 Å, and four at 3.20 Å. The four close neighbors are all on one side forming a tetragonal pyramid. This type of coordination is also found in PbO and other lead compounds. Trigonal pyramids rather than tetragonal pyramids are found in the rhombohedral perovskites. As the Pb^{2+} ions move along [111] they move closer to three oxygens. Trigonal pyramids are also common for lone-pair ions. Ammonia is a good example.

Polarization takes place along [110] in the orthorhombic ferroelectric which creates problems regarding the cubo-octahedral cation. The cubo-octahedral cation displaces directly toward an oxygen neighbor which leads to one very short Pb–O distance, a most uncomfortable configuration for lead. Therefore the orthorhombic ferroelectric phase is not found among the Pb-containing perovskites, the same systems in which morphotropic boundaries occur.

D. Phase Transformations under Pressure

The Gibbs free energy ($G = U - TS + PV$) governs the stability of solids, with the stable phase possessing the minimum free energy. At high pressures, the third term becomes very important, which means that structures with small volume or high density become more stable.

As examples, consider compounds of chemical formula RX. There are three important structures in this group: sphalerite (ZnS), rock salt (NaCl), and cesium chloride (CsCl), with coordination numbers 4, 6, and 8, respectively. At low temperature and low pressure the free energy of these phases depends primarily on the internal energy *U* which, for many inorganic solids,

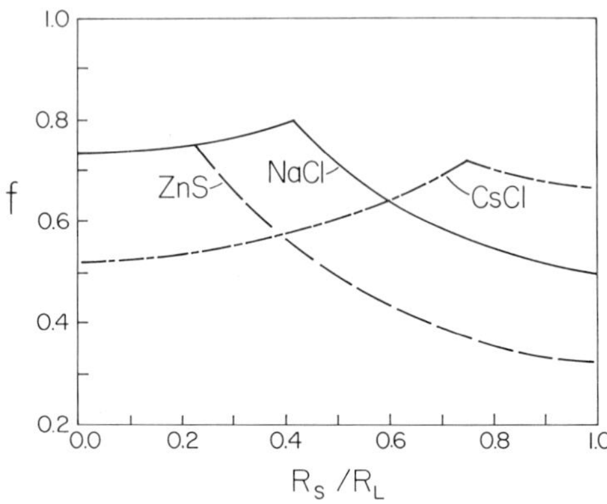

Fig. 24. Packing fractions for the zincblende, rock salt, and cesium chloride structures for various radius ratios. Here f is the fraction of space filled by large and small spheres of radius R_L and R_S. CsCl has the densest packing for cations and anions of equal size while ZnS and NaCl share honors when one ion is much larger than the other. The most efficient packing is achieved for the rock salt structure in the intermediate range accounting for the great stability of this structure.

is largely electrostatic energy. The variation of internal energy with radius ratio has been discussed previously in regard to Pauling's first rule. The ZnS structures have the lowest lattice energy for radius ratios around 0.2–0.3, rock salt structures predominate over the important middle range 0.4–0.7, and CsCl is slightly favored at larger ratios.

Phase transitions often occur under pressure, and among the RX compounds the transformation is usually to a phase with higher coordination number. Figure 24 shows the reason why. The volume V per mole appearing in the free energy function is inversely proportional to the packing fraction f, the ratio of the volume occupied by spheres to the total volume. Denser packing leads to smaller volumes, which in turn lead to lower free energies and greater stability. The packing fraction f depends on radius ratio, as shown in Fig. 24. It is important to realize that structures based on close packing do not always have the highest density. The large anions in sphalerite and rock salt are cubic close packed but when the radius ratio exceeds 0.6, the CsCl structure has a greater density. For smaller radius ratios, rock salt has the greatest packing fraction for ratios between 0.2 and 0.6, while ZnS is equally dense for radius ratios smaller than 0.2.

High pressure transformations have been reported in a number of RX compounds (Klement and Jayaraman, 1967). Most of the III–V and II–VI compounds with sphalerite or würtzite structures at low pressures, transform to rock salt structures in the 10–100 kbar range. AgI, InP, InAs, CdS, CdSe, and CdTe convert to the rock salt structure while AlSb, GaSb, and InSb transform to the white tin structure, a distorted NaCl arrangement. Near

10 kbar, the mercury salts HgTe and HgSe transform from sphalerite to cinnabar, another rock salt variant with higher density than four-coordinated compounds.

At room pressure and temperature, all the alkali halides except CsCl, CsBr, and CsI possess the rock salt structure. Under pressure, CsF and most of the sodium, rubidium, and potassium salts transform to the CsCl structure. The lithium salts, with smaller radius ratios, retain the rock salt structure to the experimental limit (40 kbar). No transformations are observed for CsCl, CsBr, and CsI since there is no RX structure with higher coordination numbers.

Similar phenomena occur in the derivative structures based on the ZnS, NaCl, and CsCl structure types. $LiGaO_2$ has four-coordinated würtzite superstructure, and under pressure transforms to a trigonal distortion of rock salt with six coordination (Marezio and Remeika, 1965).

The reason why the coordination numbers increase with pressure is related to the compressibility of the ions. In most ionic structures anions are larger than cations. Anions are also more compressible because of the low electron density in the outer part of the ion. As a result the radius ratio of cation to anion increases as pressure increases, and the coordination number increases in accordance with Pauling's rules.

Pressure sometimes alters the electron configuration of transition-metal ions, causing a transformation from high-spin state to low-spin state. The ionic radius of the low-spin configuration of Fe^{2+} is about 0.16 Å smaller than that of the high-spin configuration. Hence there is a tendency for divalent iron compounds to transform to low-spin states under pressure.

Most elements exhibit pressure-induced polymorphic transformations. In accordance with the PV term in the free energy function, those of low density convert to denser phases under pressure (Fig. 25). Among the elements, the densest and most stable structures belong to the transition-metal families.

Pressure tends to smooth out differences in crystal structure, density, and compressibility in the elements. At high pressures alkali metals such as cesium have properties similar to those of typical transition metals. The ratio of the bulk modulus of tungsten to that of cesium is reduced from 200:1 at 1 atm to 3:1 at 120 kbar. With increasing pressure many insulators are transformed into metals and are often superconducting at low temperatures. None of alkali or alkaline earth metals are superconductors under ordinary conditions, but both Cs and Ba are high-pressure superconductors. In all cases the appearance of superconductivity is accompanied by a change in crystal structure.

Ice VII, the stable form of H_2O at 25°C and 25 kbar, is an interesting example of a high-pressure phase. Ordinary ice I (Fig. 21) is a very open

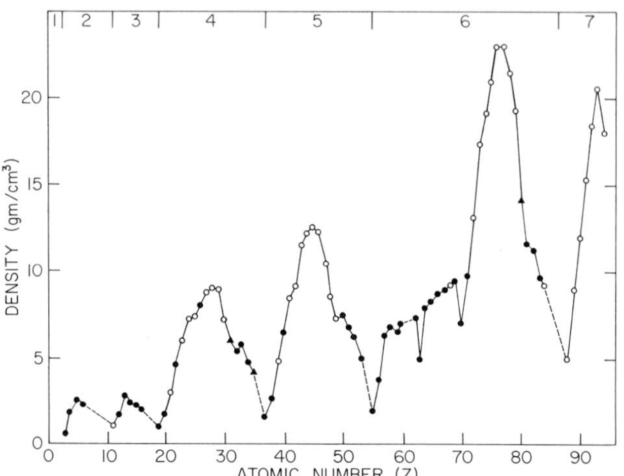

Fig. 25. Density plotted against atomic number for solid and liquid elements. Elements represented by solid circles show pressure-induced phase changes while the others do not. Dense elements seldom transform under pressure (Liu, 1973).

structure which is why it floats on water. Ice VII is more stable at high pressures because of its greater density, 1.57. The structure of ice VII consists of *two* interpenetrating frameworks, each of which is similar to the framework found in ice I. This not only greatly increases the density but offers some interesting possibilities regarding electronic transport properties, providing the two sublattices could be independently doped. There are relatively few crystals made up of interpenetrating sublattices unbonded to one another. Cuprite, Cu_2O, is another well-known inorganic example.

Pressure transformations of silicates are of geophysical interest. The dominant phases at depths around 500 km are believed to be denser high-pressure polymorphs or disproportionation products of the silicate minerals found in the outer crust. Feldspars constitute about 60% of the outer crust, and potash feldspar ($KAlSi_3O_8$) illustrates how temperature and pressure effect structure. The low temperature phases, microcline and orthoclase, convert to sanidine on heating with Al and Si randomized in the tetrahedra. At high pressure (120 kbar), potash feldspar converts to the hollandite structure in which Al and Si occupy octahedra rather than tetrahedra. As a result, the density increases from 2.55 to 2.84 g/cm^3 (Ringwood, 1967). Along with SiO_2 (stishovite) and SiP_2O_7, this is one of the few examples of silicon in sixfold coordination with oxygen.

Our knowledge of minerals is largely confined to the varieties encountered here on earth, but interesting crystalline solids are found elsewhere in the universe as well. Moon rocks resemble basalts with some compositional differences caused by vaporization losses. Jupiter and other large planets have atmospheres of hydrogen, helium, methane, and ammonia, which together with large gravitational fields, suggest the existence of some very unusual high-pressure minerals.

Even more unusual minerals may exist in solid stars. A typical white dwarf is about the size of the earth, but its mass is comparable to that of the sun. Densities are about 10^6–10^8 g/cm^3. In superdense matter, electrons are distributed nearly uniformly in space: a homogeneous sea of electrons with nuclei embedded in it. The nuclei are strongly repelled by one another and are therefore held to rigidly fixed positions. Theoretical calculations predict the stability of simple structure types like rock salt in superdense matter (Dyson, 1971).

V. METASTABLE PHASES AND KINETICS

Many compounds have not been discovered because they require unusual starting materials or unusual preparation techniques. Most have eluded detection because of the slow rate of formation from the usual starting materials: the crystal structure of a solid is often controlled by the molecular structure of the starting material. Phosphorus has white, red, and black polymorphs. The black and red forms consist of graphitelike layers, while white phosphorus is a molecular solid made up of P_4 groups, similar to those found in gaseous phosphorus. Condensing P_4 gas invariably gives white phosphorus, even though it is less stable than the red and black varieties. Catalysts, heat, or irradiation are needed to synthesize the red and black forms of phosphorus.

A great variety of unusual molecular species are found in high-temperature gases, species which sometimes condense to form unusual solids (Brewer, 1958). Amorphous silicon monoxide is produced from chilled SiO gas and persists indefinitely at room temperature. Silicon monoxide coatings are used to protect optical lenses. High-temperature gases will undoubtedly prove important at starting materials in the synthesis of other stable and metastable solids. It is generally expected that in a vapor the cation coordination will be a minimum and hence solids derived directly therefrom will have a tendency to reproduce this minimum coordination. An example is fibrous silica (W–SiO_2) which has very low density.

High pressure is useful in other systems. Several dense polymorphs of SiO_2 with unusual properties have been prepared in this manner. The keatite form is metastable at ordinary pressures and yet it can be boiled in hydrofluoric acid without attack, unlike the other forms of silica.

A phase transformation in hydrogen was recently reported at pressures of 2.8 Mbar with a density increase from 1.08 to 1.3 g/cm^3 at the transition. The high-pressure phase may be the long-sought metallic phase of hydrogen. If metallic hydrogen turns out to be metastable under ambient conditions, it may be useful but possibly dangerous. Under ambient conditions the stored energy in metallic hydrogen would be very high, approximately 40 times greater than TNT. It has been speculated that metallic hydrogen is a room-temperature superconductor, with possible applications in power transmission, though it seems doubtful if it could be produced in large volumes at low cost. Metastable solid deuterium could be used in controlled fusion, in nuclear weapons, and as a weight-saving fuel for spacecraft. Some astrophysicists have predicted that metallic hydrogen will be found as a mineral in the interior of Jupiter and Saturn.

Disproportionation makes it difficult to produce certain compounds by conventional means. Wüstite (FeO) is stable at high temperatures but disproportionates to Fe and Fe_3O_4 on cooling. Many unusual oxides are stable only at high temperatures, including the metalliclike phases Fe_3Ti_3O and Mn_2Mo_4O. Phases stable only at elevated temperatures sometimes play an unexpected role in high-temperature reactions. In bonding metals to ceramics, for instance, intermediate phases aid in sealing the join between oxide and metal.

Compounds which are stable at low temperature but disproportionate on heating are often difficult to prepare because of slow reaction rates at low temperatures. The rate of formation is fastest just below the disproportionation temperature. Furnaces with large temperature gradients have been used to study such systems. Under these conditions the sample is subjected to a wide range of temperatures, making it easy to locate the temperature at which the low-temperature phase forms at a reasonable rate.

A. Topotaxy

Topotaxy is the transformation of one crystal into another of different chemical composition without any major change in axial directions. Shape and orientation of the product crystal bears a recognizable and reproducible relation to that of the initial crystal. The definition of topotaxy is sometimes broadened to include reactions in which the composition does not change: polymorphic transitions and exsolution phenomena may also be topotactic. Epitaxy—the crystallization of one substance in regular order on another—is closely related to topotaxy. Epitaxy requires a two-dimensional accord between film and substrate, whereas the interconversion occurring during topotaxy requires three-dimensional resemblance. Despite this requirement, topotaxy appears to be widespread, and may be the commonest method of transformation in solid state reactions. Many mineral pseudomorphs result from topotactic reactions.

The practical significance of topotaxy is related to morphology. Suppose compound RX is to be prepared in a prescribed shape, perhaps as platelets, or needles, or as a reactive powder. One way is to react R and X, say crystals of R and a gas of X. The reaction may proceed by topotaxy with the crystalline product RX showing a resemblance to the original crystals R, both in shape and orientation. If this procedure does not give the desired product, other topotactic reactions may be attempted decomposing a double salt RMX_2 into crystalline RX and gaseous MX is sometimes successful. The production of acicular γ-Fe_2O_3 particles for magnetic tape illustrates the usefulness of topotactic reactions. In this case, the starting material consists of tiny needle-like crystals of FeOOH.

The way in which topotactic reactions proceed depends to a large extent on crystal chemistry (Glasser *et al.*, 1962). Size and electronegativity determine which ions move and which do not. In many oxides, the cations migrate while the larger oxygen ions remain fixed. Even strongly bonded cations such as Si^{4+}, Al^{3+}, Fe^{3+}, and Mg^{2+} migrate during the transformation, which means that SiO_4 groups are broken up. Only cations migrate during the exsolution process which occurs near 1300°C in alumina-rich spinel crystals. $MgAl_2O_4$ and Al_2O_3 form a solid solution at high temperatures but on cooling exsolve into corundum and spinel lamellae with the close-packed planes parallel. A somewhat different example of topotaxy is found in the Al_2O_3–TiO_2 system. Oriented rutile needles precipitated out of solid solution give rise to asterism in star rubies and star sapphires.

Anions may diffuse more when the cations are larger in size. In the redox reactions of lead oxides, O rather than Pb migrates. Topotactic reactions in which both anion and cation moves, or in which neither moves are also possible. Only protons and electrons move during the dehydration of manganese oxyhydroxides.

One of the major questions concerning the topotactic process is that of homogeneity. Do reactions proceed the same way in each unit cell, or do some regions develop differently than others? Two models describing the dehydration of magnesium hydroxide are shown in Fig. 26. In the homogeneous model, water molecules are evolved uniformly throughout the crystal and then migrate to the surface, leaving a contiguous MgO crystal behind. The inhomogeneous model develops into crystalline regions and pores. The evolution of water vapor is preceded by cation migration; magnesium migrates toward the crystalline regions and hydrogen toward the

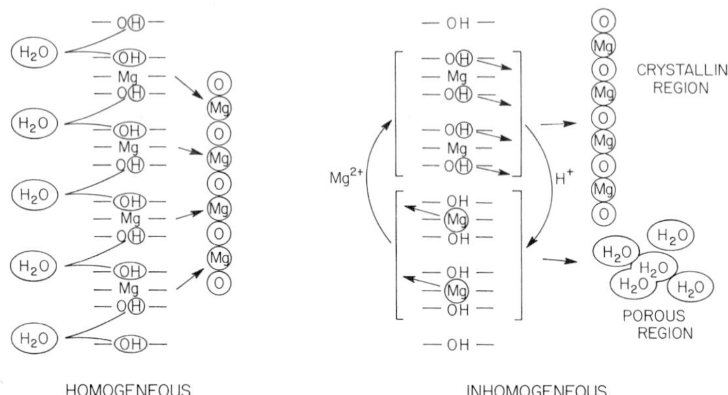

Fig. 26. Homogeneous and inhomogeneous mechanisms for the dehydration of brucite, $Mg(OH)_2$. Mg^{2+}, and $(OH)^-$ ions occur in layers as indicated by horizontal lines. In the homogenous model, water evolves uniformly without migration to pore spaces (Glasser *et al.* 1962).

pores. Both processes are topotactic but the inhomogeneous reaction leads to poorer crystallization in the final product. This is often the case in topotactic reactions. The dehydration of kaolinite to metakaolin, and that of gypsum to plaster of paris are examples. When the loss of water occurs there is no shrinkage in external dimensions, but only an increase in porosity and a deterioration in crystallinity.

The rosenhahnite–wollasonite transformation is an excellent example of a topotactic reaction (Jeffery and Lindley, 1973). A single crystal of rosenhahnite ($Ca_3Si_3O_9 \cdot H_2O$) transforms into a single crystal of wollastonite ($CaSiO_3$) on heating to 500°C, where a molecule of water is given off. There is little change in morphology and no visible cracks appear, although the crystals are less transparent following the transformation.

Crystal structure analyses provide an explanation of the topotactic relation. Wollastonite contains silicate chains similar to those in pyroxene minerals. Rosenhahnite contains linear groups of three silica tetrahedra, the two outer silicon atoms each sharing a common oxygen with the central silicon. The structure does not contain molecular water: rather, there are hydroxyl groups attached to the outer silicon atoms, terminating the short $Si_3O_8(OH)_2$ chains. During the dehydration process, the $Si_3O_8(OH)_2$ groups polymerize into infinite Si_3O_9 chains with each join producing a water molecule. Hydroxyl groups in neighboring segments are less than 3 Å apart and almost certainly coalesce at the transition, liberating a water molecule. Other small rearrangements provide channels through which the water molecules diffuse to the surface and evaporate. Orientational relations are maintained in the two structures with the length of the $Si_3O_8(OH)_2$ groups in rosenhahnite parallel to the silicate chains in wollastonite.

Similar processes may occur in cement chemistry which involves a number of hydrated calcium silicates. If two Si–OH groups on the surface of adjacent crystallites come in contact, they may form a chemical bond by the elimination of water, bridging the crystallites, and strengthening the cement.

A principal concern in low-temperature geochemistry is with equilibrium relations between minerals and the coexisting liquids and gases in or near the earth's crust. Observed mineral assemblages and their present environment are used to deduce the conditions under which the minerals or their precursors were formed, taking diagenetic and metamorphic processes into account. Field evidence is supplemented with laboratory data such as phase equilibrium diagrams and reaction rates. Some minerals can be relied upon to equilibrate quickly providing a guide to past conditions. A number of rate-determining processes are involved in the drive toward equilibrium in various geochemical systems. The splitting out of water from iron oxide hydrates and the persistence of certain metastable carbonate species are two examples.

Crystal chemistry sometimes provides an explanation of the rate-determining process. The hydrated sodium borate minerals borax ($Na_2B_4O_7 \cdot 10H_2O$), tincalconite ($Na_2B_4O_7 \cdot 5H_2O$), and kernite ($Na_2B_4O_7 \cdot 4H_2O$) are an interesting case in point (Christ, 1972). Borax and tincalconite are rapidly and reversibly interconverted by changing the water vapor pressure. But dehydration of tincalconite results in an amorphous dihydrate instead of kernite. When kernite crystals are exposed to high humidities, a crust of borax forms on the surface which subsequently reverts to tincalconite. Crystal structure analyses of the three minerals explains the reluctant conversion behavior of kernite. Borax contains isolated polyanions of composition $[B_4O_5(OH)_4]^{2-}$ so that the structural formula is $Na_2(B_4O_5(OH)_4) \cdot 8H_2O$. Tincalconite contains identical borate polyanions, and differs from borax only in the number of waters of crystallization, $Na_2(B_4O_5(OH)_4) \cdot 3H_2O$. The two minerals transform to one another readily because of the small activation energy involved in changing the number of water molecules. Only relatively weak chemical bonds are broken during the borax–tincalconite transformation but kernite has a very different borate complex. The structural formula is $Na_2(B_4O_6(OH)_2) \cdot 3H_2O$ with infinite borate chains parallel to the *b* axis. Kernite is hydrated only with great difficulty because the conversion to borax or tincalconite requires the breaking of B—O bonds.

Knowledge of the structural formulas of kernite, tincalonite, and borax also explains the peculiar nature of the temperature-solubility curve for the $Na_2B_4O_7$–H_2O system. Plotting concentration against temperature ordinarily shows two breaks when three hydrates are present. In this case there is only one break because tincalconite and kernite both contain three waters of hydration.

B. Stranded Phases and Stuffed Derivatives

Sluggish phase transitions take time and are easily bypassed, producing metastable phases, sometimes referred to as stranded phases. Stranded phases can be produced in several ways. Hexagonal $BaTiO_3$ is stable only above 1460°C but is easily quenched to room temperatures where it remains indefinitely without converting to the stable ferroelectric polymorph. Complete reconstruction is required to effect the transition, creating a large energy barrier. This is an example of *thermal stranding* by rapid temperature change. *Barically stranded* phases such as coesite, a high-pressure polymorph of SiO_2, persist to room pressure when the pressure is lowered rapidly. A third type—represented by dehydrated zeolites—is the *compositionally stranded* phase.

Differences in composition sometimes stabilize unstable phases, which is simply another way of saying that the stable phase field is a function of composition as well as temperature and pressure. Thus the transition temperature between polymorphs is often sensitive to impurities.

Silica-based structures illustrate the principle nicely. The high-temperature forms of SiO_2 (glass, cristobalite, tridymite, β-quartz) are stabilized by the presence of aluminum and alkali- or alkaline-earth ions. Compounds of formula $MAlSiO_4$ are stuffed silica derivatives in which half the silicons are replaced by aluminum, and alkali ions M^+ occupy voids (Perrotta and Smith, 1963). Void spaces are largest in high tridymite, next largest in high cristobalite, and smallest in β-quartz. Tridymite contains both large and small cavities whereas cristobalite contains many medium-sized cavities, giving it a similar density. Kalsilite ($KAlSiO_4$) is a tridymite derivative because of the space requirements for K^+ (Fig. 27). Smaller alkali ions prefer other SiO_2 structures. Carnegieite (α-$NaAlSiO_4$) is a cristobalite derivative

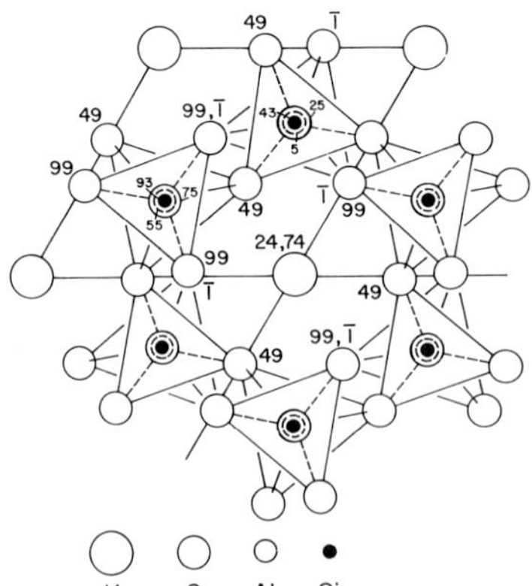

Fig. 27. Crystal structure of kalsilite ($KAlSiO_4$), a stuffed-derivative of tridymite. Tetrahedra are occupied alternately by aluminum and silicon, with potassium atoms in cavities within the aluminosilicate framework (Perrota and Smith, 1963).

while the structure of high $LiAlSiO_4$ is based on quartz. Ionic radii decrease in going from K^+ to Na^+ to Li^+.

Even small amounts of alkali ions help to stabilize high-temperature forms. High-temperature phases are usually open structures which transform to denser forms on cooling. Open cavities within the structure collapse during the process unless buttressed by impurities. Buttressing lowers the transition temperatures, stabilizing the more open forms. Mineral samples of tridymite and cristobalite are seldom as pure as quartz, indicating that their stability is at least in part due to buttressing.

Buttressing is not the only way in which impurities stabilize apparently unstable forms. The wurtzite form of zinc sulfide, normally unstable below 1020°C, can be precipitated from slightly acidic solutions. Alkaline solutions invariably precipitate sphalerite, the stable low-temperature polymorph. Similar peculiarities are found in other sulfide systems such as FeS_2 and HgS, but the reasons are not yet clear. Hydrogen or other impurities may be incorporated in the structures, or possibly the defect concentration may be controlled by pH. There is evidence that sphalerite and würtzite differ slightly in stoichiometry.

Yet another mechanism has been proposed to explain the existence of aragonite under ambient conditions. Strontium is often found in natural aragonite, but not in sufficient amounts to alter the calcite–aragonite boundary significantly. If the impurity is sufficiently insoluble, the initial $CaCO_3$ nucleus may be rich in $SrCO_3$, causing aragonite to form in preference to calcite. Epitaxial growth then proceeds on the aragonite seed.

VI. LIQUIDS AND NONCRYSTALLINE SOLIDS

The *Bernal model* of a liquid consists of a homogeneous, coherent, irregular assemblage of molecules. There are no holes large enough to admit another molecule. The irregularity results from a small concentration of atoms with a slight departure from their normal coordination. Interatomic angles and distances differ slightly from those in the crystal; but the liquid is free of major flaws. Both anions and cations are mobile, though they generally do not exchange places with ions of opposite charge. A Bernal liquid resembles the crystal from which it was formed, and is therefore often viscous and easy to supercool to a glassy state.

The *Frenkel model* assumes the liquid is permeated by a large number of ruptured surfaces, dynamic fissures continually opening and closing. At any given instant, the liquid contains a system of holes and cavities, making it a heterophase system. The Frenkel model of a liquid bears little resemblance to a crystal. Ideal fissured liquids are highly fluid and cannot be supercooled to form glasses.

The *Stewart model* pertains to molecular liquids showing some degree of orientational order. Liquid crystals provide visible evidence of this phenomenon.

Real liquids can be classified according to the Bernal, Frenkel, and Stewart models, depending on whether they are flawless, fissured, or orientable liquids. Vitreous silica resembles a flawless Bernal liquid, while molten NaCl is close to the fissured model. High polymers and vitreous B_2O_3 show orientation effects. Weyl and Marboe (1962) have discussed the importance of liquid structure in the constitution of glasses and other noncrystalline solids.

A. Melting

For all known substances, the entropy of the liquid state is greater than that of the solid at the melting point. Structural theories of melting attempt to describe the various ways in which the entropy of a crystal can be increased on melting (Ubbelohde, 1965). To a rough approximation the total entropy of fusion S_f, can be written as the sum of several contributions, each representing a different kind of disorder.

$$S_f = S_p + S_o + S_c + S_e.$$

Entropy due to positional disorder (S_p) is found in all materials. Orientational disorder (S_o) occurs in molecular crystals with nonspherical molecules, as well as inorganic materials with polyatomic ions. Configurational disorder (S_c) arises in certain molecular solids made up of flexible molecules, molecules with several different atomic arrangements, each corresponding to a minimum in potential energy. The three principal contributions to entropy refer to the position, orientation, and shape of individual molecular units. Changes in electronic configuration give rise to an entropy term S_e.

Positional randomization is usually pictured in terms of a quasi-crystalline model in which the crystal structure is expanded and disordered on melting. The most common types of defects are vacancies, interstitials, and cooperative defects involving coupled displacements of several neighboring atoms. The latter type has a smaller energy of formation, but also contributes less to the entropy change. The melting of inert gas crystals provide the best examples of positional melting. Measured values of the entropy of fusion for Ne, Ar, Kr, and Xe lie between 3.3 and 3.4 cal/mole °C, with a 15% increase in volume on melting. For most metallic elements S_f is about 1.5–2.5, and for simple ionic crystals between four and six.

Sometimes positional randomization occurs before the crystal melts. In several silver salts, the cations behave like a liquid while the anions remain fixed to their crystallographic positions. These materials are extremely good ionic conductors.

Orientational randomization occurs on heating crystals containing small nonspherical molecular units. The heats of fusion of $TiCl_4$, $SnCl_4$, and $SiCl_4$ are all about 9 cal/mole °C, considerably larger than the positional entropy change alone. Similar values are found for CO_2, Cl_2, SO_2, and N_2O, but not all solids of this type show significant orientational randomization at the melting point. For some it occurs during solid-state transitions at lower temperatures, the so-called lambda transitions. This is true for CH_4, CCl_4, CO, N_2, and HCl where the entropy of fusion is only 2.5–3.0, typical of positional entropy.

The normal paraffins are a class of flexible molecules with a sizable configurational contribution to the entropy of fusion. Configurational isomerism in butane is illustrated in Fig. 28. The number of configurational isomers increases rapidly with the chain length. When the potential energies differ by less than a few kT_f, both configurations are found in the melt, whereas the crystal generally contains only the lower energy stretched form. On melting, the crumpled isomers are created by rotation about the C—C bonds, producing a mixture of crumpled and stretched molecules. This increases the diversity of species and also raises the entropy since there is more disorder. The randomization process is limited by the potential energy term which favors the stretched forms since intermolecular attraction is maximized when the stretched molecules lie parallel to one another.

Aluminum trichloride shows a dramatic change in properties on melting which has been attributed to a change in bonding. The volume expands by 83% and the entropy of fusion is nearly an order of magnitude larger than those of other halides. For solid $AlCl_3$, the electrical conductivity rises steeply on approaching the melting point, and then drops to near zero on melting. This strange behavior has been interpreted as a change in bonding. Crystalline aluminum chloride is ionic but the melt contains Al_2Cl_6 molecules with intramolecular covalent forces and intermolecular van der Waals bonds. In this type of melting, there is an electronic contribution to the entropy of fusion.

CH_2—CH_2 / CH_3 CH_3 (a)

CH_3 / CH_2—CH_2 / CH_3 (b)

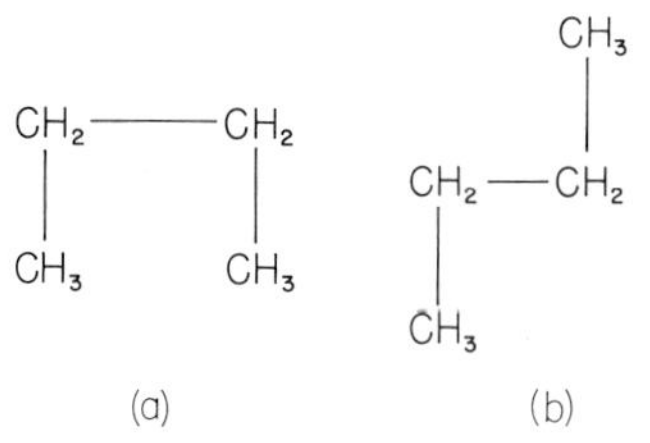

Fig. 28. The (a) syn- and (b) anticonfigurations of butane. Both configurational isomers are found in the melt.

Bonding changes also occur in semiconductors where the conductivity generally increases on melting, producing a metallic molten state. In germanium and silicon the four-coordinated diamond structure collapses to give a coordination number of about 8 in the melt. A different type of bonding change occurs in ionic crystals with polyatomic ions. Some groups such as NH_4^+ or SO_4^{2-} persist in the melt, but others dissociate. The pyrophosphate $P_2O_7^{4-}$ anion breaks up into PO_4^{3-} and PO_3^- ions, while the AlF_6^{3-} group in cryolite (Na_3AlF_6) converts to AlF_4^- tetrahedra and two F^- ions on melting.

Melting almost always involves an increase in volume. The number of crystalline solids which shrink on melting is surprisingly small. Some of the best known exceptions include the semimetals antimony, bismuth, and gallium; the semiconductors Si and Ge; and ice, which shows an 8% change in volume on melting, the largest decrease of any common material. All crystals displaying this unusual behavior have small coordination numbers and relatively open structures which partially collapse on melting.

B. Melting Points and Boiling Points

Boiling points correlate closely with bond strength because when bonds break, molecules evaporate. The relationship between bond strength and melting point is less obvious but some interesting generalizations apply. In structures containing both weak and strong bonds it is sometimes the weak bonds which determine melting while the strong bonds govern vaporization.

Substances with extended covalent bonding generally have high melting points. Crystalline diamond, graphite, silicon carbide, and boron nitride are stable to temperatures exceeding 3000°C. Fusion of such materials necessarily involves extensive disruption of strong covalent bonds, and high temperatures are required to achieve this. In molecular compounds, however, where only the molecular units are covalently bonded the melting points are much lower. Titanium tetrachloride consists of tetrahedral $TiCl_4$ molecules weakly held together by van der Waals forces. $TiCl_4$ melts at -23°C and boils at 137°C. The weak intermolecular forces constitute the points of structural weakness in such a crystal, and fusion occurs at relatively low temperatures. Inert gases with only van der Waals bonding exhibit some of the lowest melting points. The melting points of ionic crystals are generally higher than molecular solids but lower than solids with extended covalent bonding. Sodium chloride melts at 801°C and boils at 1465°C.

Among simple ionic crystals, the melting point is a measure of the strength of bonding. Comparing the melting points of RX compounds, for example, divalent substances have higher melting points than monovalent materials.

Melting points in the alkali halide family range from a low of 450°C for LiI to a maximum of 988° for NaF, far less than the isostructural alkaline-earth oxides. The melting points of MgO, CaO, SrO, and BaO are 2800, 2580, 2430, and 1923°C, respectively. The oxides also have high boiling points in the 2000–4000° range, whereas the halides all boil at less than 1700°C. Divalent ions have twice the charge and four times the Coulomb force of a monovalent ion, giving rise to larger heats of fusion and higher melting and boiling points.

Many of the best high-temperature materials are from the light elements in the first row of the periodic system. Graphite, boron nitride, beryllium oxide, and boron carbide have melting points in the 2300–3700°C range, and boiling points above 3500°C. Since nearly all the electrons in these compounds are involved in bonding, they have higher melting points than similar materials of higher atomic number. For the same reason, oxides have higher melting points than do sulfides, selenides, and tellurides. The heavier elements contain additional electrons arranged in closed shells which contribute little to the attractive forces between atoms and much to the repulsive forces. When the closed shells of neighboring atoms overlap, there is Coulomb repulsion, increasing the interatomic distance and lowering the melting point.

Oxides, borides, carbides, and other ceramic materials can generally be used at higher temperatures than metals. However, metals are generally stronger than ceramics, so that ceramic-coated metals are a good compromise. The coatings not only improve the high-temperature characteristics, providing erosion, corrosion, and oxidation protection, but are also used to modify the electrical and optical properties of the surface.

Some of the coating materials employed in the aerospace industry are listed in Table V. Coatings are generally applied by flame-spraying molten or partially molten particles onto a cold metal surface. Impacting droplets flatten, interlock, and overlap to provide a dense coherent coating of the desired thickness. Surface materials are selected to have suitable characteristics for absorbing, reflecting, and emitting radiation in the required amounts. Coatings must also be sufficiently durable to withstand harsh environments.

There are several relations between melting and structure, and between melting and other physical properties. Two of these are with thermal expansion and elasticity. Perhaps the simplest concept of melting is Lindemann's observation that melting occurs when the amplitude of vibration exceeds a certain critical fraction of the interatomic distance.

If x_m is the vibration amplitude at melting and we assume simple harmonic motion around the equilibrium site with a spring constant k, the energy is

$$\tfrac{1}{2}kx_m^2 = k_B T_m$$

TABLE V

APPROXIMATE MELTING POINTS OF FLAME-SPRAYED CERAMICS USED IN AEROSPACE APPLICATIONS[a]

Oxides		Carbides		Borides and Nitrides	
ThO_2	3300 °C	HfC	3900 °C	HfB_2	3250 °C
UO_2	2880	TaC	3900	TaB_2	3100
MgO	2800	ZrC	3500	ZrB_2	3060
HfO_2	2780	NbC	3500	NbB_2	3000
CeO_2	2730	TiC	3250	TiB_2	2980
ZrO_2	2600	VC	2800	CrB_2	2760
Y_2O_3	2400	Al_4C_3	2800	SiB_6	1950
BeO	2350	WC	2770	BN	3000
Cr_2O_3	2260	MoC	2700	TiN	2940
$MgAl_2O_4$	2130	SiC	2700	AlN	2200
Al_2O_3	2070	B_4C_3	2450	Si_3N_4	1900

[a] Love *et al.*, 1965.

where k_B is Boltzmann's constant and T_m is the melting point. The force constant k is related to elastic stiffness c and the interatomic distance l by $k = cl$ so that

$$T_m = clx_m^2/2k_B = cl^3\beta^2/2k_B$$

where $\beta = x_m/l$, the relative vibration amplitude at melting. Substituting measured values from T_m, c, and l for a wide range of materials—including metals as well as ionic crystals—it is found that β is surprisingly constant, about $\frac{1}{7}$. Thus many materials melt when the thermal vibration amplitude is about 15% the interatomic distance.

Because of the close relation with vibration amplitude, it is perhaps to be expected that melting and thermal expansion are correlated. Materials with low melting points have large thermal expansion coefficients because all materials expand by about the same amount (3–10% in volume) before melting. Therefore αT_m is nearly constant. Fortunately there is enough scatter so that materials with different melting points can be bonded together—enamel and iron, for example.

Materials with anisotropic molecular units behave like tangled threads, increasing viscosity and the melting and boiling points (Slater, 1939). Methane, ethane, propane, butane, and the other chain hydrocarbons illustrate the effect nicely (Fig. 29). Note that the separation between boiling and melting increases with chain length. This is because in the long hydrocarbons many more strong bonds must be ruptured in boiling the materials. Melting points increase less rapidly because the van der Waals bonding between molecules does not change much.

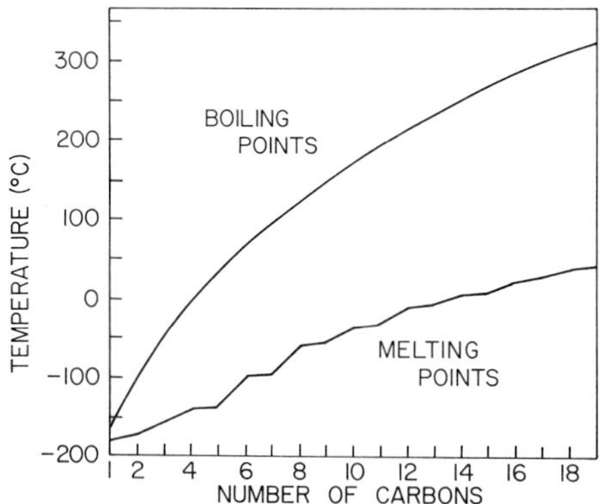

Fig. 29. Melting points and boiling points of long-chain hydrocarbons plotted as a function of backbone length. Melting points increase with length because of the tendency of chains to entangle. Boiling points increase even faster because of the large number of covalent bonds to be broken before vaporization is possible (Slater, 1939).

The importance of crystal chemistry in melting processes can be illustrated with Bowen's reaction series which described the crystallization sequence for igneous rocks. Molten magma is a complex solution of aluminosilicates under high pressure and high temperature. By weight it consists of about 47% O, 28% Si, 8% Al, 5% Fe, and 2–3% each of Ca, Na, K, and Mg. As it cools, the magma crystallizes and recrystallizes, beginning with olivine and anorthite, which occur together in mafic igneous rocks rich in Mg, Ca, and Fe, and ending with pegmatites. The crystallization sequence is as shown in the accompanying diagram.

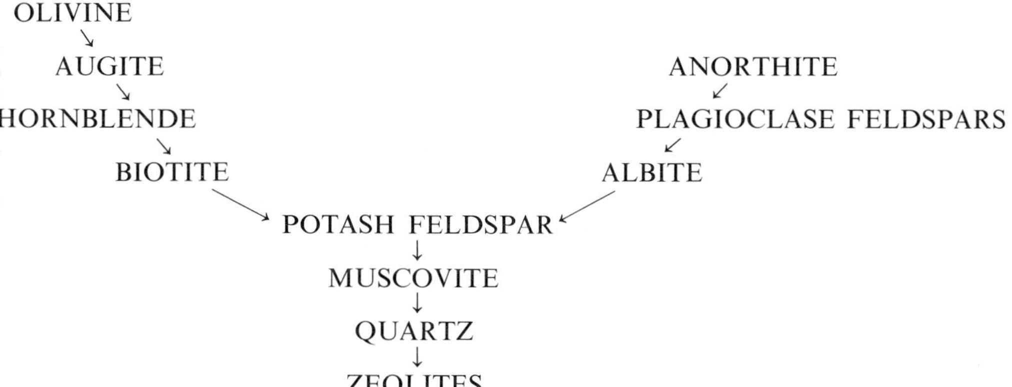

Two general trends may be noted from Bowen's series. In the left series the cooling progression is from orthosilicate (olivine) to single-chain silicate (augite) to double-chain silicate (hornblende) to sheet silicate (biotite) to framework silicate (feldspar, quartz, zeolite). The second trend is obvious from the right-hand progression, the crystallizing plagioclase feldspars become more silica rich as the magma cools.

To explain these trends it is helpful to visualize the effect of temperature on linked silicate tetrahedra. Under thermal agitation the degradation sequence is framework, sheet, multiple chain, single chain, pairs, and discrete tetrahedra. Except for the absence of paired tetrahedra this corresponds to the Bowen crystallization series. From a bonding viewpoint, melting is chiefly determined by the weakest link, which in silicates are the cations holding the silicate groups together. The number and strength of these bonds influence the melting point greatly. Using Pauling's rules, it is instructive to examine the bond strengths per oxygen to atoms *other* than silicon. In anorthite and olivine it is one and the value decreases toward zero down the reaction series. Thus the effect of aluminum substitution for silicon is to increase the disintegration temperature, by promoting bonding between the silicate groups.

C. Noncrystalline Solids

There is no sharp dividing line between crystalline and amorphous solids. Crystals possess long-range order, i.e., atomic arrangements with three-dimensional translational periodicity, while noncrystalline solids usually possess short-range order with recognizable atomic groupings in the first coordination sphere. About ten unit cells (~ 50 Å) have been suggested as an arbitrary boundary between the two. The distinction between the two is further blurred by the existence of materials with one- or two-dimensional order such as liquid crystals. Thixotropic materials, gels, plastics, clay minerals with extensive stacking faults, and crystals with extensive radiation damage also fall in this borderline region.

One approach to the study of noncrystalline solids is the link between structure and preparation technique: as a genetic classification, noncrystalline solids are always metastable because there is always a crystalline assemblage with lower free energy. Preparation methods can therefore be grouped according to the way in which free energy is added to produce the metastable state (Roy, 1973). Crystals can be amorphized by mechanical working (shear stresses are particularly effective), by radiation through radiation damage, or by chemical reaction as in dessicated gels. Usually, however, the excess energy is added thermally; beginning from a high-temperature state, perhaps vapor or melt, excess thermal energy is frozen into the solid state by rapidly lowering the temperature.

Contrary to many textbook statements almost all materials can be rendered noncrystalline. Rapid quenching rates of 10^6 deg/sec have been achieved by the "splat cooling" method in which molten droplets collide with a cold metallic surface and heat is rapidly conducted away. A number of amorphous metals, especially tellurium- and palladium-based alloys, have been prepared in this way. More examples will undoubtedly follow as superior experimental techniques are developed, leading to a new class of glassy solids. Currently there is a good deal of interest in amorphous Gd–Co films which show promise as magnetic bubble domain devices.

Among metals it has been observed that the alloys which form noncrystalline solids most easily are mixtures of transition metals with metalloids from columns IV, V, or VI. The glassy state forms best when the metal to metalloid ratio is approximately 4:1. This composition corresponds to a deep eutectic in the binary system, and can be explained from structural considerations. The liquid structure of metals is dense random packing in which the smaller metalloid atoms would occupy the holes and interstices. With a 4:1 ratio the metalloid just fill the larger holes and there are no strong metalloid–metalloid interactions to disrupt the dense random packing of the liquid.

1. Glasses

The relatively few compounds which are easily supercooled to an amorphous state are classified as molecular and ionic glasses. Both contain extensive covalent bonding giving rise to the strong directional bonding responsible for polymerization in the liquid state from which glasses are formed. Metals are seldom amorphous because the melts possess low viscosities and the crystal structures are very simple.

If a melt contains molecular groups with strong internal forces and if these groups are so large or so irregular that it is difficult to add them to a crystal structure, such a melt will show a tendency to supercooling and glass formation. In adding a spherically symmetric atom or ion to a simple crystal, the orientation of the atom is unimportant and there are generally a number of closely spaced equivalent sites for the atom to occupy, so that crystal growth proceeds readily. If, however, the molecular group is large or asymmetric, then it must be carefully positioned and oriented precisely when incorporated in a crystal structure.

The structure of the melt is therefore crucial to glass formation. As discussed earlier, Bernal and Stewart liquids form glasses easily but Frenkel liquids do not. Furthermore, the liquid structure is a function of temperature with higher temperatures favoring the Frenkel model. Glasses form when the liquid is cooled fast enough to prevent nucleation, the critical rate being determined primarily by the structure of the melt.

Among molecular compounds, the tendency for glass formation is particularly noticeable for large molecules such as glucose which become entangled as the melt cools, and are therefore unable to crystallize. Crystallization of long chain polymers is often hampered by insufficient stereoregularity. Prior to the discovery of stereospecific catalysis many polymers could not be crystallized at all. Branching and cross linking reduce the mobility of the molecules producing glasslike systems.

Similar phenomena occur in ionic glasses such as SiO_2, B_2O_3, P_2O_5, and GeO_2, and in closely relatedly chalcogenide glasses (Se, Te, As_2S_3). A lively interest in amorphous semiconductors and ovonic switches has prompted investigation of the chalcogenide glasses.

The cations in ionic glasses have been classified as network formers (e.g., Si^{4+}), potential network formers (e.g., Zn^{2+}), and modifiers such as Na^+. Random networks form most easily when the cations are triangularly or tetrahedrally coordinated and are flexibly linked through their vertices. In silicate structures no more than two SiO_4 tetrahedra share a given corner since two silicon neighbors are sufficient to satisfy the bonding requirements of oxygen. Crystal structures with shared corners possess a great deal more flexibility than those with shared edges or shared faces. Flexibility is absent in MgO, for example, where the octahedrally coordinated Mg^{2+} ions share edges in the crystalline state and are unable to polymerize in the melt so that the viscosity and tendency toward glass formation is small. In silica there is no marked restriction on the Si–O–Si angle, as evidenced by the many polymorphic forms of SiO_2.

Na_2O and other fluxes are used to modify the properties of glass formers like B_2O_3. Fluxes are useful in lowering freezing points, reducing viscosities, and assisting reluctant network formers such as Al_2O_3 in forming glasses. Large cations (K^+, Ca^{2+}, Na^+) fit into large interstices breaking some of the Si–O linkages, thereby changing the properties. Tetrahedral ions like Zn^{2+}, Al^{3+}, and Be^{2+} become glass network formers in the presence of sodium ions for the following reason. In pure ZnO each Zn is surrounded by four O, and each O by four Zn. Therefore there is little flexibility in Zn—O—Zn bonds. Thus when ZnO is added to SiO_2, Zn_2SiO_4 crystallizes readily, giving opal glasses. However when Na_2O flux is added simultaneously in equal amounts with ZnO, the tetrahedral cation to oxygen ratio becomes 1:1, greatly increasing the flexibility and glass formation.

Crystal structures often persist into the molten phase even though long-range ordering disappears. DeWys (1960) has shown that chainlike structures exist in glass made from rapidly melted diopside. He also demonstrated that melt viscosities could be grouped according to the crystal structures of the raw materials from which the melt was derived. Melts obtained from crystals containing silicate chains show similar viscosity–temperature relationships. Isoviscosity lines for glasses held at equal temperature intervals above their melting points exhibited abrupt changes with composition when different silicate groups produce different liquid structures.

Alkali- and alkaline-earth silicate melts are important in glass production. It has been postulated that the melts contain discrete SiO_4^{2-} tetrahedra at the orthosilicate composition. As the silica content is increased, chains of increasing length form because of tetrahedral polymerization. Above 50 mole % SiO_2, both short chains and rings are present. Still higher SiO_2 contents result in polymerization of these chains and rings into larger anions. At 80–90 mole %, SiO_2 a continuous three-dimensional network structure forms and a large change in properties occurs.

Alkali ions tend to surround themselves with nonbridging O^{2-} ions, opposing the random distribution of these nonbridging oxygens. This principle favors the segregation into regions of higher and lower alkali concentration. The smaller the alkali size the greater the tendency to develop submicroscopic segregation. This influence of the alkalies on nonbridging oxygens accounts for the immiscibility observed in certain alkali silicate melts.

2. Immiscible Liquids

Liquid immiscibility (Fig. 30) is very common in glass forming systems and a number of interesting two-phase glass and glass–ceramic products are formed from two-liquid regions of the phase diagram. Pyroceram is shaped first as a glass and then converted to a strong glass–ceramic composite by prolonged heat treatment. Liquid immiscibility is also important in the development of improved silica refractories and in the differentiation of molten magmas.

Levin (1970) has showed that bond strength determines the presence or absence of immiscibility, but that geometrical considerations are needed to explain the extent of the effect. Unmixing is found in oxide melts when cations of vastly different field strengths are present. As explained earlier, field strength is defined as the cation valence Z divided by the square of the interatomic distance d. Immiscibility occurs in ionic melts when three criteria are met: (1) one of the cations possesses a field strength greater then

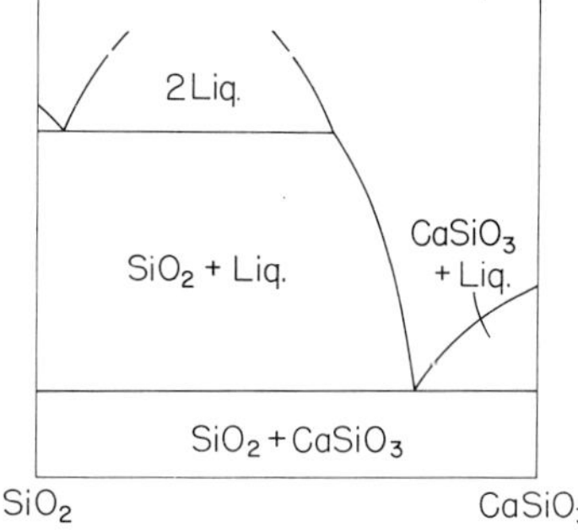

Fig. 30. The formation of two liquid regions is intimately related to glass formation. Unmixing occurs in systems such as CaO–SiO_2 in which cation field strengths do not differ by more than 1.0.

1.0, (2) the differences in field strength for the two cations is not large, less than 1.3, and (3) the two cations have different coordinations. At least one of the cations should be very polarizing (high charge and small size) to enforce its polyhedral requirements, thereby determining its surroundings.

The extent of unmixing is semiquantitatively related to the difference in field strengths of the cations involved. For example, across the series SiO_2–Na_2O, SiO_2–Li_2O, SiO_2–BaO, SiO_2–SrO, SiO_2–CaO, SiO_2–MgO the difference in field strengths progressively diminishes while unmixing increases. In the SiO_2–K_2O series the difference is $1.57 - 0.13 = 1.44$, a very large difference so that Si^{4+} dominates and K^+ fits in interstices, yielding a single phase. For SiO_2–MgO the field strength difference is only $1.57 - 0.45 = 1.12$ and two liquids form because Si^{4+} no longer dominates. Unmixing occurs not only for silicates but for many other systems when $\Delta(Z/d^2)$ is less than 1.0. A few transition-metal cations such as Ti^{4+} and Cr^{3+} appear to be exceptions to the above scheme.

The conditions governing the separation of two liquids have been discussed by Charles (1967). Consider the mixing of two different liquids, liquid A consisting of long intertwined molecular groups weakly held together by van der Waals forces, and liquid B made up of cations and shorter molecular groups, negatively charged. Ionic forces are important in liquid B.

When one A molecule is transferred to B, it straightens, relieving strain, and heat is given off because of the relaxation of bent bonds. Disorder increases both because of the heat involved and because of mixing. The process is repeated many times as more A is added to B. Cations become increasingly scarce in the melt and the molecular groups grow longer and more entangled. Soon the relaxation energy decreases until it just matches the ionic energy taken up fluxing ions and negatively charged molecular groups. Heat is no longer evolved at this point but disorder tends to increase because of mixing. Further additions draw heat from the system in lengthening the chains and separating positive and negative ions. Heat loss results in ordering, and when the ordering is sufficient to overcome the disordering due to mixing, phase separation begins. Adding more liquid A leads to the formation of A-rich regions. Two liquid separation is aided by strong ionic forces between the fluxing ions and matrix, and by low temperatures.

D. Plastic Crystals and Liquid Crystals

There are two mesomorphic states of matter between solids and liquids: solidlike liquids and liquidlike solids. Solidlike liquids are referred to as *liquid crystals*, and the liquidlike solids are called *plastic crystals*.

Mesomorphic states exist because two types of randomization occur during melting: translational and rotational. Ordinarily both translational and rotational disorder set in at the same temperature but there are some materials in which one is much easier than the other. *Two* transitions occur in this case. Which transition is melting? By definition, crystals have translational periodicity, liquids do not. Therefore, melting is associated with the loss of translational order. This is the lower transition for liquid crystals and the upper transition for plastic crystals.

Molecules of very anisotropic shape tend to maintain parallel orientation. As shown in Fig. 31, elongated molecules need considerable room and kinetic energy to rotate freely. As a result their rotations are restricted to rotation about the axis of the molecule. The molecules remain parallel but displace from their crystallographic positions on melting. At a higher temperature, the molecules acquire sufficient energy to rotate freely and a liquid crystal-to-liquid transition takes place. The nematic liquid crystal state just described is found in *p*-azoxyanisole ($C_2O_3N_2H_6$) and other chain molecules.

The common types of liquid crystals are called nematic, smectic, and cholesteric. Nematic liquid crystals contain rodlike organic molecules resembling matches in a box. The molecules maintain a parallel arrangement but can rotate about the rod axis and can move to some extent. Smectic and cholesteric liquid crystals also retain orientational order but have layer and spiral structures, respectively. Soaps sometimes form smectic liquid crystals with layered arrangements in which the molecules are oriented perpendicular

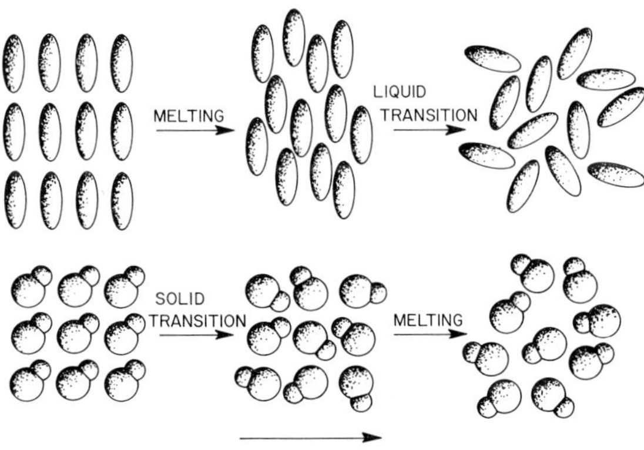

Fig. 31. Melting involving the onset of two types of disorder, positional and orientational disorder. In some materials the two commence separately. On heating, liquid crystals lose positional order before orientational order, while the sequence is reversed in plastic crystals. The latter generally consist of near-spherical molecular groups, whereas anisotropic molecules sometimes show liquid crystal behavior.

to the layers. Cholesteric liquid crystals also form layers, but the molecules in each layer lie parallel to the plane of the layer. Molecular orientations in neighboring layers usually differ, producing spiral arrangements. The most important property is the extremely large optical rotatory power and light-scattering effects that have been exploited in electrooptic applications.

Plastic crystals resemble plastics in their mechanical properties but possess the three-dimensional periodicity of a crystal. The molecules tend to be near spherical as illustrated in Fig. 31, with weak nondirectional bonding between molecules. The molecules are ordered at low temperatures and then undergo a solid-state transition to the plastic crystal region in which the molecules rotate freely but retain translational periodicity. At still higher temperatures plastic crystals melt and all long-range order disappears. In plastic crystals the energy barrier hindering rotation is less than that for diffusion but because of rotation, molecules slip by one another easier, making the crystal more plastic.

Camphor, $C_{10}H_{16}O$, is one of the better known plastic crystals. Below $-23°C$, camphor is a normal crystal, and above 179°C a normal liquid. In between camphor is a plastic crystal. Soft and waxy when cut with a knife, plastic crystals spread like butter. Some, like perfluorcyclohexane, flow under their own weight. Other molecular plastic crystals are H_2, H_2S, CCl_4, adamantane, and cyclohexane.

Plastic crystals are characterized by a small entropy of fusion, which is compensated by one or more solid-state transitions. Other properties include high triple-point pressure and temperature, high symmetry (usually hexagonal or cubic), transparency, tackiness, and relatively high plasticity. Microscopically, the materials are made up of globularlike molecules which undergo order–disorder transitions before melting. Abrupt changes in molecular freedom occur at the solid-state transitions.

Such properties are not restricted to organic crystals. The transition-metal hexafluorides undergo transitions to the plastic-crystal state a few degrees below fusion. PtF_6 transforms at 276°K and melts at 335°K, and WF_6, ReF_6, OsF_6, and IrF_6 behave in a similar manner (Pitzer and Gwinn, 1942). In the vapor phase the molecules are regular octahedra but no primary bonds formed on crystallization, hence PtF_6 is a molecular solid made up of near-regular octahedra. A closely packed orthorhombic structure is observed at low temperatures. Before melting, the solid undergoes a transition to the plastic state in which the octahedral molecules reorient freely.

Plastic crystals have yet to find an important application but some of their properties are sufficiently different to warrant consideration. Because of their butterlike consistency, plastic crystals may be of use as a medium for transmitting low hydrostatic pressures at low temperatures where a solid is required. In addition to their mechanical properties, plastic crystals are noted for their very narrow liquid ranges; perfluorocyclohexane melts at 51°C and boils at 52°C. Another distinguishing feature is the low entropy of melting, ΔS_m is very small because there is little difference in order between the plastic crystal and liquid phases. This has led to their use in determining molecular weights since the freezing point of a plastic crystal is extremely sensitive to additives because the entropy of mixing is sizeable compared to ΔS_m. The depression of the melting point is directly proportional to the number of molecules added. Knowing the weight added, and the number of molecules gives the molecular weight of the unknown.

Study of the mesomorphic states helps in understanding melting. The melting of any substance depends on the ease of rotation compared to the ease of diffusion. Materials with molecules that rotate easily have high melting points, low volume changes on melting, low entropy of melting, and a solid-state phase transition. The transition is near second order in plastic crystals when the molecules rotate very easily. Solid rare gases where the atoms are completely spherical show a gradual transition to a plastic crystal range near the melting point. Materials with more shape anisotropy tend to show first-order transitions. When the shape is very nonspherical, translational disorder occurs before rotational disorder and the material behaves as a liquid crystal.

References

Anderson, J. S. (1971). *Rev. Pure Appl. Chem.* **21**, 67.

Anderson, J. S., and Burch, R. (1971). *J. Phys. Chem. Solids* **32**, 923.

Berkes, J. S., and Roy, R. (1970). *Z. Kristallogr.* **131**, 60.

Brewer, L. (1958). *J. Chem. Educ.* **35**, 153.

Brewer, L. (1967). *Acta Metall.* **15**, 553.

Brown, A. F. (1972). *Proc. R. Soc. Edinburgh* **A70**, 27.

Buerger, M. J. (1971). *Kristallografia* **16**, 1084.

Charles, R. J. (1967). *Sci. Am.* **217**(7), 127.

Christ, C. L. (1972). *J. Geol. Educ.* **20**, 235.

Darken, L., and Gurry, R. W. (1953). "Physical Chemistry of Metals." McGraw-Hill, New York.

DeVries, R. C., and Roy, R. (1953). *J. Am. Chem. Soc.* **75**, 2481.

DeWys, E. C. (1960). *Min. Mag.* **32**, 825.

Dietzel, A. (1942). *Z. Elektrochem.* **48**, 9.

Dyson, R. J. (1971). *Ann. Phys.* (NY) **63**, 1.

Engel, N. (1949). *Kem. Maanesbl. Nord. Handelsb. Kem. Ind.* Nos, 5, 6, 8–10.

Farrell, E. F., and Newnham, R. E. (1967). *Am. Mineral.* **52**, 380.

Fine, M. E. (1972). *Bull. Am. Ceram. Soc.* **51**, 510.

Glasser, L. S. D., Glasser, F. P., and Taylor, H. F. W. (1962). *Q. Rev.* **16**, 343.

Goldschmidt, V. M. (1926). *Skr. Nor. Vidensk. Akad. Oslo, 1:* No 8.

Greenwood, N. N. (1968). "Ionic Crystals, Lattice Defects and Nonstoichiometry." Butterworth, London.
Hoermann, F. (1928). *Anorg. Allgem. Chem.* **177**, 154.
Houlihan, J. F., and Roy, R. (1974). *J. Am. Ceram. Soc.* **57**, 234.
Hume-Rothery, W. (1936). "The Structure of Metals and Alloys." Inst. Metals, London.
Hume-Rothery, W. (1967). *Prog. Mat. Sci.* **13**, 229.
Jaffe, B., Cook, W. R., and Jaffe, H. (1971). "Piezoelectric Ceramics." Academic Press, New York.
Jamieson, J. C. (1957). *J. Geol.* **65**, 338.
Jeffery, J. W., and Lindley, P.F. (1973). *Nature* **241**, 42.
Kapustinskii, A. F. (1956). *Q. Rev.* **10**, 283.
Kitaigorodskii, A. I. (1961). "Organic Chemical Crystallography." Consultants Bureau, New York.
Levin, E. M. (1970). *In* "Phase Diagrams: Materials Science and Technology" (A. M. Alper, ed.), Vol. II. Academic Press, New York.
Levin, E. M., Robbins, C. R., and McMurdie, H. E. (1964). "Phase Diagrams for Ceramists." Amer. Ceram. Soc., Columbus, Ohio.
Liu, L. (1973). *J. Appl. Phys.* **44**, 2470.
Love, R. W., Esty, C. C., and Wheildon, W. M. (1965). *In* "Materials Science Research" (H. M. Otte and S. R. Locke, eds.), Vol. 2. Plenum, New York.
Ludwiczek, H., and Zemann, J. (1973). *Tschermaks Mineral Petrogr. Mitt.* **19**, 133.
Marchese, B., Mascolo, G., and Sersale, R. (1972). *J. Am. Ceram. Soc.* **52**, 463.
Marezio, M., and Remeika, J. P. (1965). *J. Phys. Chem. Solids* **26**, 1277.
McCarthy, G. J., White, W. B., and Roy, R. (1969). *J. Am. Ceram. Soc.* **52**, 463.
Megaw, H. D. (1965). *In* "The Molecular Designing of Materials and Devices" (A. R. von Hippel, ed.). MIT Press, Cambridge, Massachusetts.
Moss, S. C., and Newnham, R. E. (1964). *Z. Kristallogr.* **120**, 359.
Muller, O., and Roy, R. (1974). "The Major Ternary Structural Families." Springer-Verlag, Berlin and New York.
Newnham, R. E. (1961). *Min. Mag.* **32**, 683.
Newnham, R. E., and Megaw, H. D. (1960). *Acta Crystallogr.* **13**, 303.
Pauling, L. (1929). *J. Am. Chem. Soc.* **51**, 1010.
Payne, W. H., and Tennery, V. J. (1965). *J. Am. Ceram. Soc.* **48**, 413.
Pedone, C., and Benedetti, E. (1972). *Acta Crystallogr.* **B28**, 1970.
Perrotta, A. J., and Smith, J. V. (1963). *Acta Crystallogr.* **16**, A13.
Pitzer, K. S., and Gwinn, W. D. (1942). *J. Chem. Phys.* **10**, 428.
Ramberg, H. (1954). *Am. Mineral.* **39**, 79.
Ringwood, A. E. (1967). *Acta Cryst.* **23**, 1093.
Roy, R. (1973). *In* "Phase Transitions and Their Applications in Materials Science" (L. E. Cross, ed.) Pergamon, New York.
Schneer, C. J. (1955). *Acta Crystallogr.* **8**, 279.
Shannon, R. D., Chenavas, J., and Joubert, J. C. (1975). *J. Solid State Chem.* **12**, 16.
Sinha, A. K. (1972). *Prog. Mater. Sci.* **15**, 79.
Slagle, O. D., and McKinstry, H. A. (1966). *Acta Crystallogr.* **21**, 1013.
Slater, J. C. (1939). "Introduction to Chemical Physics." McGraw-Hill, New York.
Spear, K. (1975). *In* "Phase Diagrams: Materials Science and Technology" (A. M. Alper, ed.), Vol. IV. Academic Press, New York.
Tosi, M. P. (1964). *Solid State Phys.* **16**, 1.
Ubbelohde, A. R. (1965). "Melting and Crystal Structure." Clarendon, Oxford.
Vorres, K. S. (1965). *J. Am. Ceram. Soc.* **48**, 113.

Waber, J. T., Gschneider, K., Larson, A. C., and Prince, M. Y. (1963). *Trans. Metall. Soc. AIME.* **227**, 717.
Wells, A. F. (1958). *Solid State Phys.* **7**, 425.
Weyl, W. A., and Marboe, E. C. (1962). "The Constitution of Glasses." Wiley (Interscience), New York.
Young, R. A., Van der Lugt, W., and Elliott, J. C. (1969). *Nature* **223**, 729.

II

Thermodynamics and Structure of Nonstoichiometric Binary Oxides

O. TOFT SØRENSEN

METALLURGY DEPARTMENT
RISØ NATIONAL LABORATORY
DENMARK

I. INTRODUCTION

At higher temperatures many of the binary transition metal oxides, rare-earth oxides and actinide oxides have nonstoichiometric phases extending over a considerable composition range. Previously, these phases were considered as grossly nonstoichiometric and their thermodynamic properties were described in terms of randomly distributed and noninteracting point defects. Thorough structural studies by x rays, by neutron diffraction and particularly by high resolution electron microscopy have, however, shown that many nonstoichiometric oxide systems are ordered to a much higher degree than previously visualized, and today it is generally realized that

ISBN 0-12-053205-0

point defect theories do not give a realistic description of grossly nonstoichiometric systems. As the thermodynamics and structure of a solid are closely interrelated, the thermodynamic properties of a nonstoichiometric system should therefore reflect the actual structures observed, i.e., they should be expressed in terms of extended defects, defects ordered into superstructures, or in some systems in terms of intermediate phases formed by a crystallographic shearing mechanism. It is the purpose of the present review to discuss the most important thermodynamic theories of nonstoichiometry in binary oxide systems and to compare the structures assumed in or predicted from these theories with those actually observed. In this way the areas where further thermodynamic studies are necessary should become clear.

Thermodynamic properties of nonstoichiometric transition-metal oxides have recently been reviewed by Navrotsky (1974), and the thermodynamic properties of nonstoichiometric actinide oxides have also previously been reviewed (Rand and Kubaschewski, 1963; IAEA, 1967). It is not the intention of the present author to duplicate these reviews, but rather to consider a few typical oxide systems of each type as examples. Finally, thermodynamic theories of nonstoichiometric systems have previously been reviewed by Anderson (1970, 1974) and Fender (1972), and these reviews are recommended for further details.

II. GENERAL THERMODYNAMIC CONSIDERATIONS OF NONSTOICHIOMETRIC PHASES

For a two-component system such as a binary oxide, the phase rule dictates that bivariant behavior ($\mu_i = \mu_i(T, x)$) should be expected for a single nonstoichiometric phase of variable composition, whereas a mixture of two phases should show univariant behavior ($\mu_i = \mu(T)$). This is illustrated in Fig. 1 that shows a hypothetical μ, T, x diagram for a two-component system. At lower temperatures this system consists of two-phase ranges separated by the line phases A_2B and AB_2, but at some critical temperature, T_c, the envelope of the two-phase range $A_2B + AB_2$ closes and a nonstoichiometric phase is formed. This is also indicated on the phase diagram shown in the figure, which is obtained by projecting the boundaries between the μ–T–x surfaces onto the T–x plane.

In studies of nonstoichiometric oxides the oxygen activities are often expressed as log p_{O_2} or $\Delta\bar{G}_{O_2}$ ($= RT \ln p_{O_2}$)—relative partial free energy of oxygen—both values being related to the chemical potential of oxygen of the oxides. In order to evaluate the phase relationships of the system these parameters are plotted at constant temperature as a function of composition,

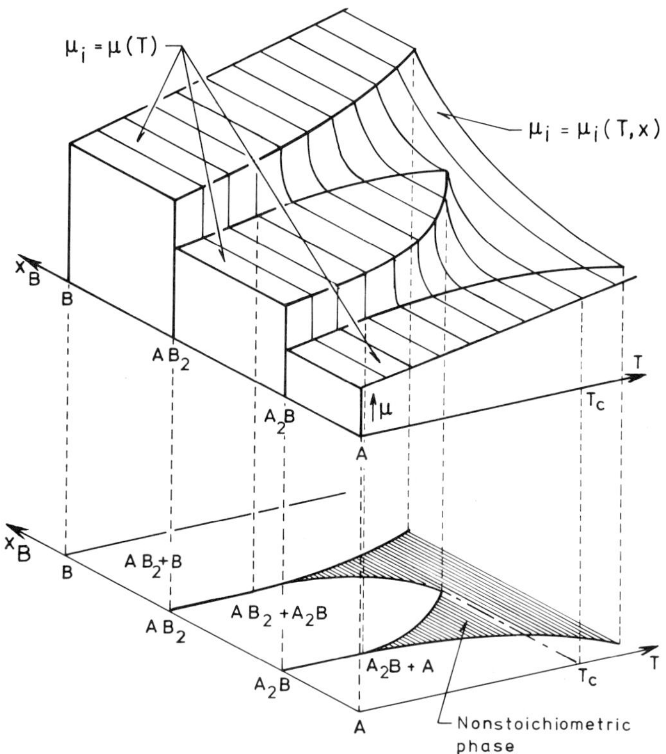

Fig. 1. Hypothetical μ, T, x diagram for a two-component system.

and from Fig. 1 it is clear that the isotherms for the different types of phase regions must show the following behavior:

(i) For a two-phase range a horizontal line should be observed

(ii) For a line phase (discrete compound) a vertical line should be observed, and

(iii) For a nonstoichiometric phase range a line of intermediate slope should be observed. As will be shown later, the appearance of such a line is, however, in no way a proof that a phase is nonstoichiometric since this can also be obtained by a succession of ordered or partly ordered phases.

Although $\Delta\bar{G}_{O_2}$ is an important property from which the phase relationship of a system can be deduced as shown above, the relative stabilities of the different phases can only be evaluated by considering their total free energies, $G\,(= x_A\bar{G}_A + x_B\bar{G}_B)$—the most stable phase will be that with the lowest free energy. It is, however, important to realize that the free energy of an oxide phase at a given composition and temperature and its corresponding

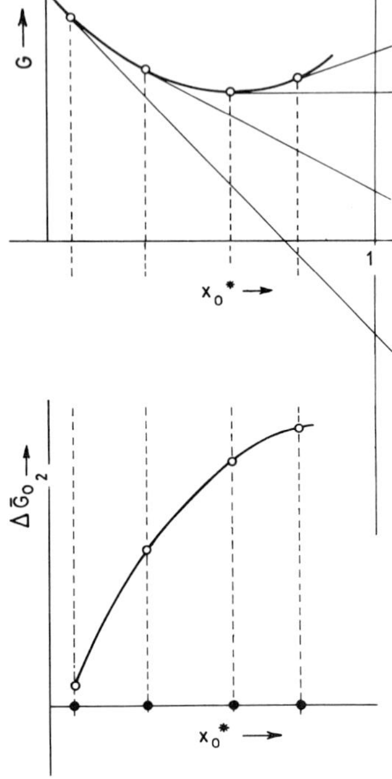

Fig. 2. Construction of a $\Delta\bar{G}_{O_2} - x_O^*$ (mole fraction of oxygen) curve from a G–x_O^* curve.

$\Delta\bar{G}_{O_2}$ are closely related. As shown in many standard thermodynamic textbooks [see for instance Darken and Gurry (1953)], $\Delta\bar{G}_{O_2}$ can be found by extrapolating the tangent to a G versus x_O^* (mole fraction of oxygen) curve to $x_O^* = 1$, and, as indicated in Fig. 2, the shape of the G–x_O^* and $\Delta\bar{G}_{O_2}$–x_O^* curves must closely correspond to each other.

The form of the G–x curves to be expected for a line phase and a nonstoichiometric phase is shown in Fig. 3. In a system where the energy associated with the formation of defects is high, the line phase will show a sharply pointed curve in contrast to the broad and asymmetric (Brebrick, 1967; Kröger, 1968; Thorn and Winslow, 1967) G–x curve expected for a nonstoichiometric phase. In a G–x plot a common tangent between curves for different phases indicates that these phases are in equilibrium and, besides the two hypothetical line phases A_2B and AB_2 and the nonstoichiometric phase AB indicated in Fig. 3, phase regions where a line phase and a nonstoichiometric phase coexist should also be possible. Whether this is observed in a real system, however, depends on how easily the line phase (discrete compound) can nucleate from a nonstoichiometric phase, and in the absence of such a nucleation the nonstoichiometric phase might very well exist in the whole composition range between the line phases.

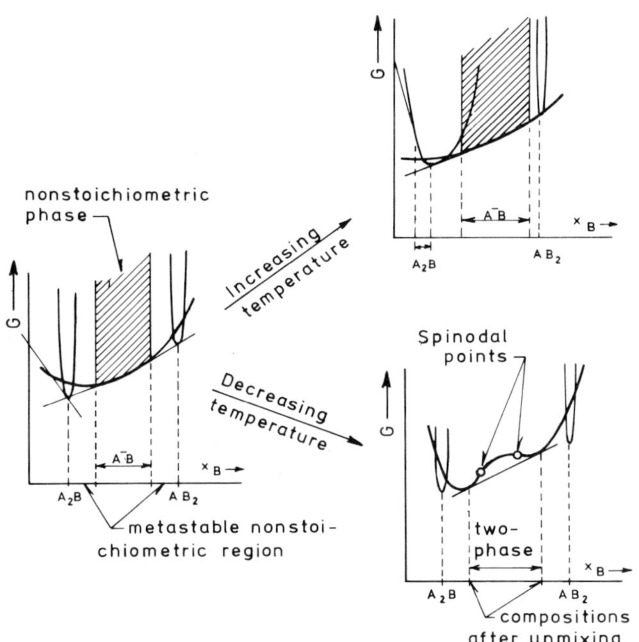

Fig. 3. Shape and relative positions of *G–x* curves for line phases and a nonstoichiometric phase with increasing and decreasing temperature.

The relative stabilities of the different phases can be changed in several ways with increasing temperature (Fig. 3):

(i) the *G–x* curve for the nonstoichiometric phase becomes broader and the region for this phase will be extended,
(ii) the nonstoichiometric phase becomes relatively more stable than the neighboring line phases and the metastable region becomes smaller,
(iii) the *G–x* curve for a line phase may broaden and a new nonstoichiometric phase is formed ($\overline{A_2B}$).

Another situation that can be imagined with increasing temperature, and which is not covered in Fig. 3, is that a line phase becomes completely unstable with respect to a nonstoichiometric phase, i.e., its *G–x* curve will lie well above the *G–x* curve for the nonstoichiometric phase. In this case a peritectic decomposition will take place in which the line phase (A_2B for instance) is transformed into the two adjacent phases (A and $\overline{AB}$), as has been observed, for instance, in the PrO_x–O_2 (Hyde *et al.*, 1966) and CeO_x–O_2 systems (Bevan and Kordis, 1964; Blank, 1967).

Whereas the range of existence of nonstoichiometric phases is expected to increase with increasing temperature, instabilities may become important at lower temperatures. If the G–x curve takes the spinodal form shown in Fig. 3, small composition fluctuations between the spinodal points will lead to a spontaneous unmixing into a pair of coexisting phases—between the spinodal points $d^2G/dx_B{}^2 < 0$—and a fluctuation in the composition will be irreversible and self-propagating since it gives a decrease in the total free energy as shown by Cahn (1961). The two-phase range observed at lower temperatures for many nonstoichiometric oxide systems is probably formed in this way, and the observation by Manenc *et al.* (1962) and Herai *et al.* (1964) of an unmixing into iron-rich and oxygen-rich lamellae (probably consisting of defect clusters ordered into superstructures (Koch and Cohen (1969)) in quenched wüstite (FeO_x) has also been explained by spinodal decomposition reactions.

Detailed structural studies by x-ray and high resolution electron microscopy have revealed that many oxide systems previously considered grossly nonstoichiometric in fact consist of a succession of ordered intermediate (line) phases (Magnéli phases). This has, for instance, been observed for the TiO_x system (Bursill and Hyde, 1972) where crystallographic shear can produce a whole series of phases with compositions following the general formula Ti_nO_{2n-1}. As shown in Fig. 4, the free energy of the successive phases in such a system differ only slightly and the coexistence tangents

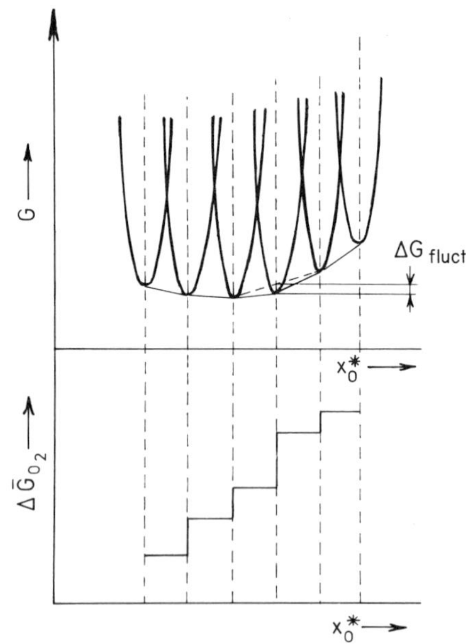

Fig. 4. G–x_O^* and $\Delta\bar{G}_{O_2}$–x_O^* curves for a sequence of intermediate phases. ΔG_{fluct} represents a change in free energy due to a local fluctuation is composition.

between the phases are very close to the continuous curve expected for a grossly nonstoichiometric system. In a $\Delta\bar{G}_{O_2}$–x representation, the presence of these line phases would be indicated by a step curve, as also shown in Fig. 4, but unless closely spaced and accurate experimental points are available, this curve could also easily be mistaken for a continuous curve. For many oxide systems it is, however, questionable whether a thermodynamic study can give sufficiently accurate $\Delta\bar{G}_{O_2}$ values to judge whether a phase is grossly nonstoichiometric, or consists of successive line phases; hence in order to evaluate the real nature of a solid phase, thermodynamic and structural studies should be combined.

III. DEFECT STRUCTURE AND THERMODYNAMIC THEORIES OF NONSTOICHIOMETRIC OXIDES

A. Classification

As a first broad and general classification nonstoichiometric oxide systems can be divided into:

(i) essentially stoichiometric oxides with a very narrow composition range
(ii) nonstoichiometric oxides with a limited composition range, and
(iii) grossly nonstoichiometric oxides with broad composition ranges.

Each of these systems has a characteristic defect structure and in the next sections the thermodynamics of defect formation will be considered for each type.

Nonstoichiometry is primarily observed in oxide systems where the cations can exist in several oxidation (valence) states, i.e., in the transition-metal oxides, rare-earth oxides, and actinide oxides. Considering these oxides it is further apparent that the type and extent of nonstoichiometry also depend on the structure; generally, nonstoichiometry has been observed in oxides with the following structures:

(1) NaCl structure: monoxides of the first and second transition series, i.e., TiO, VO, (CrO?), MnO, FeO, CoO, NiO (first series), and NbO (second series).

(2) Fluorite structure: rare-earth and actinide dioxides, e.g., CeO_2, PrO_2, TbO_2 (rare-earth series), and UO_2, PuO_2, AmO_2, and other *trans*-plutonium oxides (actinide series).

(3) Rutile structure: transition-metal dioxides, e.g., TiO_2, VO_2, MnO_2 (first series), and ZrO_2 (second series).

(4) ReO_3 structure: transition-metal trioxides, e.g., CrO_3 (first series), MoO_3 (second series), and WO_3 (third series).

(5) Nb_2O_5 structures: transition-metal oxides as Nb_2O_5 (second series) and Ta_2O_5 (third series).

For the NaCl and fluorite structures, nonstoichiometry can be described in terms of defects—point defects or defect clusters depending on the extent of deviation from stoichiometry. For the other structures, the defect concentrations are considered to be rather small and for these oxides the nonstoichiometric phases are considered to consist of a series of discrete phases.

Finally, nonstoichiometric and grossly nonstoichiometric oxides can also be classified according to the direction taken by a deviation from stoichiometry. Generally, a change in composition should correspond to a change in the oxidation state of the cations toward the next stable state, and nonstoichiometric oxides can further be divided into two main groups (Kofstad, 1972):

(1) Oxygen-deficient oxides in which oxygen vacancies ($V_O^{\cdot\cdot}$) are predominantly formed, e.g., CeO_{2-x}, PuO_{2-x}. Alternatively, the oxygen deficiency can also in these oxides be described as the presence of excess metal, in which case the predominating defects are considered to be interstitial metal atoms or ions (M_i).

(2) Metal-deficient oxides in which metal vacancies (V_M') are formed, e.g., $Fe_{1-y}O$, $Mn_{1-y}O$. Alternatively, the metal deficiency may reflect the presence of excess oxygen in the form of interstitial oxygen ions (O_i'') as in UO_{2+x}.

B. Thermodynamics of Defect Formation in Essentially Stoichiometric Oxides

By statistical thermodynamic calculations it can readily be shown that the free energy of a crystal is lowered by the introduction of vacancies or other point defects above the absolute zero temperature (Wagner and Schottky, 1930; Swalin, 1972). The reason for this is that the energy expenditure required to form such defects is more than compensated by the entropy increase resulting from the defect formation ($\Delta G = \Delta H - T\,\Delta S$). Even at perfect stoichiometry each compound will thus contain an equilibrium concentration of atomic point defects, which are formed inherently within the crystal (internal or thermal disorder) and not by a reaction with a surrounding atmosphere, like the defects in nonstoichiometric oxides described in the next sections. The atomic point defects formed in a stoichiometric compound are termed primary or native defects and characteristically they are formed in pairs such that the simple ratio between the constituents can be maintained. Generally the following five types of defects have been considered:

(1) Schottky disorder: Equal concentration of anion and cation vacancies.

(2) Frenkel disorder on cation sublattice: Equal concentrations of cation vacancies and interstitial cations.

(3) Frenkel disorder on anion sublattice: Equal concentrations of anion vacancies and interstitial anions.

(4) Anti Schottky disorder: Equal concentrations of interstitial cations and anions.

(5) Antistructure disorder: Equal concentrations of cations on anion sites and anions on cation sites.

Although the concentration of primary defects is small, properties such as electrical conductivity and self-diffusion strongly depend on the type and number of defects formed. Deviations from the stoichiometric composition have also been considered to be the result of an unbalance between the primary defects; as an introduction to nonstoichiometry it is thus worthwhile to consider in some detail the thermodynamics of the formation of these defects and the equilibrium between them.

Assuming that the defect pairs are randomly distributed and noninteracting and that they form an ideal solution with the crystal, the formation of primary defects can be described by quasi-chemical equations. For a Schottky type of disorder, for instance, such an equation would be

$$\mathrm{O} = (\mathrm{V_M V_O}) \tag{1}$$

where O represents the perfect crystal and $(\mathrm{V_M V_O})$ an imperfection consisting of a metal vacancy and an oxygen vacancy. According to Kröger and Vink (1956), this equilibrium can be treated according to the general laws of chemical thermodynamics and the defect concentrations can therefore be expressed in terms of an equilibrium constant:

$$[(\mathrm{V_M V_O})] = K_{(\mathrm{VM})} = C_{(\mathrm{VM})} \exp(-W_{(\mathrm{VM})}/RT) \tag{2}$$

In this equation $C_{(\mathrm{VM})}$ and $W_{(\mathrm{VM})}$ are constants determined by the change in entropy and by energy involved in the formation, respectively.

At increasing temperature the defect pairs will tend to dissociate into single defects according to:

$$(\mathrm{V_M V_O}) = \mathrm{V_M} + \mathrm{V_O}. \tag{3}$$

The law of mass action can also be applied to this equilibrium and

$$\frac{[\mathrm{V_M}]\cdot[\mathrm{V_O}]}{[(\mathrm{V_M V_O})]} = K_\mathrm{D} = C_\mathrm{D}\exp(-W_\mathrm{D}/RT), \tag{4}$$

where W_D is the dissociation energy. If $W_\mathrm{D} > 0$ the fraction of single defects will increase with increasing temperature and the presence of defect pairs is usually neglected when equilibria are considered at high temperatures. The equation, however, is important since it emphasizes that there is a tendency for the dissociated defects to become associated during cooling, which might perhaps explain the formation of superstructures observed in many nonstoichiometric oxides at lower temperatures. Neglecting the formation of defect pairs, the reaction describing the Schottky disorder should then be written as

$$\mathrm{O} \rightleftarrows \mathrm{V_M} + \mathrm{V_O} \tag{5}$$

and the equilibrium constant will be

$$[\mathrm{V_M}][\mathrm{V_O}] = K_\mathrm{S}. \tag{6}$$

The equilibrium constants for the formation of the other types of primary defects can be written in the same way as for the Schottky disorder, but before the concentrations of the different types of defects can be evaluated the electronic disorder must also be taken into account. This disorder, which is especially important for nonmetallic systems (insulators and semiconductors), and which was not considered in the original statistical thermodynamic treatments of primary defects, arises in the following way: At low temperatures the electrons occupy low-energy states and no electron transfer to or from the defects will take place. At increasing temperature, however, some of the electrons can be transferred to higher energy states because of the entropy effects, as described for the atomic defects, and there will be an increasing tendency toward either ionization or uptake of electrons by the defects.

According to Kröger and Vink, the basic process of creating point defects is the displacement of neutral atoms so that every lattice site bears its normal charge relative to the perfect lattice. In the basic process of formation of oxygen vacancies, for instance, two electrons are trapped in the neutral vacancy (neutral relative to the surrounding lattice), but, depending on the temperature, these may be excited and liberated from the vacancy according to:

$$\mathrm{V_O}^{\mathrm{x}} \rightleftarrows \mathrm{V_O^{\cdot}} + \mathrm{e}' \tag{7}$$

$$\mathrm{V_O} \rightleftarrows \mathrm{V_O^{\cdot\cdot}} + \mathrm{e}' \tag{8}$$

where V_O^x designates a neutral vacancy, and $V_O^{\cdot}$ and $V_O^{\cdot\cdot}$ a single positively charged and double positively charged vacancy, respectively.

For the transition-metal oxides, where the cations can exist in several oxidation states, the electrons liberated by the vacancies can be considered to be associated with cations on normal lattice sites, which are then reduced to a lower oxidation state:

$$M_M + V_O^X \rightleftarrows V_O^{\cdot} + M_M' \tag{9}$$

$$M_M + V_O^{\cdot} \rightleftarrows V_O^{\cdot\cdot} + M_M'. \tag{10}$$

In order to understand the equilibria involved in electronic disorder the band model of solids should be considered. According to this model, which, for instance, has been described for typical transition metal and nontransition-metal oxides by Adler (1971) and in general in many standard textbooks (Kofstad, 1972; Kittel, 1966; Azaroff, 1960), the electron orbitals of the single atoms overlap and, depending on the symmetry, are split when the atoms are brought together in a crystal, and the energy levels for the electrons may be considered to form energy bands as shown in Fig. 5. The energy bands of importance for the electronic properties are termed the valence and conduction bands, and for a pure insulator or semiconductor these bands are separated by a forbidden energy gap E_g. The forbidden energy gap is much greater for a nontransition-metal oxide than for a transition-metal oxide, as also indicated in Fig. 5, and much less electronic disorder must be expected

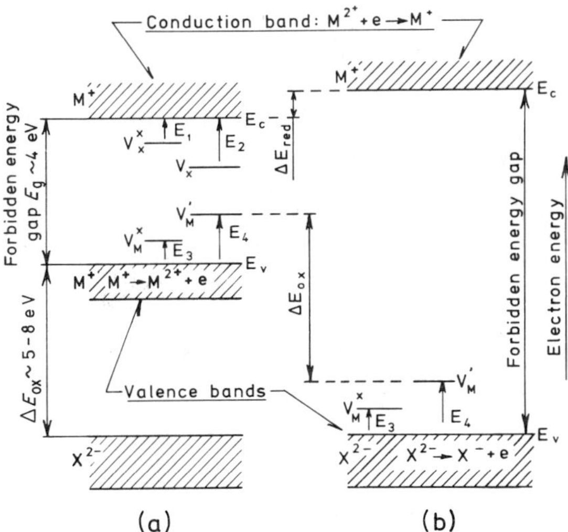

Fig. 5. Schematic illustration of electron energy bands for $M^{2+}X^{2-}$: (a) transition-metal oxide; (b) nontransition-metal oxide.

for the former type of oxide than for the latter. Two types of electronic disorder can be envisaged: (i) instrinsic ionization and (ii) ionization of imperfections. In the former case, the electrons in the valence band may be excited across the forbidden energy gap to the conduction band when the temperature is increased, leaving a positive hole ($h^{\cdot}$) in the valence band. This excitation can be described by the equation:

$$O \rightleftarrows e' + h^{\cdot} \tag{11}$$

and applying the law of mass action to this equilibrium, the relation between the concentration of electrons and holes will be

$$[e'] \cdot [h^{\cdot}] = K_i = C_i \exp(-E_g/RT), \tag{12}$$

where C_i is a constant determined by the entropy change of the intrinsic ionization and E_g is the forbidden band gap energy.

The equilibria involved in ionization of imperfections can also be described in terms of excitation energies, which, as shown in Fig. 5, are considered to be smaller than the forbidden band gap energy. For neutral and singly charged oxygen vacancies, the energy levels are considered to lie below but relatively close to the conduction band with excitation energies E_1 and E_2, in contrast to the levels for metal vacancies, which are close to the valence band. Other types of imperfections will give a different scheme of levels, but generally the law of mass action can again be applied to the equilibria involving ionization of the imperfections, as described for the intrinsic ionization, and relations between the concentrations of free electrons or holes and the imperfections in their various states of ionization can be obtained.

By the application of the mass action law to the various equilibria describing the formation of neutral imperfections and their subsequent ionization, as described above, a set of equations is obtained relating defect concentrations to the corresponding equilibrium constants. Using the conditions of electrical neutrality and constant stoichiometry, which give two extra equations, it should in principle be possible to calculate the concentrations of the different types of imperfections provided that the equilibrium constants are known. Although this is seldom the case, an approximate solution can be obtained by an iterative method described by Kröger and Vink (1956).

In the classical treatment outlined above it is assumed that the host lattice is not affected when defects are created. However, in the vicinity of a defect, local rearrangements of the host lattice atoms should be expected, as shown in Fig. 6 for an anion vacancy. After ionization the vacancy will carry a positive charge and the surrounding atoms will be displaced in the directions indicated by the arrows. This results in relaxation processes and using the

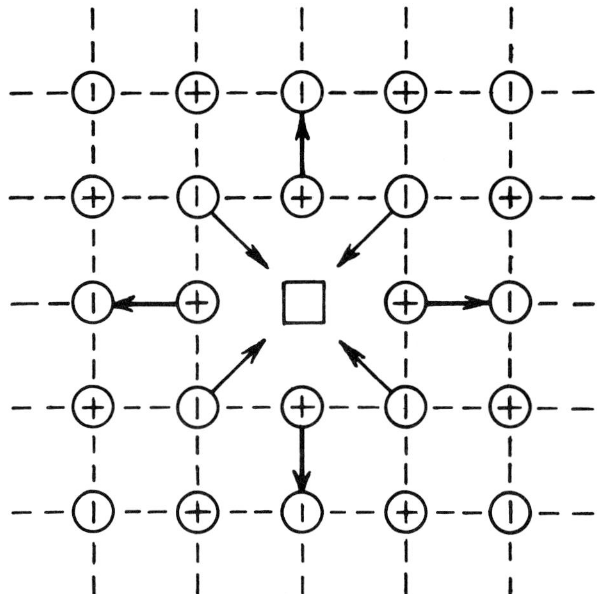

Fig. 6. Relaxation of ions around an anion vacancy in an ionic crystal.

classical dielectric theory it has been shown that the energy of vacancy formation is considerably reduced when this effect is taken into account. Early calculations (Mott and Littleton, 1938) for ionic crystals gave formation energies slightly larger than those observed experimentally, but in later refinements (Scholtz, 1964) where interaction energies between defect pairs, as well as the energy of lattice strain in the neighborhood of the defects, were also taken into account complete agreement with experimental data was obtained. Relaxation phenomena probably also play an important role in structure changes in nonstoichiometric oxide systems, although these effects have not yet been fully evaluated.

C. Thermodynamics of Nonstoichiometric Oxides with Narrow Composition Ranges

Small deviations from perfect stoichiometry can be considered to arise from an unbalance between the primary defects as described in the previous section. Take, for instance, a Schottky defect consisting of a pair of metal and oxygen vacancies in an oxide with the composition MO_{1+x}—e.g., TiO_x (Banus and Reed, 1970)—then metal vacancies will predominate when $x > 0$, whereas oxygen vacancies will be present in the greatest concentration when

$x < 0$. Similar to stoichiometric oxides, defect formation in nonstoichiometric systems has been treated by a statistical thermodynamic approach (Swalin, 1972) that gave the following relationship between the oxygen pressure (p_{O_2}) and the deviation from stoichiometry $-x$ in MO_{1+x}:

$$\frac{p_{O_2}(x)}{p_{O_2}(x=0)} = 1 + \frac{x^2 \pm x(x^2 + 4n_i^2)^{1/2}}{2n_i^2}, \tag{13}$$

where n_i designates the intrinsic disorder at the stoichiometric composition, i.e., $n_i = N_S/N$ (N_S is the number of Schottky pairs and N the number of cation or anion sites). $p_{O_2}(x)$ and $p_{O_2}(x = 0)$ designate the oxygen pressures in equilibrium with the oxide when it has a nonstoichiometric and a stoichiometric composition, respectively. From the equation, which also applies if the deviation from stoichiometry is described in terms of an unbalance between defects of the Frenkel type (Libowitz, 1965), i.e., vacancies plus interstitial ions, it is clear that the intrinsic disorder plays a significant role in determining the extent of deviation obtained for a given oxygen pressure. Generally, a greater intrinsic disorder results in a larger deviation from stoichiometry and the largest extension of the nonstoichiometric range should thus be expected for systems in which primary defects are easily formed.

In the original treatment by Wagner and Schottky, ionization of the defects and electronic disorder were not taken into account. If, however, the electronic disorder is so great that the activities of electrons and holes are independent of composition, then, as recently shown by Iyengar (1973), the higher the state of ionization of the defects, the greater the deviation from stoichiometry for a given oxygen pressure; whereas the opposite effect is observed if the electronic disorder is small; i.e., all defects are in the same state of ionization. The statistical approach leading to Eq. (13) is, however, incomplete as it only takes the electronic disorder of the defects into account and ignores the electronic properties of the host lattice atoms. As shown in the previous section, electrons liberated, for instance, during ionization of an oxygen vacancy in an ionic crystal may be considered to change the oxidation state of the cations, and the processes leading to nonstoichiometry can only be described in a realistic way if this electronic disorder is also included. The importance of such equilibria can be seen from the fact that the largest deviation from stoichiometry is always observed in systems where there is a small energy difference between the oxidation states of the cations (transition-metal oxides for instance).

For small concentrations of randomly distributed and noninteracting points defects, the various equilibria involving the formation of defects and the electronic disorder can, similar to the essential stoichiometric oxides according to the approach taken by Kröger and Vink, be described by quasi-chemical equations. In the case of an oxygen-deficient oxide, for instance, the formation and ionization of oxygen vacancies will involve the total reaction:

$$O_O + 2M_M = V_O^{\cdot\cdot} + 2M_M' + \tfrac{1}{2}O_2 \tag{14}$$

where, according to Kröger's notation, O_O and M_M represent oxygen and metal atoms on their respective sites in the crystal lattice, $V_O^{\cdot\cdot}$ represents a double, positively charged oxygen vacancy, and M_M' a negatively charged metal ion (e.g., M^{3+}, which is negative relative to a normal lattice with M^{4+}) in the normal cation lattice, which has taken up one of the electrons liberated during ionization of the neutral oxygen vacancies primarily formed.

By assuming that the law of mass action is valid for this equilibrium, by introducing the neutrality condition

$$[M_M'] = 2[V_O^{\cdot\cdot}] \tag{15}$$

and finally by expressing $[V_O^{\cdot\cdot}]$ as the fraction of unoccupied sites in the oxygen lattice—x in MO_{2-x}—i.e.,$[V_O^{\cdot\cdot}] = x/2$, it can be shown that the composition of the oxide should depend on the oxygen pressure according to

$$x \propto p_{O_2}^{-1/6}. \tag{16}$$

In deriving this equation it was assumed that double charged oxygen vacancies ($V_O^{\cdot\cdot}$) were formed. If, however, single charged or neutral vacancies are formed, direct proportionality between x and p_{O_2} will also be observed, but the exponents will in these cases be $-\frac{1}{4}$ and $-\frac{1}{2}$, respectively. Other types of defects (e.g., metal vacancies) or defect clusters would give still other exponents, but generally

$$x \propto p_{O_2}^{-1/n}. \tag{17}$$

If this treatment is valid, the important thermodynamic quantity $\Delta\bar{G}_{O_2}$ (relative partial free energy of oxygen) should depend on the composition in the following way:

$$\Delta\bar{G}_{O_2} = RT \ln p_{O_2} \propto -nRT \ln x, \tag{18}$$

and a straight line should thus be expected if experimental $\Delta\bar{G}_{O_2}$ values obtained at a constant temperature are plotted as a function of ln x, provided that only one type of defect is formed $-n$ constant. This representation is very similar to that used by Brouwer (1954), and by Kröger and Vink, for analysis of the type of defects predominating at different oxygen pressures, and in which the logarithm of defect concentrations is plotted as a function of log p_{O_2}. In studies of oxides, however, $\Delta\bar{G}_{O_2}$ is an important thermodynamic quantity, as shown in Section B, and the results obtained in a recent investigation of the CeO_2–Ce_2O_3 system (fluorite structure), which was analyzed according to Eq. (18), will be described in the next section.

Under the assumptions made in the classical defect theory described above—randomly distributed and noninteracting defects—the following relations between composition and $\Delta\bar{H}_{O_2}$ (relative partial enthalpy of oxygen) or $\Delta\bar{S}_{O_2}$ (relative partial entropy of oxygen) can be derived for the formation of $V_O^{\cdot\cdot}$ in an oxygen-deficient oxide (Kofstad, 1972; Panlener, 1975) (Eq. 14)):

$$\Delta\bar{H}_{O_2} = -2\Delta H_{V_O^{\cdot\cdot}} \tag{19}$$

and

$$\Delta\bar{S}_{O_2} = -(S^0_{O_2} + 2\Delta S_f^{\ v} - 6R\ln x + 4R\ln 2) \tag{20}$$

or

$$\partial\,\Delta\bar{S}_{O_2}/\partial\ln x = 6R, \tag{21}$$

if the vibrational entropy is considered to be independent of composition. In these equations $\Delta H_{V_O^{\cdot\cdot}}$ is the enthalpy of formation of oxygen vacancies, S_{O_2} the entropy of oxygen gas (~ 60 eu at 1000°C), and $\Delta S_f^{\ v}$ the total vibrational entropy change. According to these equations, $\Delta\bar{H}_{O_2}$ should thus be independent of composition whereas $\Delta\bar{S}_{O_2}$ should vary linearly with composition if the assumptions made in the classical defect theory are valid.

A typical example of nonstoichiometric oxide systems, which at first sight appear to conform to the classical defect theories outlined above, are the transition-metal monoxides with the NaCl- structure. The composition ranges and the predominant types of defect observed for these oxides are given in Table I, which has been compiled from a similar table and phase diagrams published by Navrotsky (1974), and from phase diagrams published elsewhere (Fender and Riley, 1969, 1970; Watanabe *et al.*, 1970).

Comparing the data summarized in the table, the following important generalizations about the behavior of these oxides can be made:

(1) The nonstoichiometric range for the metallic oxides in the first part of the transition-metal series (TiO, VO, and NbO) is generally much wider than the range covered by the essential ionic oxides (MnO, FeO, CoO, and NiO) in the last part of the series. Especially for $TiO_{1\pm x}$ and $VO_{1\pm x}$, the nonstoichiometric range becomes broader with increasing temperature, although this is also observed to some extent for MnO_{1+x} and FeO_{1+x}.

(2) For the metallic oxides, deviations from stoichiometry are found both on the oxygen-rich side (metal deficiency) and the metal-rich side (oxygen deficiency), whereas the ionic oxides only show deviations on the oxygen-rich side. The reason for this difference is that both lower and higher oxides exist for the metallic oxides (Ti_2O/Ti_2O_3 and V_2O/V_2O_3), whereas only higher oxides exist for the ionic oxides.

(3) The predominant defects formed in the metallic oxides are cation or anion vacancies depending on the composition. In all three oxides a strong

TABLE I

COMPOSITION RANGES AND DEFECT TYPES OF TRANSITION-METAL MONOXIDES

Oxide	Temp. (°C)	Nonstoichiometric composition range	Predominant defect type
$TiO_{1\pm x}$	1400	0.72–1.27[a]	Strongly interacting cation and anion vacancies
	1000	0.89–1.22[a]	
	below 900		Ordered vacancies[a,b]
$VO_{1\pm x}$	1400	0.85–1.27[c]	Strongly interacting cation and anion vacancies
	1000	0.85–1.22[c]	
	below 700		Ordered vacancies[d]
MnO_{1+x}	1500	1.00–1.13[e]	Cation vacancies + Mn^{3+} ions
	1000	1.00–1.01[e]	Probably change in defect structure with increasing defect concentration
FeO_{1+x}	1200	1.05–1.17[c]	Cation vacancies + Fe^{3+} ions
	800	1.05–1.13[c]	Defect clustering. Probably change in defect structure with increasing defect concentration
	below 350	metastable	Superstructure ordering in metastable existence range[f]
CoO_{1+x}	1200	1.00–1.01[c]	Cation vacancies + Co^{3+} ions. Some interactions between defects at higher defect concentrations[g]
NiO_{1+x}	1000	1.00–1.0002[c]	Cation vacancies + Ni^{3+} ions
$NbO_{1\pm x}$	1025	0.95–1.02[c]	Strongly interacting cation and anion vacancies

[a] Watanabe *et al.* (1970).
[b] Watanabe *et al.* (1967).
[c] Navrotsky (1974).
[d] Westman and Nordmark (1960).
[e] Fender and Riley (1970).
[f] Manenc *et al.* (1962).
[g] Bransky and Wimmer (1972).

interaction between the defects has been observed and at lower temperatures ordered vacancy structures (superstructures) are formed.

(4) The predominant defects formed in the ionic oxides are singly or doubly ionized cation vacancies (V_M' or V_M''). The positive holes formed by this ionization are associated with lattice metal ions which are then transformed into a formal higher oxidation state. The defect structure of FeO_{1+x}, and probably also of MnO_{1+x}, is very complex—defect clustering and changes in types of defect with increasing defect concentration—whereas

independent and randomly distributed defects are believed to be present in CoO_{1+x} and NiO_{1+x} to a great extent.

The nature and extent of nonstoichiometry in the transition-metal monoxides MnO, FeO, NiO, and CoO have been analyzed by Kröger (1968) in terms of their electronic properties. As explained in Section B, a wide, asymmetric curve is generally observed when the free energy G of a nonstoichiometric phase is plotted against its composition (mole fraction of oxygen x_O*, for instance). For a binary compound, MX, the slope of the tangent to this curve at the stoichiometric composition, $(dG/dx)_O$, will be related to the chemical potentials of the components by the relation

$$\mu(\mathrm{X})_O - \mu(\mathrm{M})_O = (dG/dx)_O, \tag{22}$$

as pointed out by Brebrick (1967, 1961) and Kröger, and the sign and magnitude of $(dG/dx)_O$ can thus be taken as a measure for the ability of a stoichiometric phase to form a nonstoichiometric phase. If $(dG/dx)_O$ is negative, for instance, then there is a tendency for achieving deviations from stoichiometry by incorporation of excess nonmetal atoms, whereas there is a tendency to incorporate excess metal if $(dG/dx)_O$ is positive.

The values of $\mu(\mathrm{X})_O$ and $\mu(\mathrm{M})_O$ depend on the energetics of the processes leading to nonstoichiometry, which for incorporation of excess nonmetal in a transition-metal compound can be written as

$$\mathrm{X(g)} + 2\mathrm{M_M}^{\mathrm{x}} \rightarrow \mathrm{X_X}^{\mathrm{x}} + \mathrm{V_M''} + 2\mathrm{M_M}, \tag{23}$$

whereas the reaction for a nontransition-metal compound will be

$$\mathrm{X(g)} + \mathrm{X_M}^{\mathrm{x}} \rightarrow 2\mathrm{X_X}^{\cdot} + \mathrm{V_M''}. \tag{24}$$

For both types of compound, neutral metal vacancies are primarily formed, but the subsequent ionization of these vacancies is different for the two types of compound. As shown in Fig. 5 (Section III.B), the valence band for the transition-metal compounds is considered to be a high-lying M^{2+} band, whereas the valence band for a nontransition-metal compound will be a low-lying X^{2-} band. For the former compounds, then, the electrons necessary for the ionization of the neutral metal vacancies will be provided by the lattice metal atoms, which become oxidized, but for the latter compounds these electrons must be supplied by the lattice nonmetal atoms, which then become positively charged (relatively).

By expressing the chemical potentials of the components for nontransition- and transition-metal compounds, respectively, in terms of the virtual potentials of the defects formed according to Eqs. (23) and (24), Kröger showed that for transition-metal compounds,

$$\{\mu(\mathrm{X})_O - \mu(\mathrm{M})_O\}_T \equiv \left(\frac{dG}{dx}\right)_O = 2(\Delta E_{OX} - \Delta E_{red}), \tag{25}$$

where ΔE_{OX} and ΔE_{red}, respectively, are the energy difference between the valence bands and conduction bands for the two types of compound, as also indicated in Fig. 5. Using this equation, Kröger further calculated the values of $(dG/dx)_O$ for the oxides MnO, FeO, NiO, and CoO. In all cases he obtained strongly negative values indicating a strong preference for forming nonstoichiometric compounds by incorporation of excess oxygen for these oxides in close agreement with the experimental observations summarized in Table I. Similar calculations were also carried out by Kröger for transition-metal oxides in intermediate oxidation states (Mn_3O_4, Fe_3O_4, etc.) and in the highest oxidation state (Mn_2O_3, Fe_2O_3, etc.). For the intermediate states, strongly negative values of $(dG/dx)_O$ were still obtained, but for the highest oxidation states this calculation gave positive values of $(dG/dx)_O$ indicating that deviation from stoichiometry in these oxides will take place preferentially by incorporation of excess metal, as also observed experimentally (Gardner *et al.*, 1963). Finally, it should also be mentioned that by comparing the $(dG/dx)_O$ values obtained for the three iron oxides with the slopes of the common tangents between their G–x curves, Kröger was also able to explain the width of the nonstoichiometric ranges observed for these oxides.

As shown in Table I, the nonstoichiometric range for many of the transition-metal monoxides is quite extensive and according to the classification given in Section III.A most of these oxides should be considered as grossly nonstoichiometric oxides rather than nonstoichiometric oxides. Furthermore, strong interactions between the defects were observed for many of these oxides, and the characteristics of a truly nonstoichiometric system—randomly distributed and noninteracting defects—are probably only approached in the CoO_{1+x} and NiO_{1+x} systems. Only these two oxides will therefore be treated in this section, whereas the nature of the nonstoichiometry of some of the other transition-metal monoxides will be discussed in the next section.

Previous investigations of the nature and extent of nonstoichiometry in the metal-deficient oxides $Co_{1-y}O$ and $Ni_{1-y}O$ have been thoroughly reviewed by Libowitz (1965) and Kofstad (1972), and for further details these reviews should be consulted. Generally, these investigations have shown that the composition (y) as well as the electrical conductivity (σ) observed for these oxides is proportional to $p_{O_2}^{1/6}$ near the stoichiometric composition, whereas y or $\sigma \propto p_{O_2}^{1/4}$ at greater deviations from stoichiometry. These observations have been interpreted in terms of a metal vacancy model, in which singly and doubly charged vacancies are formed according to

$$\tfrac{1}{2}O_2 \rightleftarrows O_O + V_M' + h^{\cdot} \qquad [V_M'] = K_{V_M'} \cdot p_{O_2}^{1/2}/p, \tag{26}$$

$$V_M' \rightleftarrows V_M'' + h^{\cdot} \qquad [V_M''] = K_{V_M''} \cdot p_{O_2}^{1/2}/p^2, \tag{27}$$

where h˙ is a free hole, $K_{V_M'}$ and $K_{V_M''}$ are the equilibrium constants for the two reactions and p is the concentration of free holes. The electroneutrality condition requires $[V_M'] = p$ and $[V_M''] = p/2$ and from Eqs. (26) and (27)

$$y = V_M' \propto p_{O_2}^{1/4} \qquad \text{for} \quad V_M' \tag{28}$$

and

$$y = V_M'' \propto p_{O_2}^{1/6} \qquad \text{for} \quad V_M'', \tag{29}$$

indicating that V_M'' is predominantly formed near the stoichiometric composition, whereas V_M' is considered to be formed at greater deviations from stoichiometry. For $Ni_{1-y}O$, which only shows small deviations from stoichiometry, only V_M'' is formed, but for $Co_{1-y}O$ both types of defect are important.

In deriving the ideal equations for the oxygen pressure dependence, the effect of impurities and interactions between the defects has been neglected. Careful and accurate thermogravimetric and electrical conductivity measurements have, however, shown that neither $Co_{1-y}O$ nor $Ni_{1-y}O$ near the stoichiometric composition ($y \propto p_{O_2}^{1/6}$ region) exactly follows the ideal behavior predicted by these equations and, as shown by Bransky and Wimmer (1972) for $Co_{1-y}O$, a consistent description can only be obtained for this region if the effect of impurities is also taken into account.

At larger defect concentrations, as observed for $Co_{1-y}O$ (near the CoO/Co_3O_4 phase boundary), deviations from the ideal behavior have also been observed ($y \infty p_{O_2}^{\sim 1/3}$). According to Kofstad, these deviations can be explained by the formation of neutral metal vacancies V_{Co}^x, which are formed according to

$$\tfrac{1}{2}O_2 \rightleftarrows V_{Co}^x + O_O, \qquad V_{Co}^x = K_{V_{Co}^x} \cdot p_{O_2}^{1/2}, \tag{30}$$

but according to the recent work by Bransky and Wimmer the steeper pressure dependence observed can probably also be explained in terms of interactions between the vacancies. These interactions, which are very important for the grossly nonstoichiometric oxides described in the next section, thus also seem to be of some importance for the nonstoichiometric oxides with limited composition ranges although this has not yet been fully clarified.

D. Thermodynamics of Grossly Nonstoichiometric Oxides

For the small concentration of point defects formed in the nonstoichiometric oxides with narrow composition ranges considered in the previous section, the assumptions made in the classical defect theories—noninteracting and randomly distributed defects—were considered to be approximately fulfilled. For the grossly nonstoichiometric oxides, which extend over wide composition ranges, the apparent point defect concentration, however, is so high—for $Fe_{0.85}O$, for instance, the apparent cation vacancy concentration will be 15%—that these assumptions can no longer be fulfilled. The defect structure for these oxides, therefore, will be dominated by extended defects in the form of defect clusters, long-range ordering of the defects into superstructures, or in some systems by shear structures formed by an elimination of the defects by a crystallographic shearing mechanism. The first step in clustering, formation of superstructures, and shearing requires some interactions between the single point defects, and before the thermodynamics of grossly nonstoichiometric compounds is reviewed the nature of such interactions will be considered.

1. Defect Interactions in Grossly Nonstoichiometric Oxides

As already indicated in Table I, most of the transition metal monoxides form grossly nonstoichiometric phases at higher temperatures, which, upon cooling, are transformed into ordered phases (superstructures). This behavior has also been observed in many other nonstoichiometric oxide systems—for the systems CeO_{2-x} and UO_{2+x}, for instance, the grossly nonstoichiometric high-temperature phases are transformed into two-phase mixtures of ordered phases upon cooling, as indicated in the phase diagrams shown in Fig. 7.

For a compound of perfect stoichiometry and for a true nonstoichiometric system with randomly distributed and noninteracting defects, the high endothermicities (ΔH positive) connected with the formation of defects were more than compensated by the increase in entropy associated with the formation of a random defect structure. The same is probably true for a grossly nonstoichiometric compound at very high temperatures, but at intermediate and lower temperatures the endothermicity can be decreased considerably by interactions between the defects, which exist in much larger concentrations in these compounds, although this will be offset to a small extent by the decrease in entropy associated with these interactions. Supporting evidence for the importance of such interactions has been given by Bertaut (1953), who calculated the electrostatic lattice enegery for $FeS_{1.14}$ (pyrrhotite) supposing that the defects (Fe^{3+} ions and iron vacancies) were either randomly distributed over all cation sites or the vacancies were ordered into alternate cation sheets. The calculation showed that the energy liberated by ordering of the vacancies was very large (320 kcal/mole), and it is probably correct to conclude that generally a nonstoichiometric phase with random distribution of vacancies or interstitials should be highly endothermic compared to ordered phases of fixed composition. Using Bertaut's

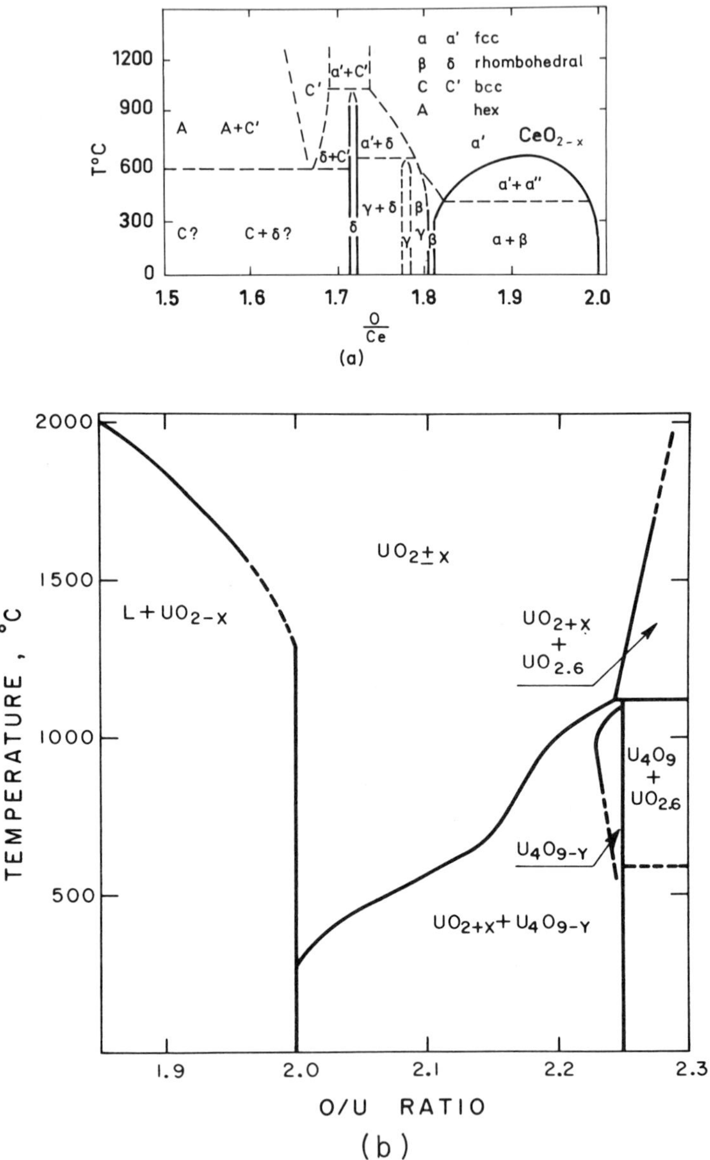

Fig. 7. Phase diagrams of nonstoichiometric fluorite-oxide systems showing two-phase regions and line phases at lower temperatures and grossly nonstoichiometric regions at higher temperatures [Ce–O diagram after Blank (1967) and U–O diagram after Roberts and Markin (1967)].

method, electrostatic lattice energy calculations were also carried out by Anderson and Burch (1971) for a random distribution of vacancies and for the observed shear structures in the TiO_x system. In accordance with Bertaut's results these calculations also showed that the electrostatic energy is lowered by 10–15% for the ordered shear structures compared to that of the random distribution of vacancies.

The forces operating during ordering of defects can be divided into long-range and short-range forces, which can be either attractive or repulsive. Considering first the attractive long-range forces between single defects of similar and equal charges, which operate over distances larger than the unit cell, these can be considered to give

(i) superstructures of large unit cells,
(ii) a collection of vacancies or interstitial ions along certain directions or crystallographic planes (precursor for shear structures), and
(iii) nucleation of a new phase of different stoichiometry.

Characteristically the formation of these structures requires some movement of the defects and they can therefore only be formed at intermediate temperatures where the diffusional mobility is sufficiently high. There is also substantial evidence that ordering is a very rapid process. In studies of the CeO_{2-x} system, for instance (Bevan and Kordis, 1964), it was impossible to freeze the defective fluorite phase, which exists at moderate temperatures, by quenching, and O'Keeffe and Stone (1962) showed that even by extremely rapid quenching of $Cu_2O_{1.0034}$ only a small percentage of the copper vacancies present at 1000°C could be retained at room temperature.

Some energy is gained by long-range ordering, but much higher gains should be expected when short-range ordering within a single or a few unit cells takes place. As explained in the previous sections, defect pairs are formed that consist, for instance, of metal vacancies and electron holes associated with the host lattice cations (e.g., Fe^{2+}, $h^{\cdot}$, or Fe^{3+}) or oxygen vacancies and electrons, which are also associated with cations (e.g., Ce^{4+}, e', or Ce^{3+}). Especially in ionic compounds a strong attraction between these oppositely charged point defects should be expected, and for these systems it has been postulated that defect complexes are formed at intermediate temperatures by short-range interactions both between the two types of defect and between these defects and the surrounding crystal lattice. For the metal-deficient ferrous oxide, $Fe_{1-x}O$, for instance, Koch and Cohen (1969) inferred from their x-ray crystallographic examinations on quenched samples with the composition $Fe_{0.9}O$ that the defect complexes in this oxide consist of four tetrahedral Fe^{3+} ions surrounded by 13 vacant cation sites, as shown in Fig. 8. Ferrous oxide is usually regarded as a defective rock salt structure with all cations in octahedral sites, and the Fe^{3+} ions

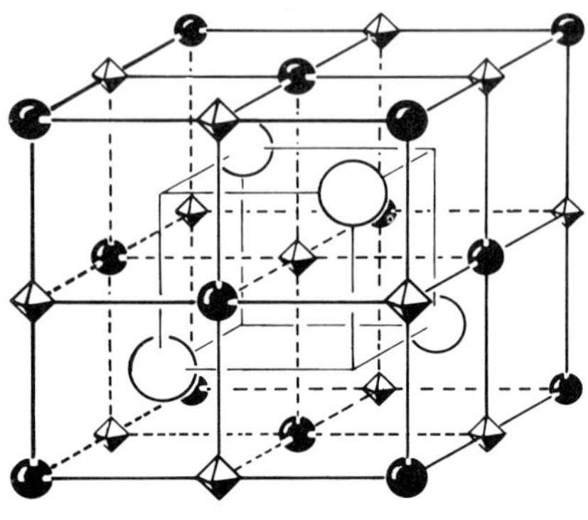

Fig. 8. Structure of Koch–Cohen cluster for $Fe_{1-x}O$.

formed in the defect reactions:

$$\tfrac{1}{2}O_2 \rightleftarrows O_O + V_{Fe}^{x} \tag{31}$$

$$Fe_{oct}^{2+} + V_{Fe}^{x} \rightleftarrows V_{Fe}' + Fe_{oct}^{3+} \tag{32}$$

are considered to be displaced to a tetrahedral site to form an additional vacancy according to

$$Fe_{oct}^{3+} \rightleftarrows Fe_{tet}^{3+} + V_{Fe}. \tag{33}$$

The same type of complex has also been inferred from the diffuse neutron-scattering measurements, and the actual existence of this complex at higher temperatures was shown by the neutron diffraction studies of Cheetham *et al.*, (1971) in the temperature range 800–1200°C. The Mössbauer measurements by Greenwood and Howe (1972) also confirm that defect complexes are formed in ferrous oxide, although they suggest that the complex consists of five tetrahedral Fe^{3+} ions surrounded by 16 vacancies. Finally it should be noted that the total charge of the vacancies (negative) cannot be fully compensated by the four tetrahedral Fe^{3+} ions. The complex will therefore be negatively charged and a compensating atmosphere of Fe^{3+} ions is considered formed in the surrounding matrix. Because of the net negative charge of the complexes, these will tend to order under the action of Coulombic forces, and the development of superstructures observed in the metastability region at lower temperatures has been explained by such long-range ordering phenomena, which in principle could generate a family of discrete phases.

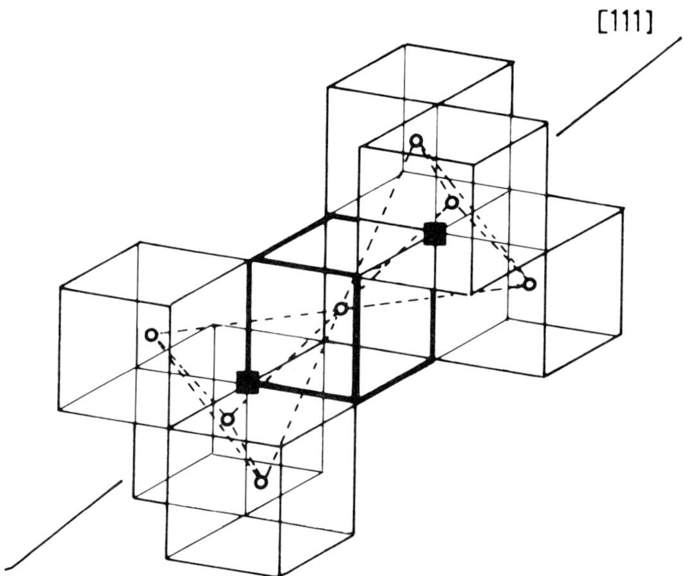

Fig. 9. Structure of defect complex proposed for oxygen-deficient fluorite-related oxides by Thornber and Bevan (1970). O, cation; ■, empty anion site.

For oxygen-deficient oxides, the defect structure has also been explained in terms of defect complexes. In ionic oxides of the fluorite structure, for instance, the positively charged oxygen vacancies and the negatively charged cations have been proposed by Thornber *et al.* (1970) to associate in a defect complex of the type shown in Fig. 9, in which a pair of vacancies is situated on the diagonally opposite corners of the cubic coordination cell of a cation of lower charge. Long-range ordering of such complexes has been proposed to take place preferentially along the [111] directions of the fluorite structure, and in this way the structure of the ordered intermediate phases observed in the CeO_{2-x}, PrO_{2-x}, and Y_2O_3–YOF systems, as well as the structure of the cubic (type C) L_2O_3 phase (L = Ce, Pr, Tb) has been described. In the intermediate phases the defect complexes are considered to order linearly along a single ⟨111⟩ direction whereas nonintersecting strings of complexes along all four ⟨111⟩ directions have been proposed for the L_2O_3 phases.

Recently Martin (1974) and Hoskins and Martin (1975) introduced another type of defect cluster to describe the structure of intermediate phases of oxygen-deficient oxides. In these phases, whose compositions can all be described by the formula M_nO_{2n-2}, the basic defects are considered to be octahedrally coordinated oxygen vacancies formed by attractive interactions between a positive oxygen vacancy and the surrounding six oxygen ions in the oxygen lattice. These defect clusters are considered to gather on oblique

{213} planes and, depending on the thickness of the defect-free layers separating these planes, the structure of the different intermediate phases of the series can be described. Whether the defect structure of oxygen-deficient oxides is best described by coordinated vacancies, or by defect complexes consisting of vacancy pairs, as described above, cannot definitely be decided before structure determinations have been carried out on at least some of the members of the homologous series, M_nO_{2n-2}. One drawback of the model involving coordinated vacancies is, however, that it neglects the role of the reduced cations and thus the electronic properties of the crystal. Intuitively, therefore, the model involving pairing of vacancies around a cation of lower charge appears to be more correct.

In less ionic oxygen-deficient oxides, such as PuO_{2-x}, yet another defect complex consisting of one oxygen vacancy and two Pu-ions has been proposed. Originally this defect complex was inferred from measurements of the Pu-self-diffusion in mixed (U, Pu)O_{2-x} by Schmitz and Marajofsky (1974), but later Manes and Manes-Pozzi (1975) showed that the thermodynamic properties of PuO_{2-x} can also be described in terms of such defect complexes. In contrast to the complexes postulated for the ionic oxides, the electrons supplied by the ionization of the oxygen vacancy are in this complex considered to form bonds with adjacent cations, and this complex can thus be characterized as a quasi-molecular complex (Pu_2O_3 nucleus) with a well-defined energy of formation. In Manes' statistical thermodynamic treatment, the interactions between these molecular complexes were considered to be long-range electrostatic forces, which arise from dipolar moments established in the complexes due to the transfer of electronic charge from the vacancy to the neighboring cations. Taking these interactions as well as the energy of formation of the complexes into account, Manes showed that this type of defect can exist isolated only within a limited composition range ($2.00 \leq O/M \leq 1.91$), whereas it orders into other types of defect complexes containing more oxygen vacancies at larger deviations from the stoichiometric composition.

That the same type of complex cannot exist throughout a nonstoichiometric phase extending over a considerable range has also been postulated for other systems. For the CeO_{2-x} system, for instance, Sørensen (1976) showed by thermogravimetric measurements in atmospheres of well-defined oxygen pressures that the nonstoichiometric α'-phase (see Fig. 10) can be divided into several subregions, some of which are nonstoichiometric and some of which apparently consist of a succession of intermediate phases separated by two-phase regions. For the nonstoichiometric subregions, linear isotherms but with different slopes were obtained for the subregions when the experimental $\Delta\bar{G}_{O_2}$ values were plotted against log x (x in CeO_{2-x}), indicating that different types of defect complex exist in the different sub-

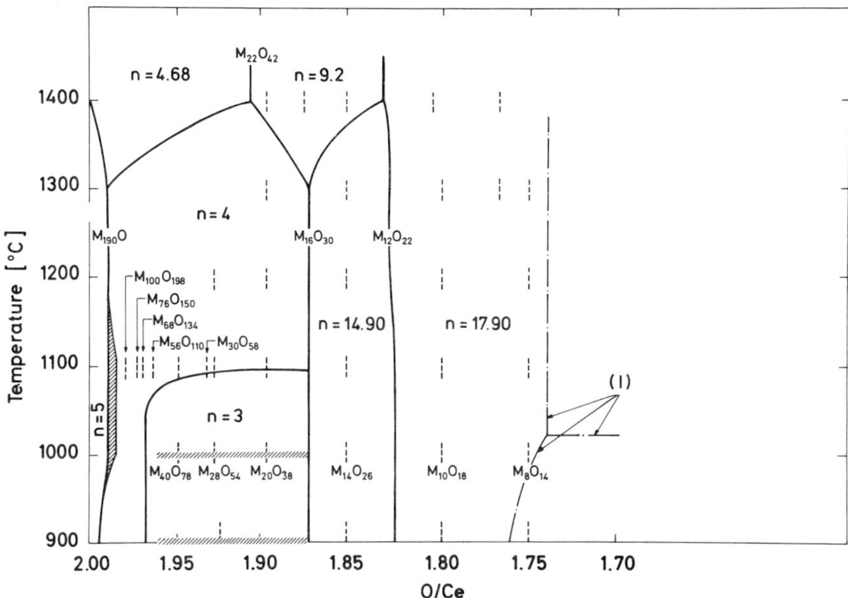

Fig. 10. Diagram of subregions with possible ordered intermediate phases in the α' phase for the Ce–O system. The n values given in the different regions were calculated from the slope of a $\Delta\bar{G}_{O_2} - \log x$—$x$ in CeO_{2-x}-plot ($\Delta\bar{G}_{O_2}\alpha = -nRT \ln x$). Signatures: ______, Phase boundaries; -----, Possible intermediate phases; [hatched], Possible two phase regions; —·—; phase boundaries from Bevan and Kordis diagram (1964).

regions. By a similar analysis Raccah and Vallet (1965) showed that also the single nonstoichiometric $Fe_{1-x}O$ phase can apparently be divided into several subregions (three). Supporting evidence for this behavior was also obtained by Fender and Riley (1969) from solid-state electrochemical measurements for $Fe_{1-x}O$, as well as for a similar behavior of the $Mn_{1-x}O$ system (Fender and Riley, 1970). In these oxides, different patterns of defect ordering were also postulated for the three subregions, and the boundaries between the regions were taken to signify second-order phase transitions. The recent direct measurements by high-temperature calorimetry of the relative partial enthalpy of oxygen ($\Delta\bar{H}_{O_2}$) in wüstite by Marucco *et al.* (1970), which probably are more reliable than the values derived from equilibrium data by Fender *et al.*, do not, however, support the idea of a subdivision of the nonstoichiometric phase field of $Fe_{1-x}O$, whereas good agreement between derived and directly measured $\Delta\bar{H}_{O_2}$ values was obtained for the subregions observed in the CeO_{2-x} system.

Also in anion-excess oxides, such as UO_{2+x}, convincing evidence has been provided for the existence of defect complexes. As shown by Willis

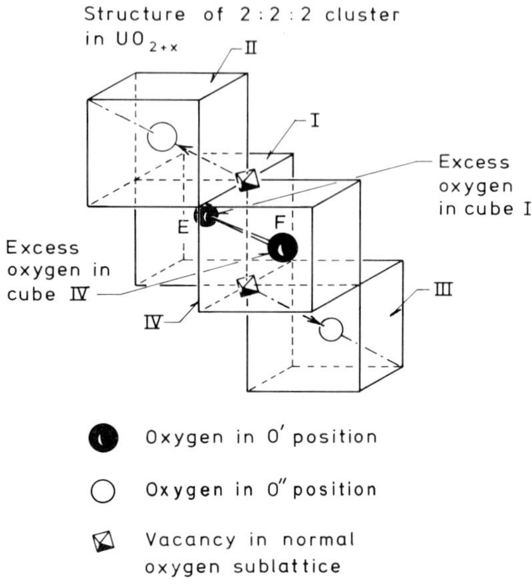

Fig. 11. Structure of defect complex proposed for UO_{2+x} by Willis (1964).

(1964) by neutron diffraction studies on U_4O_9 and $UO_{2.12}$, excess oxygen does not simply enter the high-symmetry interstitial site $\frac{1}{2}, \frac{1}{2}, \frac{1}{2}$ of the fluorite lattice but is paired with oxygen vacancies to form a defect complex consisting of two kinds of interstitial oxygen atom ($O_i{}^a$ and $O_i{}^b$) and one oxygen vacancy (V_O). The basic unit can be described as a $O_i{}^a : O_i{}^b : V_O = 2:2:2$ complex, i.e., it consists of two interstitial atoms of each type and two vacancies (Fig. 11). At low temperatures these complexes are considered to order completely to form the superstructure of U_4O_9 (Ishii *et al.*, 1971; Naito *et al.*, 1967), which can nucleate and grow as a separate phase, whereas at intermediate temperatures these complexes are randomly distributed but anchored by the immobile compensating cations (U^{5+}). At higher temperatures, however, electrical conductivity measurements by Matsui and Naito (1975) on the UO_{2+x} and U_4O_{9-y} phases seem to indicate that these defect complexes dissociate into doubly charged oxygen interstitials (O_i'') and doubly charged oxygen vacancies ($V_O^{\cdot\cdot}$). This is an important principle that probably also applies at high temperatures to many other nonstoichiometric systems in which the defect structure can be described in terms of defect complexes.

A characteristic feature of the defect complexes described above is that they are capable of coherent intergrowth in three dimensions with the parent structure. For the complexes considered for the oxygen-deficient oxides, for instance, only the oxygen lattice is affected, whereas the parent cation lattice retains its original structure. The type of complex proposed for the metal-deficient oxides is also dimensionally compatible with the matrix, and generally both types of complex can be formed without considerable strain or interphase energy being involved. Another characteristic property of the complexes, which it is important to note, is that empty cation or anion sites are so tightly bound that they cannot be regarded as defects, but should rather be considered as structure elements. The properties of the oxides depending on such defects—e.g., diffusion—therefore involve a dissociation of the complexes as discussed by Greenwood and Howe (1972b).

So far the effect of temperature on the formation of defect complexes has been described, but undoubtedly there is also a concentration effect. At low concentrations of defects near the stoichiometric composition the defects are probably still randomly distributed and noninteracting, as in the real nonstoichiometric compounds treated in the previous section. At intermediate temperatures, however, the tendency for defect complex formation will soon become important and the composition range in which single point defects exist is probably very narrow. The thermodynamic properties of a grossly nonstoichiometric compound close to the stoichiometric composition should thus be described in terms of single defect–defect complex equilibria. At large concentrations of defect complexes, the crystal lattice becomes saturated and a new phase will start to nucleate. The composition at which this will happen depends on the form of the free energy versus composition curve for the grossly nonstoichiometric phase, and the relative position of this curve with respect to the free energy curves for adjacent phases. As discussed in Section B, a general requirement for thermodynamic equilibrium between a nonstoichiometric and an adjacent phase is that a common tangent can be drawn to the free energy curves for the two phases, but in some systems nucleation of a second phase can be kinetically hindered and the possibility of an extension of the nonstoichiometric phase into a metastable region must also be taken into account.

2. Thermodynamic Models for Grossly Nonstoichiometric Oxides

In the previous section it was shown that defect interactions are considered to be of great importance in grossly nonstoichiometric systems, and that the original model of Schottky and Wagner, which considers randomly distributed and noninteracting defects, cannot be applied at appreciable defect concentrations. The first approach taking interactions into account was the statistical thermodynamic treatment by Anderson (1946), Libowitz (1962), and Rees (1954), which considers pairwise interactions between randomly distributed but like defects on nearest neighbor sites. In this model only one type of defect is considered to be present, and it is further assumed that the interaction energy is constant and independent of the degree of deviation from stoichiometry. For high concentrations of defects this is obviously not a good assumption, and this model can therefore only be expected to be useful at relatively small deviations from stoichiometry. In the case of a compound MX_s, in which nonstoichiometry arises because of the formation of X vacancies, Libowitz, by including both the energy of formation of the vacancies (E_v) and the interaction energies (E_{vv}) in the partition function, showed that

$$\ln p = \ln p_O + 2\ln[n/(s-n)] + (2E_{vv}/skT)(s-2n), \tag{34}$$

where p is the pressure of the volatile component (X_2) in equilibrium with MX_s, s is the X/M ratio at the stoichiometric composition, n is the X/M ratio at the nonstoichiometric composition, and z is the number of nearest neighbor X sites around each X vacancy. If the interaction energy E_{vv} is considered as the energy liberated when a vacancy pair is formed, then negative energies will signify that the interactions are attractive and that clustered configurations are favored. Below a critical temperature, which depends on the magnitude of E_{vv} according to

$$T_c = -zE_{vv}/4k, \tag{35}$$

there is thus a tendency for clustering, which can also be illustrated by plotting p from Eq. (34) as a function of composition (n/s). As shown in Fig. 12, the composition is a continuous function of p for $T > T_c$, but for $T < T_c$ three compositions ($n/s = 0.5$, and two other values) are obtained for a given pressure, p_O. For a given compound, a single nonstoichiometric phase extending over the entire composition range is thus predicted from this model at higher temperatures, whereas formation of a two-phase range consisting of defect-rich and defect-poor phases is predicted at lower temperatures. The magnitude of the interaction energies can also be taken as a measure of the ability of a given system to form nonstoichiometric compounds. As also indicated in Fig. 12, high interaction energies not only

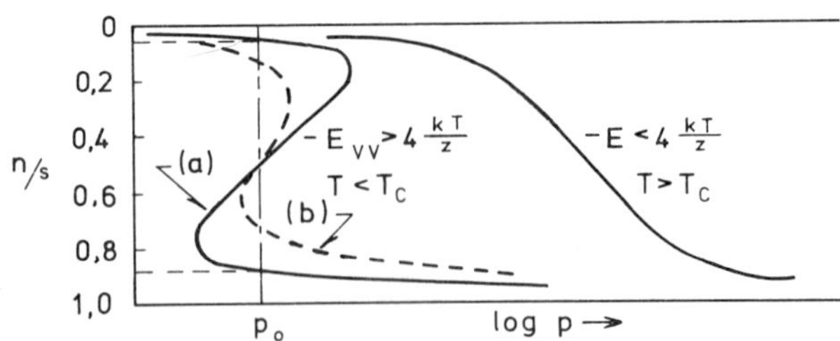

Fig. 12. Theoretical pressure–composition curves according to Anderson and Libowitz's model. Interaction energy larger (more negative) for curve (a) than for curve (b).

tend to favor clustering at a given temperature, but also influence the shape of the curves so that the composition range in which clustering takes place also becomes broader at large interaction energies, as pointed out by Libowitz (1965).

Although the model by Anderson and Libowitz has been applied with some success on nonstoichiometric lanthanide hydrides, and it reproduces the general behavior observed in many nonstoichiometric systems—two-phase systems or ordered phases at low temperatures and grossly nonstoichiometric phases at higher temperatures—it is obviously too crude to describe the detailed behavior of any system. One limitation of this model is that it only considers pairwise interactions between like defects, whereas a complete treatment would require that interactions of any defect with all other defects and with the surrounding lattice ions should be taken into account. Mathematically this would, however, be very difficult and so far this has only been tried for near-stoichiometric systems with very small defect concentrations by Allnatt and Cohen (1964a,b) by using cluster theory. Hoch (1962; 1963a,b, 1964) also applied a pairwise interaction model in his calculation of vacancy interaction energies in transition metal monoxides. For TiO_x, VO_x, and NbO_x, which contain oxygen and metal vacancies (see Table I), however, pairwise interactions both between like defects and between unlike defects were taken into account. In this way reasonable interaction energies could be obtained and for TiO, for instance, the critical temperatures for ordering calculated from the interaction energies agree well with experimental data.

Another limitation of the Anderson–Libowitz model is that it neglects the temperature dependence of the energy of formation of the defects. According to Thorn and Winslow (1966), who considered the $UO_{2\pm x}$ system, such temperature dependencies can be introduced by taking the vibrational partition function into account. Assuming that the occupation of the cation sites in the fluorite crystal was constant, vibrational partition functions for interstitial oxygen and for oxygen vacancies were derived and equations describing the behavior of both hyperstoichiometric (UO_{2+x}) and hypostoichiometric urania (UO_{2-x}) could be obtained by an extension of Anderson's theory. For hypostoichiometric material near the lower phase boundary between $U + UO_2$ and UO_{2-x}, thermodynamic quantities calculated from Thorn and Winslow's model agree fairly well with experimentally determined quantities (Tetenbaum and Hunt, 1968; Wheeler and Jones, 1972; Markin *et al.*, 1968; Javed, 1972), but near the stoichiometric composition the fit is poor. The reason for this probably is that this model only takes oxygen vacancies into account, whereas optical measurements (Ackerman *et al.*, 1959) have shown the presence of at least two types of defect—interstitials and vacancies—near the stoichiometric composition. At the very high temperatures at which UO_{2-x} exists it is also possible that uranium vacancies are formed, and in order to treat this system in a realistic way many more energy parameters should be taken into account. Finally, the relative partial entropies of oxygen, $\Delta\bar{S}_{O_2}$, calculated from experimental results (Tetenbaum and Hunt, 1968) have also shown that there is an entropy minimum in the hypostoichiometric region at O/U = 1.98, indicating that partial ordering of the lattice defects takes place (Thorn, 1966). The appearance of such minima, however, seems to be in accordance with predictions based on defect theory (IAEA (1965)).

In the case of hyperstoichiometric urania the Thorn and Winslow model assumes that there is one interstitial site per uranium atom available for the excess oxygen atoms. Compared to the Willis defect complex described in the previous section, this is an oversimplification and in fact a comparison with experimental data (Thorn and Winslow, 1966; Tetenbaum and Hunt, 1968; Markin *et al.*, 1968) has shown that this model is not valid for oxygen contents larger than $UO_{2.08}$. Except for compositions very close to the stoichiometric composition, where the defect structure is very complex as explained for the UO_{2-x} system, the model does, however, give a fairly good description of the experimental data up to this limit at higher temperatures (well above 1100°C) where the Willis defect complexes probably are dissociated into single defects (O_i'' and $V_O^{\cdot\cdot}$), as shown by Matsui and Naito (1975).

At oxygen contents larger than $UO_{2.08}$, the assumption of one available interstitial site per uranium atom in the lattice cannot apply, and at high temperatures the experimental data have been explained by site exclusion effects. As shown in Fig. 13, excess oxygen can enter the fluorite lattice in the interstitial positions at the center of the unit cell ($\frac{1}{2}, \frac{1}{2}, \frac{1}{2}$) and at all equivalent positions at the middle of each edge of the unit cell ($\frac{1}{2}$, 0, 0), (0, $\frac{1}{2}$, 0) and (0, 0, $\frac{1}{2}$). Each unit cell contains four uranium atoms and four interstitial sites—the composition can be written as $U_4O_8\square_4$—and if all interstitial positions were filled with oxygen, UO_3 would be obtained. At small deviations from stoichiometry and at high temperatures all four interstitial sites in the unit cell are considered to be equally available for occupany ($\alpha = 1$, i.e., one site per uranium atom), but at higher oxygen contents it seems unlikely, for steric and energetic reasons, that nearest neighbor interstitial sites are occupied by oxygen. Introduction of an oxygen atom into an interstitial position thus prevents a certain number of neighboring interstitial positions from being occupied. Moreover, in the limit where each unit cell has received one oxygen atom ($\alpha = \frac{1}{4}$), 12 nearest neighbor sites will be excluded, as proposed by Hagemark and Broli (1966)—the composition at the limit will be U_4O_9, corresponding to the introduction of one extra oxygen atom into the center of the unit cell. The fraction of available sites per uranium

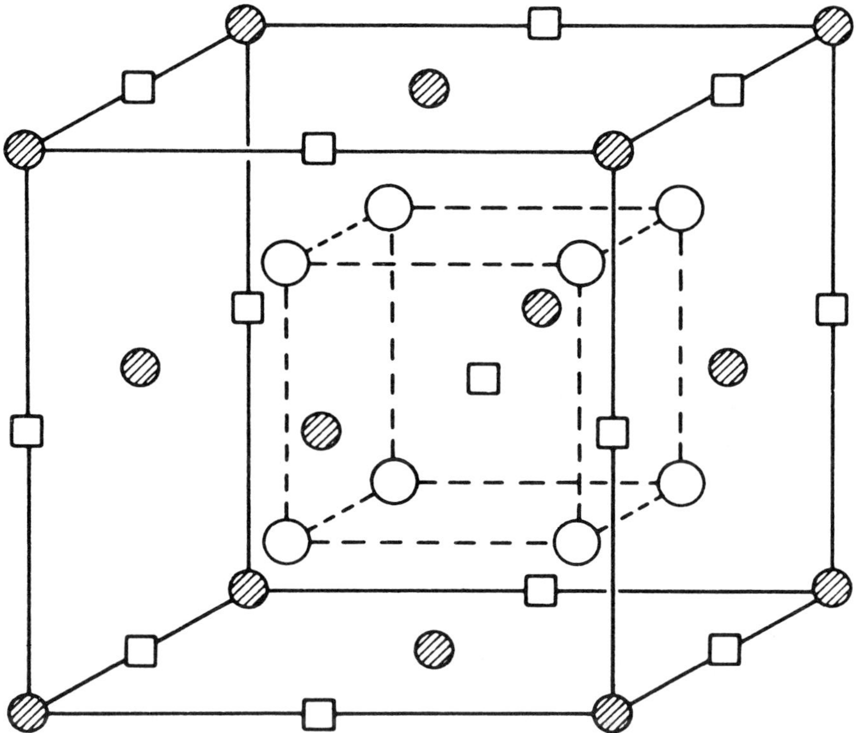

Fig. 13. Interstitial sites in fluorite lattice available for incorporation of excess oxygen.

atom that depends on the oxygen content (x in UO_{2+x}) has been expressed by

$$\alpha \approx 1/(1 + 12x), \tag{36}$$

where α can change smoothly from $\alpha = 1$ at $x = 0$ to $\alpha = \frac{1}{4}$ at $x = 0.25$. By introducing this factor Hagemark derived an equation for the relative partial molar entropy of oxygen ($\Delta\bar{S}_{O_2}$) that fairly well describes his experimental data (1200 and 1400°C) up to the composition $UO_{2.20}$. Thorn and Winslow also considered the effect of site exclusion and for $x \geq 0.125$ they obtained a good agreement with experimental data when a constant value of α ($\alpha = \frac{1}{4}$) was used. The site exclusion principle, which was originally used by Rees (1954) in his very general statistical treatment of nonstoichiometric binary compounds, thus appears to be an important principle for nonstoichiometric systems with large concentrations of interstitial atoms, but it has also been applied to interpret the thermodynamic behavior of metal-deficient oxides such as $Fe_{1-x}O$ (Kofstad and Hed, 1968; Seltzer and Hed, 1970).

As described above, the Thorn and Winslow model does not take the formation of defect complexes into account, and for the UO_{2+x} system it cannot be used at temperatures below 1100°C where Willis complexes are formed. Furthermore, the model neglects the electronic disorder arising from the incorporation of excess oxygen—each interstitial oxygen atom creates two positive holes that are associated with two uranium atoms which are then oxidized to U^{5+}. Taking the formation of defect complexes and the electronic disorder into account, however, Gerdanian (1974) recently proposed an improved statistical model for the UO_{2+x} system, which for small departures from stoichiometry ($x < 0.03$) gives a good description of relative partial enthalpies ($\Delta\bar{H}_{O_2}$) measured by microcalorimetry at 1100°C. With this model it is still not possible to treat the system at larger deviations from stoichiometry, but by using the site exclusion principle in connection with the formation of 2:2:2 Willis complexes and U^{5+} ions, Saito (1974) obtained an expression for $\Delta\bar{S}_{O_2}$ which describes his experimental data for UO_{2+x} fairly well over an extended composition range.

Another and more advanced model, also based on the exclusion principle, has been proposed by Atlas for oxygen-deficient oxides (e.g., CeO_{2-x}) (Atlas, 1968a) and for the UO_{2+x} system (Atlas, 1968b, 1970). In this model, as in the models discussed above, the energy of formation of an isolated defect is considered to be constant, independent of composition, whereas the interaction energies are considered to arise from Coulombic forces—i.e., attractive forces between all unlike defect pairs and repulsive forces between all like pairs. For the CeO_{2-x} system, Atlas assumes that the predominating defects consist of singly charged oxygen vacancies ($V_O^{\cdot}$) and reduced cations (Ce^{3+}). The central idea in his model for this system is that the repulsive energies in vacancy–vacancy or reduced cation–reduced cation interactions can be divided into discrete levels—$u_0, u_1, \ldots, u_{LV}$ for the vacancies and $\varepsilon_0, \varepsilon_1, \ldots, \varepsilon_{Le}$ for the reduced cations, where u_0 or ε_0 are the smallest significant and u_{LV} or ε_{Le} are the largest values to be included. Taking a particular defect as center, a sphere with a radius equal to the distance corresponding to the interaction energy level u_0 or ε_0 can be constructed, and the net energy of this particular defect will depend on (i) the number of other vacancies and reduced cations within the sphere (interaction volume), and (ii) the distance between this defect and the surrounding defects. Instead of considering the real distribution of defects in a nonstoichiometric crystal, which is very complicated, a hypothetical distribution with the same total energy is devised. Consider, for instance, that a local excess of the vacancies and reduced cations is produced within a small volume around a given vacancy due to partial ordering, and that the energy in this volume is lowered by a small amount ΔE through this accumulation of defects. The same lowering of the energy could also be achieved by distributing the cationic charge over all cation sites into a uniform distribution in the crystal, and at the same time by decreasing the number of surrounding vacancies. In this hypothetical distribution, which gives a negative image of the real distribution in the respect that when the latter has a local high density of defects the former will have a low density and vice versa, an interaction volume can again be constructed around the vacancy. In this case, however, each interaction volume is now constructed so that it only contains one oxygen vacancy at the center, which can then be said to exclude all other vacancies within this volume. For a vacancy–vacancy pair with a repulsion energy of u_i corresponding to a distance of d_i between the defects, a spherical envelope with a radius $r_i = \frac{1}{2}d_i$ can further be constructed around each vacancy and the configurational entropy will now be given by the number of ways these envelopes containing C_A, $C_A - 1$, $C_A - 2, \ldots, C_A - i$ sites can be distributed over all the anion sites in the crystal (N_A). If there are $n_0, n_1, n_2, \ldots, n_i$ vacancies with the interaction energies $u_0, u_1, \ldots, u_i$, and if the number of anion sites available for n_i vacancies of state i is restricted to those not already occupied by vacancies of lower energies, the number of configurations in the ith level will be

$$\Omega_{v(i)} = \frac{(C_A - i)^{n_i} W_i!}{n_i!(w_i - n_i)!} \tag{37}$$

where w_i is the number of available sites for n_i vacancies. The total number of vacancy configurations is then

$$\Omega_v = \prod_{i=0}^{L_v} \Omega_{v(i)}. \tag{38}$$

By using the same arguments, similar expressions for Ω_e can be obtained for the reduced cations and the total degeneracy used in the partition function can now be calculated from $\Omega_{ve} = \Omega_v \cdot \Omega_e$. Furthermore, the magnitudes of the Coulombic repulsive energies u_i and ε_i, which also must be used in the partition function, can be evaluated in terms of the dimensions of the unit cell of the fluorite lattice. Finally, the attractive energies between vacancies and the counter ions can be found from the number of the uniformly distributed reduced cations within each spherical envelope.

Using the reasoning outlined above, Atlas obtained a rather complex partition function, which, however, could be handled numerically. In order to compare the model with experimental observations, the thermodynamic functions $\Delta\bar{G}_{O_2}$, $\Delta\bar{H}_{O_2}$, and $\Delta\bar{S}_{O_2}$ were computed from this partition function for CeO_{2-x} in the composition range $0.03 < x < 0.35$. Compared to the experimental results reported by Bevan and Kordis (1964), the computed values agree moderately well with the experimental ones in the range $0.03 < x < 0.2$, and a significant feature is that the computed functions, especially $\Delta\bar{G}_{O_2}$, show instabilities in the ranges where a two-phase region has been observed both at lower temperatures (below 900°C) and at higher defect concentrations ($x > 0.2$). Although the model thus seems to reflect the main trends in the experimental observations, there are, however, significant differences between the calculated and observed values of $\Delta\bar{H}_{O_2}$ and $\Delta\bar{S}_{O_2}$, and it is doubtful whether the true defect structure in CeO_{2-x} really can be simulated by Atlas's model. A serious limitation seems to be that the distributions of vacancies and reduced cations are considered to be independent of each other—in the defect complex proposed by Thornber and Bevan (1970) two vacancies and one reduced cation are considered to be associated as described in the previous section—but it is also generally believed that doubly charged oxygen vacancies ($V_O^{\cdot\cdot}$) rather than singly charged vacancies ($V_O^{\cdot}$) are formed in this system, at least near the stoichiometric composition.

In the case of UO_{2+x}, Atlas considered the predominant defects to consist of 2:1:2 Willis clusters with two oxygen vacancies and one trapped counter ion (U^{5+}) balanced by other U^{5+} ions distributed in the surrounding crystal structure. In the first approach (Atlas, 1970), the partition function for this defect structure was derived in the same way as used for the CeO_{2-x} system, but recently the model has been extended so that the dependency of the distribution of one type of defect on the other can be taken into account (Atlas, 1970). The $\Delta\bar{G}_{O_2}$ values computed from this improved model agree fairly well with experimental data in the composition range $0.04 < x < 0.20$, but significant differences still exist between calculated and observed $\Delta\bar{H}_{O_2}$ and $\Delta\bar{S}_{O_2}$ values. Thus it seems questionable whether this model, even in its improved version, can describe the real nature of the grossly nonstoichiometric UO_{2+x} phase. A reason for the discrepancies observed has been argued to be that the model considers the defect complexes to be of the 2:1:2 type, whereas it is generally believed that 2:2:2 clusters are more important. According to Atlas, however, the consideration of these complexes should not significantly change the general appearance of the computed curves, and the inadequacies of the model are probably of a more fundamental character.

While Atlas's model can only be applied to ionic oxides, the recent model proposed by Manes and Manes-Pozzi (1975) for oxygen-deficient plutonium oxides (PuO_{2-x}) appears to be more generally applicable. As described in Section III.D.1, this model considers the formation of electrically neutral, quasi-molecular defect complexes that consist of one oxygen vacancy ($V_O^{\cdot\cdot}$) and two covalently bonded reduced cations (Pu_2O_3 nucleus—type 1 complex) near the stoichiometric composition. To each complex can be assigned a characteristic energy of formation and the interaction energies are considered to arise from electrostatic forces between dipoles set up in the defect complexes. Because of the bonding between the oxygen vacancies and the cations, higher energy is required to form vacancies in the adjacent oxygen sites and, as in Atlas's model, an envelope in which no other vacancies are formed can be assigned to each complex. The configurational term in the partition function, which represents the number of possible ways the defects can be distributed in the crystal, can thus be calculated from the number of ways these envelopes can be distributed in the crystal. At a certain limiting composition, however, the crystal becomes saturated with complexes of type 1 and additional vacancies must be formed in the high energy oxygen positions surrounding the complex. By this process enlarged complexes with a higher energy of formation are formed—complex type 2 with one additional oxygen vacancy can be expressed as a Pu_2O_3–PuO nucleus—but as the electrical neutrality is considered to be maintained, the distribution and the interaction energies of the enlarged complexes can be treated in the same manner within their respective range of existence as described above for the simple complexes.

The $\Delta\bar{H}_{O_2}$, $\Delta\bar{S}_{O_2}$, and $\Delta\bar{G}_{O_2}$ values calculated from Manes' model agree fairly well with experimental data and, compared to the model proposed by Atlas for CeO_{2-x}, this model has the following advantages: (i) it considers the formation of doubly charged oxygen vacancies ($V_O^{\cdot\cdot}$) instead of the less realistic $V_O^{\cdot}$ considered in Atlas's model, (ii) it takes the formation of defect complexes into account and it considers the change in type of complex with increasing deviation from stoichiometry, and finally (iii) it takes into account the change in formation energy of the defects to be expected when the concentration of defects is increased. However, the model also has limitations and it is still uncertain whether it can be considered to be completely realistic. One of the most serious limitations seems to be that only one type of defect complex is considered to be present at the same time, thus local ordering cannot be taken into account. It is also still somewhat uncertain whether the defect complexes postulated in this model really exist and intuitively it appears that a realistic model cannot be advanced before thorough structural studies, for instance, by electron microscopy or neutron diffraction of the nonstoichiometric phase have been carried out.

3. Thermodynamics of Microdomains

In all the models discussed in the previous sections the grossly nonstoichiometric phases were considered to be solid solutions with monophasic properties that can be described in terms of point defects or defect complexes. One of the reasons for adopting this idea was that many binary nonstoichiometric phases over extended composition ranges often show an apparent bivariant behavior ($\mu_i = \mu_i(T, x)$), which, according to the classical phase rule, closely corresponds to the behavior expected for a single phase. Some doubt has, however, been raised about the validity of the classical phase rule for the often continuous transformations in the solid state as this rule assumes complete independence of the two phases involved; a condition which certainly cannot be fulfilled if the structures of these phases are closely related as in a nonstoichiometric system.

In order to take the structural relationship into account, Ubbelohde (1957, 1966) proposed that continuous (smeared) solid–solid transformations proceed via a "hybrid" crystal, in which microdomains of both phases coexist. Because of the close relationship between the structures of the two phases, the appearance of domains of one phase within a matrix of the other phase does not involve sufficient mechanical strain to lead to a breakaway of the new crystallites. Therefore the hybrid crystal survives the transformation as a unit, although it passes through a state of maximum strain energy around the transformation temperature (T_c), where regions of the two structures coexist in comparable amounts. A large internal surface will also develop in the hybrid crystal around T_c, because of the formation of domains, and for a domain of structure 1 in a matrix of structure 2

$$G_1 = f_1(p, V, T, \xi_{12}, \eta_{12}) \tag{39}$$

where ξ_{12} is the strain energy and η_{12} is the internal surface energy. Taking these extra terms into account, the phase rule must be modified to

$$F = C - P + 2 + \sum \pi, \tag{40}$$

where $\sum \pi$ signifies the additional degrees of freedom arising from the formation of microdomains—besides strain and surface energy, $\sum \pi$ can also include the effect of position and orientation of the domains relative to the matrix.

Considering now the thermodynamic consequences of the extended phase rule, it is evident that the bivariant behavior usually observed for a nonstoichiometric phase does not necessarily mean that it is a single homogeneous phase, it may equally well be considered as a pseudo two-phase microdomain system which otherwise, according to the classical phase rule, would be expected to show univariant behavior ($\mu_i = \mu(T)$). According to this concept, the structure within the single domains must be well ordered, and the defects introduced at increasing deviation from stoichiometry must now be considered as structural entities rather than interacting defects, such as considered in the defect theories described in the previous sections. In x-ray, nonstoichiometric oxides usually give monophasic diffraction patterns corresponding to a small pseudocell, often of high symmetry, which represents a common structural framework of the structures of the two types of domains. Therefore the size of the microdomains must be extremely small and they must be randomly distributed. The composition of the nonstoichiometric phase is determined by the relative proportion of the two types of domain, which in accordance with Ubbelohde's idea can be treated as discrete domains of one type dispersed in a matrix of the other or vice versa depending on the composition.

The idea of treating nonstoichiometric oxide systems according to the microdomain concept was first suggested by Ariya and his co-workers (1958, 1962). Considering the enthalpies of formation (ΔH_f°) of nonstoichiometric TiO, VO, and FeO, Ariya observed that ΔH_f° for any composition could be expressed as a linear function of the enthalpies of the neighboring oxides. Based on this observation, he proposed that the nonstoichiometric oxides consist of an ideal solution (zero enthalpy of a mixing) of discrete domains in which the local order corresponds to the perfect order in either of the adjacent phases. For example, in $TiO_{1\pm x}$, which exists within the range $TiO_{0.89}$–$TiO_{1.22}$ at 1000°C, the oxygen vacancies are in one type of microdomain considered to be arranged in a structure resembling that of Ti_2O, whereas the Ti vacancies are arranged in the other type of microdomain so that the structure resembles the Ti_2O_3 structure. $TiO_{1\pm x}$ has the cubic NaCl structure, and in order to maintain a coherent structure this author considers both types of domain to have a metastable cubic structure, although Ti_2O and Ti_2O_3 both have a hexagonal close packing. Finally, from the overall concentration of oxygen and metal vacancies in $TiO_{1\pm x}$, Ariya estimated the size of the microdomain with the composition Ti_2O to be between 9 and 750 unit cells, depending on the overall composition. For $VO_{1\pm x}$ the size of V_2O domains was found to range between 8 and 8000 unit cells.

A general requirement for stability in a structure consisting of microdomains is, as already pointed out, that the microdomains and the matrix should to a large extent be coherent. In some systems this can be achieved if either the cation sublattice, or the anion sublattice, is common to both types of domain, e.g., in oxygen-deficient or metal-deficient oxides as described in Section III.D.1. In other cases coherence can be obtained, as discussed by Wadsley (1963), if there is a close structural and dimensional similarity between the two types of domain, which is the case for the shear and block structures to be considered in the next section. For the anion-excess oxide UO_{2+x}, however, it is less clear whether the structure can tolerate the strains generated by the microdomains. If this system is considered as consisting of domains of U_4O_9 in a matrix of UO_2, the difference in the size of the common subcell for the two structures—$a_0(U_4O_9) = 5.430$ Å and $a_0(UO_2) = 5.470$ Å—will create a considerable strain. If breakaway does not take place, then at least the U_4O_9 domains must be considered to be under compression, whereas the UO_2 matrix undergoes a dilation in the nonstoichiometric phase.

Assuming that the microdomains behave as an ideal solution, Anderson (1970) examined the thermodynamic conditions, under which a dispersed microdomain state is stable, relative to a mixture of microscopic crystals of two phases. At low temperatures the formation of a two-phase mixture can be expressed by

$$\underset{\text{(structure A)}}{MX_a} + \tfrac{1}{2}bX_2 \leftrightarrows \underset{\text{(structure B)}}{MX_{a+b}}, \tag{41}$$

where MX_a and MX_{a+b} are compounds of definite compositions. If the overall composition of this mixture is $MX_n (a < n < a + b)$, then the phase composition will be:

$$(1 - y)\,MX_a + y\,MX_{a+b},$$

where $y = (n - a)/b$. At higher temperatures at which a nonstoichiometric phase is formed, the formation of MX_n can be expressed by

$$MX_a + \tfrac{1}{2}(n - a)X_2 \leftrightarrows MX_n, \tag{42}$$

and if the total system contains N_M atoms of M, this phase will contain $y\,N_M$ atoms of M in microdomains of structure B and $(1 - y)\,N_M$ atoms in the matrix (structure A). If the chemical potentials of the low- and high-temperature reactions are $\mu^*_{X_2}$ and $\mu_{X_2}(y)$, respectively, it can be shown that a dispersed microdomain state can only be stable if $\mu_{X_2}(y) - \mu^*_{X_2} = \mu^{ex} < 0$. Using this condition, Anderson obtained the following expression for the average size $\bar{N}$ of microdomain B in a matrix of A:

$$\bar{N} - N_0 + 1 = \frac{1}{1 - X_A(1 - mb\delta)} \exp\left(-\frac{\sigma(N)}{kT}\right), \tag{43}$$

where $\delta = \mu^{ex}/kT$, X_A is the mole fraction of A, m is the number of formula units of structure A required for the formation of one lattice molecule of structure B (e.g., in UO_{2+x}, $m = 4$ if the microdomains are considered to be a Willis complex), N_0 is the size of the smallest microdomain of B ($=1$ if clusters are formed), and finally $\sigma(N)$ is the effective internal surface energy (including strain) between the microdomains and the matrix. Assuming strain-free coherence—i.e., $\sigma(N) \to 0$—it was shown that the microdomain size will only be appreciable if μ^{ex} is extremely small; using the data of Bevan and Kordis (1964) the value of $\bar{N}$ was found to be only about 5 for $CeO_{1.94}$ at 1000°K. Compared to the domain sizes found by Ariya, this appears to be an astonishingly small value and the assumptions made in these calculations are probably not realistic. In particular, the behavior of a nonstoichiometric system is never ideal as assumed in this model, even for the CeO_{2-x} system as shown by Iwasaki and Katsura (1971).

The strain energy term (ξ) to be included in the free energy expressions for a nonstoichiometric phase will usually differ in the case where domains of structure 1 are distributed in a matrix of structure 2 (ξ_{12}), and in the case where 2 is distributed in 1 (ξ_{21}). As already pointed out by Ubbelohde, this

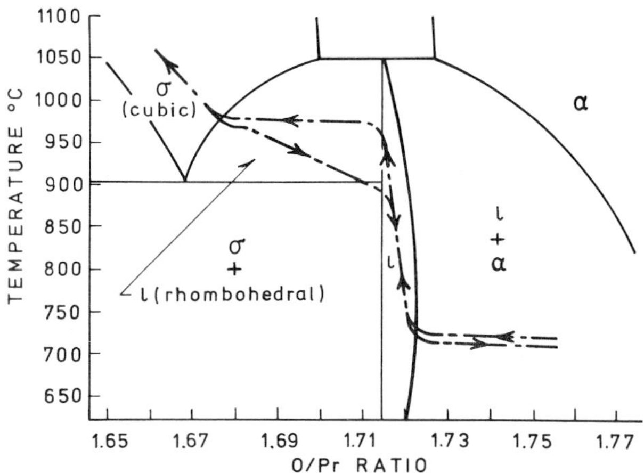

Fig. 14. Simplified partial phase diagram of the PrO_x–O_2 system with isobar obtained in thermogravimetric measurements (Hyde, *et al.* (1966)).

inevitably leads to hysteresis and in fact the oxide systems in which hysteresis has been observed—e.g., PrO_{2-x} (Hyde *et al.*, 1966, Turcotte *et al.*, 1973) and TbO_{2-x} (Faeth and Clifford, 1963)—have successfully been treated according to the microdomain concept. The hysteresis observed in the PrO_{2-x} system is illustrated in Fig. 14, which shows a partial phase diagram of the Pr–O system and the results obtained in an isobaric measurement of the composition as a function of temperature. During heating of the sample horizontal isobars are obtained for the two-phase regions $\iota + \alpha$ and $\sigma + \iota$, whereas isobars with an intermediate slope are obtained for the nonstoichiometric single phases ι and σ as expected from the classical phase rule. During cooling, however, a hysteresis effect is observed across the two-phase region $\sigma + \iota$ where a sloping isobar is obtained instead of the horizontal isobar observed during heating. The hysteresis in the PrO_{2-x} system has been explained in terms of pseudophases identical to the hybrid crystals proposed by Ubbelohde. In the σ phase the oxygen vacancies are considered to be ordered so that strings of octahedrally coordinated Pr^{3+} ions are arranged in all four $\langle 111 \rangle$ directions in the cubic, fluorite-type structure. The ι phase has a similar structure except that it is slightly distorted to a rhombohedral structure in which the strings are all parallel to only one of the $\langle 111 \rangle$ directions. In passing from the ι phase to the σ phase during heating, a randomization of the parallel strings in the ι phase apparently takes place and a relatively sharp transition is obtained as shown in the figure. During cooling from the disordered σ phase to the more ordered ι phase, however, it is possible for the strings to order along any of the $\langle 111 \rangle$ directions, and regions with different orientations of the strings can be formed. Thus microdomains of the ι phase are formed in a matrix of the residual σ phase; as the temperature further decreases these microdomains will grow at the expense of the surrounding matrix until the transformation is complete. According to the microdomain concept, this behavior should give an isobar with an intermediate slope (bivariant behavior) as also observed in the experiments. A similar hysteresis effect has also been observed at higher oxygen content where the rhombohedral ι phase is transformed into a triclinic phase and both the PrO_{2-x} and TbO_{2-x} systems probably provide some of the best examples of a successful application of the microdomain concept. Even for the high-temperature α phase in the PrO_{2-x} system, hitherto considered as a nonstoichiometric single phase, evidence has recently been presented of a domain structure (Jenkins *et al.*, 1970); and further studies (Lowe *et al.*, 1975; Lowe and Eying, 1975) supporting the microdomain concept have recently given a better understanding of the factors leading to hysteresis in these oxide systems.

Hysteresis has also been observed in other systems—e.g., in the TiO_x (Bursill and Hyde, 1972; Herritt *et al.*, 1973) and the NbO_x systems (Marucco, 1974) described in the next section—and in all cases this phenomenon has been explained in terms of a domain structure. Whether a nonstoichiometric system, which shows little or no hysteresis, can also be considered in this way is still uncertain, however. For instance, for the CeO_{2-x} system, which usually shows a fully reversible behavior (Bevan and Kordis, 1964), the small domain size found by Anderson, as described above, appears to be too small to represent a domain of a new structure although it shows that there is an appreciable interaction between the defect centers. Small domain sizes were also calculated by Anderson for the reversible UO_{2+x} system (Anderson, 1969), and before the microdomain concept can be adopted also for these systems, further and more detailed thermodynamic and structural studies are needed. A suitable tool for further structural studies appears to be high resolution electron microscopy that, by lattice imaging techniques, can show the presence of microdomains directly. This technique has successfully been applied to reveal the existence of domains in many systems with shear—e.g., TiO_x (Hyde and Bursill, 1970)—or block structures, and recently it has also been used in studies of the fluorite-related structures in the PrO_{2-x} system (Kunzmann and Eyring, 1975).

4. Thermodynamics of Shear and Block Structures

In the case of the oxides discussed in the previous sections, the defects were considered to exist in the lattice either as point defects, or in defect complexes that can order into a microdomain structure in certain systems. For the anion-deficient transition-metal oxides in high oxidation states—

Ti^{4+}, V^{4+}, V^{5+}, Nb^{5+}, Mo^{6+}, and W^{6+}—the formation of oxygen vacancies, however, leaves the high formal charges of adjacent cations unscreened, and structural adjustments that preserve the coordination of the cations are likely to occur. In these oxides there is also a considerable anion–cation orbital overlap, which should be preserved, and the structural adjustments accommodating changes in the composition that best fulfill these requirements have been shown to be a crystallographic shearing—see, for instance, Hyde *et al.* (1974) or Tilley (1972).

The parent structures of the oxides are based on the ReO_3, TiO_2 (rutile), and MoO_3 structures in which the cations are octahedrally coordinated. Taking the ReO_3 structure as an example, the formation of a shear structure preserving this coordination as well as to a certain extent the cation–anion overlap is shown in Fig. 15. According to a model proposed by Anderson and Hyde (1965, 1967), the first step in the formation of a shear plane (Wadsley defect) is the accumulation of oxygen vacancies into a planar disk in a certain orientation in the crystal. When the concentration of vacancies locally becomes too great, the crystal will collapse and shear across this disk and a shear plane bounded by a dislocation loop will be produced. As the elastic stresses in the dislocation will act as a vacancy sink, the loop will expand as more and more vacancies are produced and accumulated, and finally the shear plane will extend through the whole crystal. From the figure it will be noted that a complete sheet of anion sites has been eliminated in this process and that a lamellar structure is obtained which consists of slabs of the parent structure coherently intergrown across the shear plane. In the parent structure the octahedra share corners, but in the shear plane a closer distance between the cations, probably with an overlap of the cation orbitals, is obtained by an edge sharing of the octahedra. This results in a smaller anion to cation ratio and, depending on the orientation of the shear planes and on the thickness of the slabs between them, a whole series of intermediate phases can be envisaged. For oxides derived from the ReO_3 structure, the composition of these phases can be expressed by the formula M_nO_{3n-m}, where m depends upon the orientation of the shear planes, whereas the intermediate phases in rutile-related oxides are of the type M_nO_{2n-m} (Magnéli, 1953).

Several other models have been proposed to describe the crystallographic shearing mechanisms—see for instance von Landuyt and Amelinckx (1973) and Bursill and Hyde (1972)—but the exact mechanism has not yet been established. Especially the first step in the process—i.e., the accumulation of vacancies into a planar disk—seems to be doubtful, and, as discussed by Anderson (1972), the formation of shear structures should perhaps rather be considered as arising from fluctuations in a structure with either randomly distributed defects or defects arranged into a superstructure. In TiO_{2-x} or

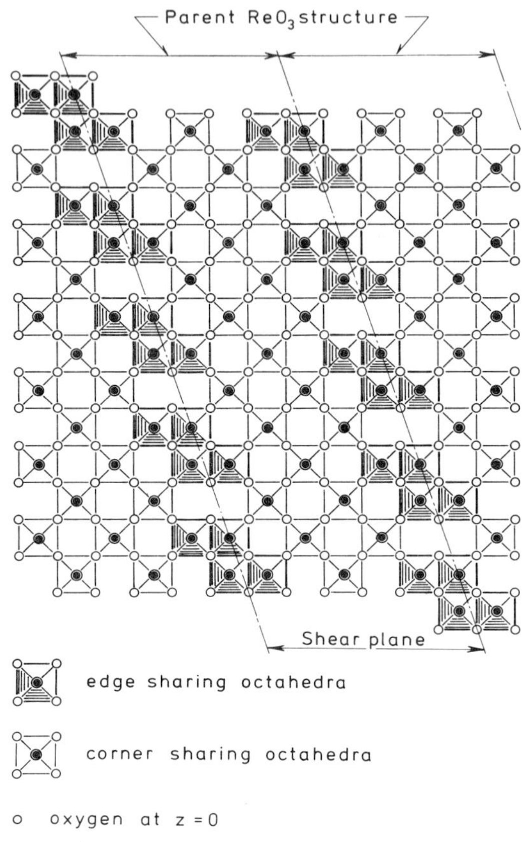

Fig. 15. Shear planes formed in oxides of ReO_3 parent structure.

WO_{3-x}, for instance, there is no experimental evidence of an ordering of the vacancies into disks, whereas superlattice ordering at very small deviations from stoichiometry ($M_{400}O_{1199}$) was observed in MoO_{3-x} by Bursill (1969). Once collapse and shear have taken place it is energetically advantageous that the shear plane extends throughout the whole crystal, and the expansion of the planes is probably very fast. Because of the high energy involved in the formation of the dislocation boundary, however, the nucleation of the shear plane is difficult and hysteresis effects are often observed when shear planes are either generated or destroyed during reduction or oxidation, as shown by Bursill and Hyde (1972).

Another characteristic feature of these oxides is that the number of defects in the parent structure between the shear planes is considered to be very small—

in fact the only defects that can be present will be those inherently formed as a result of thermal disorder (native point defects). Experimentally this has been verified by electron microscopy examinations of slightly reduced rutile (Bursill and Hyde, 1971) which showed that numerous isolated shear planes were formed even for a composition of $TiO_{1.9986}$. From the density of the shear planes it was inferred that the concentration of defects in the rutile slabs could not exceed 10^{-4}. Similar observations have been made on slightly reduced WO_3 (Allpress *et al.*, 1969) and it is clear that the thermodynamics of these systems cannot be treated according to the defect theories described in the previous sections.

Each of the intermediate phases is considered to have an ordered structure with parallel and equidistant shear planes. For each phase there is a characteristic distance between the shear planes and it must be energetically favorable for the crystal to order the shear planes into a regular pattern. The nature of the interactions between the shear planes operating during the ordering processes is not known, however, but thermodynamically it can be shown that the free energy of these phases is dominated by a decrease in enthalpy (ordering) rather than by an increase in entropy (disordering).

In the shear phases discussed so far the shear planes have been shown to be parallel. For the block (or column) structures, which play an important role for the chemistry of Nb- and Ta-oxides and which are also derived from the ReO_3 parent structure, double shear on two orthogonal sets of parallel planes takes place. As a result of this process the whole structure is divided into blocks or columns of corner- or edge-shearing octahedra, and the total composition will depend on the size of the blocks and the way in which they are connected to each other. A large number of intermediate phases also exists for these oxides, and a general formula has also been derived for the compositions of these phases.

The difference between the compositions of the intermediate phases, both in the single- and double-sheared oxides, is very small, and in order to detect the existence of these phases by, for instance, thermogravimetry, very accurate measurement is needed of samples differing only slightly in composition, as clearly demonstrated by Bursill and Hyde (1972). These conditions are probably not fulfilled in many of the measurements hitherto reported in the literature, but the very careful and accurate measurements of Merritt *et al.* (1973) on the TiO_x system merit a closer description. As shown in Table II, this study revealed that the substoichiometric range for rutile (TiO_{2-x}) can be divided into several regions, some of which show a considerable degree of hysteresis whereas others show good reversibility. By comparing their findings with the structural information obtained by the detailed electron microscopy examinations of the TiO_x system by Bursill and Hyde, it was noted that the regions showing hysteresis effects closely correspond

TABLE II

THERMODYNAMIC AND STRUCTURAL BEHAVIOR OF REDUCED RUTILE[a]

Composition range, O/M	n in Ti_nO_{2n-1}[b]	Structure	Thermodynamic behavior
2,0000–1.9999		Random point defect in parent structure	
1.9999–1.999		Widely spaced (132) shear planes in different orientations	
1.999–1.98	37 in microdomains	Microdomains of ordered (132) shear planes in a rutile matrix saturated with (132) shear planes	Hysteresis
1.98–1.93	$16 \leq n \leq 37$	Intermediate phases with ordered (132) shear planes and coherently intergrown microdomains for compositions between the intermediate phases	Hysteresis
1.93–1.90		Orientation of shear plane changing from (132) to (121). Domains with shear planes between (132) and (121)	Reversibility
1.90–1.75	$4 \leq n \leq 10$	Intermediate phases with ordered (121) shear planes and coherently intergrown microdomains for compositions between these phases	Hysteresis
1.75–1.66		Ti_3O_5 coexisting with Ti_4O_7	

[a] According to Merritt *et al.* (1973).
[b] General formula for intermediate phases.

to the composition ranges in which shear structures are formed, and the main reason for hysteresis in this system is attributed to the formation of either (132) or (121) shear planes. The reversible region observed in the composition range $1.90 \leq x < 1.93$ could in the same way be correlated with ordered or partly ordered phases in which the shear planes lie between (121) and (132). Hence it was concluded that the composition changes in this region were accommodated by a change in the shear plane orientation rather than by an increase in the concentration of the shear planes as observed for the other regions. Many of the intermediate phases existing within the different regions were also observed in this study, and the curve relating the free energy of formation and the composition of the reduced samples (mole fraction of oxygen) obtained from the experimental data could be explained entirely in terms of such phases. There is no doubt that these studies have given a much better understanding of the behavior of this type of oxide in general. Especially the combination of an accurate and detailed thermodynamic study and a comprehensive structural study by electron microscopy has proved to be rewarding; similar combined studies on other systems appear to be fertile ground for future research.

The mechanism of formation and ordering of shear planes is still unknown, as discussed above, and a thermodynamic model for oxides with shear structures has not yet been formulated. Obviously this will be a very difficult task because a realistic model must probably be based on irreversible thermodynamics which so far is largely confined to homogeneous systems (e.g., reactions in solution). The rather empirical approach taken by Kimura (1973), in which the mass action law is applied to hypothetical reactions describing crystallographic shearing in Nb-oxides, thus does not appear to be completely realistic, although it might prove useful as a first step in a formulation of a model. A treatment according to fluctuation theories might also prove to be useful because the number of possible stable intermediate phases in a system can be considered to be determined by compositional (structural) fluctuations (see Anderson, 1972). If, for instance, the G–x curve for a system consisting of a succession of line phases can be represented as shown in Fig. 4 (Section II), and if the difference in free energy between successive pairs becomes sufficiently small, then a compositional fluctuation which locally changes the free energy (ΔG_{fluct}) will disproportionate a line phase into its two adjacent phases. The probability of a fluctuation depends on the temperature and the free energy change involved according to $w = \exp[-\Delta G/kT]$; especially at high temperature the possibility of such disproportionations is important—direct evidence of a disproportionation in a shear structure is provided by the electron microscopy study of ternary W/Nb-oxides by Allpress *et al.* (1969). Another possibility which should also be taken into account is that the ordered intermediate phases become disordered at higher temperatures, either by a disordering of the shear planes themselves, or by a disordering into a nonstoichiometric phase extending over a broad composition range. The small entropy involved in a disordering of the shear planes, however, seems to exclude the possibility of the first process, and the observation of congruent melting in many of the intermediate phases in the Ti–O, V–O, Mo–O, and W–O (Anderson and Khan, 1970a,b) systems seems to exclude the possibility of the second.

Finally the question arises of whether crystallographic shearing is confined to the oxide systems considered in this section, or whether this principle is generally used in nonstoichiometric systems as a means to accommodate compositional changes. Before this question can be answered, however, a thorough characterization of the real structures of many more nonstoichiometric systems should be carried out, for instance, by electron microscopy. When this has been done, a new picture of the nature of nonstoichiometry will no doubt emerge. Recent studies by Sørensen (1976) on the CeO_{2-x} system have, for instance, shown that at least part of the nonstoichiometric phase range, even at higher temperatures, can be considered to consist of a succession of intermediate phases rather than one grossly nonstoichiometric phase. In accordance with these findings, a preliminary examination of reduced single crystals of ceria by electron microscopy showed the presence of a lamellar structure, which could be interpretated as a shear structure from the streaking observed in the diffraction pattern.

References

Ackermann, R. J., Thorn, R. J., and Winslow, G. H. (1959). *J. Opt. Soc. Am.* **49**, 1107–1112.

Adler, D. (1971). Band Structure and Electronic Properties of Ceramic Crystals. *In* "Physics of Electronic Ceramics" (L. L. Hench and D. B. Dove, eds.), Part A, pp. 29–66. Marcel Dekker, New York.

Allnatt, A. R., and Cohen, M. H. (1964a). *J. Chem. Phys.* **40**, 1860–1870.

Allnatt, A. R., and Cohen, M. H. (1964b). *J. Chem. Phys.* **40**, 1871–1890.

Allpress, J. G., Sanders, J. V., and Wadsley, A. D., (1969). *Acta Crystallogr. Sect. B* **25**, 1156–1164.

Allpress, J. G., Tilley, R. J. D., and Sienko, M. J. (1971). *J. Solid. State Chem.* **3**, 440–451.

Anderson, J. S. (1946). *Proc. R. Soc. London Ser. A* **185**, 69–89.

Anderson, J. S. (1969). *Bull. Soc. Chim. Fr.*, 2203–2214.

Anderson, J. S. (1970). The Thermodynamics and Theory of Nonstoichiometric Compounds. *In* "Problems of Nonstoichiometry" (A. Rabenau, ed.), pp. 1–76. North-Holland Publ., Amsterdam.

Anderson, J. S. (1972). Reaction Paths and Microstructure in Crystals. *In* "Proc. Int. Symp. Reactivity of Solids. 7th, (J. S. Anderson, M. W. Roberts, and F. S. Stone, eds.), p. 1. Chapman & Hall, London.

Anderson, J. S. (1974). The Real Structure of Defect Solids. *In* "Defects and Transport in Oxides" (M. S. Seltzer and R. I. Jaffee, ed.), pp. 25–48. Plenum, New York.

Anderson, J. S., and Burch, R. (1971). *J. Phys. Chem. Solids* **32**, 923–926.

Anderson, J. S., and Hyde, B. G. (1965). *Bull. Soc. Chim. Fr.*, 1215–1216.

Anderson, J. S., and Hyde, B. G. (1967). *J. Phys. Chem. Solids* **28**, 1393–1408.

Anderson, J. S., and Khan, A. S. (1970a). *J. Less Common Met.* **22**, 219–223.

Anderson, J. S., and Khan, A. S. (1970b). *J. Less Common Met.* **22**, 209–218.

Ariya, S. M., and Morozova, M. P., (1958). *J. Gen. Chem. USSR* **28**, 2647–2652.

Ariya, S. M., and Popov, Yu. G. (1962). *J. Gen. Chem. USSR* **32**, 2054–2057.

Atlas, L. M. (1968a). *J. Phys. Chem. Solids* **29**, 91–100.

Atlas, L. M. (1968b). *J. Phys. Chem. Solids* **29**, 1349–1358.

Atlas, L. M. (1970). *In* "The Chemistry of Extended Defects in Non-Metallic Solids" (L. Eyring and M. O'Keeffe, eds.), p. 425. North-Holland Publ., Amsterdam.

Azaroff, L. V. (1960). "Introduction to Solids." McGraw-Hill, New York.

Banus, M. D., and Reed, T. B. (1970). *In* "The Chemistry of Extended Defects in Non-Metallic Solids" (L. Eyring and M. O'Keeffe, eds.), p. 488. North-Holland Publ., Amsterdam.

Bertaut, E. F. (1953). *Acta Crystallogr.* **6**, 557–561.

Bevan, D. J. M., and Kordis, J. (1964). *J. Inorg. Nucl. Chem.* **26**, 1509–1523.

Blank, H. (1967). "A Comparison of the Pu–O System with the Rare Earth Oxide Systems Ce–O, Pr–O and Tb–O." *Euro. At. Community* 42 pp. EUR 3653e.

Bransky, J., and Wimmer, J. M. (1972). *J. Phys. Chem. Solids* **33**, 801–812.

Brebrick, R. F. (1961). *J. Phys. Chem. Solids* **18**, 116–128.

Brebrick, R. F. (1967). *Prog. Solid State Chem.* **3**, 213–264.

Brouwer, G. (1954). *Philips Res. Rep.* **9**, 366–376.

Bursill, L. A. (1969). *Proc. R. Soc. London Ser. A* **311**, 267–290.

Bursill, L. A., and Hyde, B. G. (1971). *Philos. Mag.* **23**, 3–18.

Bursill, L. A., and Hyde, B. G. (1972). *Prog. Solid State Chem.* **7**, 177–253.

Cahn, J. W. (1961). *Acta Metall.* **9**, 795–801.

Cheetham, A. K., Fender, B. E. F., and Taylor, R. I. (1971). *J. Phys. C* **4**, 2160–2165.

Darken, L. S., and Gurry, R. W. (1953). "Physical Chemistry of Metals," 535 pp. McGraw-Hill, New York.

Faeth, P. A., and Clifford, A. F. (1963). *J. Phys. Chem.* **67**, 1453–1457.

Fender, B. E. F. (1972). "Solid State Chemistry" (H. J. Emeleus and L. E. F. Roberts, eds.), Vol. 10, pp. 243–278 (Ser. 1 of MTP Int. Rev. of Sci., Inorganic Chem.) Butterworth, London and University Park, Baltimore.

Fender, B. E. F., and Riley, F. D. (1969). *J. Phys. Chem. Solids* **30**, 793–798.

Fender, B. E. F., and Riley, F. D. (1970). *In* "The Chemistry of Extended Defects in Non-Metallic Solids" (L. Eyring and M. O'Keeffe, eds.), p. 54. North-Holland Publ., Amsterdam.

Gardner, R. F. G., Sweett, F., Tanner, D. W. (1963). *J. Phys. Chem. Solids* **24**, 1183–1196.

Gerdanian, P. (1974). *J. Phys. Chem. Solids* **35**, 163–170.

Greenwood, N. N., and Howe, A. T., (1972a). *J. Chem. Soc. Dalton Trans.*, 110–116, 116–121, 122–130.

Greenwood, N. N., and Howe, A. T. (1972b). *In* "Reactivity of Solids—7th International Symposium" (J. S. Anderson, M. W. Roberts, and F. S. Stone, eds.), p. 240. Chapman & Hall, London.

Hagemark, K., and Broli, M. (1966). *J. Inorg. Nucl. Chem.* **28**, 2837–2850.

Herai, T., Thomas, B., Manenc, J., and Bénard, J. (1964). *C. R. Acad. Sci.* **258**, 4528–4530.

Hoch, M. (1963a). *J. Phys. Chem. Solids* **24**, 157–159.

Hoch, M. (1963b). *J. Phys. Soc. Jpn. Suppl. II* **18**, 147–151.

Hoch, M. (1964). *Trans. Metal. Soc. AIME* **230**, 138–147.

Hoch, M., Iyer, A. S., and Nelken, J. (1962). *J. Phys. Chem. Solids* **23**, 1463–1471.

Hoskins, B. F., and Martin, R. L. (1975). *J. Chem. Soc. Dalton Trans.*, 576–588.

Hyde, B. G., Bagshaw, A. N., Anderson, S., and O'Keeffe, M. (1974). *Ann. Rev. Mater. Sci.* **4**, 43–92.

Hyde, B. G., and Bursill, L. A. (1970). *In* "The Chemistry of Extended Defects in Non-Metallic Solids" (L. Eyring and M. O'Keeffe, eds.), p. 347. North-Holland, Amsterdam.

Hyde, B. G., Bevan, D. J. M., and Eyring, L. (1966). *Philos. Trans. R. Soc. London Ser. A.* **259**, 583–614.

Thermodynamic and Transport Properties of Uranium Dioxide and Related Phases. (1965). *IAEA, Tech. Rep. Ser.* No. 39; 44, Vienna 1965.

The Plutonium–Oxygen and Uranium–Plutonium–Oxygen Systems: A Thermochemical Assessment (1967). 86 pp. *IAEA Tech. Rep. Ser.* No. 79, Vienna 1967.

Ishii, T., Naito, K., Oshima, K., and Hamaguchi, V., (1971). *J. Phys. Chem. Solids* **32**, 235–241.

Iwasaki, B., and Katsura, T. (1971). *Bull. Chem. Soc. Jpn.* **44**, 1297–1301.

Iyengar, G. N. K. (1973). *J. Sci. Ind. Res.* **32**, 633.

Javed, N. A. (1972). *J. Nucl. Mater.* **43**, 219–224.
Jenkins, M. S., Turcotte, R. P., and Eyring, L. (1970). *In* "The Chemistry of Extended Defects in Non-Metallic Solids" (L. Eyring and M. O'Keeffe, eds.), p. 36, North-Holland Publ., Amsterdam.
Kimura, S. (1973). *J. Solid State Chem.* **6**, 438–449.
Kittel, C. (1966). "Introduction to Solid State Physics," 3rd. ed. Wiley, New York.
Koch, F., and Cohen, J. B. (1969). *Acta Crystallogr. Sect. B* **25**, 275–287.
Kofstad, P. (1972). "Nonstoichiometry, Diffusion and Electrical Conductivity in Binary Metal Oxides." Wiley, New York.
Kofstad, P., and Hed, A. Z. (1968). *J. Electrochem. Soc.* **115**, 102–104.
Kröger, F. A. (1964). "The Chemistry of Imperfect Crystals." North-Holland Publ., Amsterdam, and Wiley, New York.
Kröger, F. A. (1968). *J. Phys. Chem. Solids* **29**, 1889–1899.
Kröger, F. A., and Vink, H. J. (1956). *Solid State Phys., Adv. Res. Appl.* **3**, 307–435.
Kunzmann, P., and Eyring, L. (1975). *J. Solid State Chem.* **14**, 229–237.
Libowitz, G. G. (1962). *J. Appl. Phys.* **33**, 399–405.
Libowitz, G. G. (1965). *Prog. Solid State Chem.* **2**, 216.
Lowe, A. T., and Eyring, L. (1975). *J. Solid State Chem.* **14**, 383–394.
Lowe, A. T., Lau, K. H., and Eyring, L. (1975). *J. Solid State Chem.* **15**, 9–17.
Magnéli, A. (1953). *Acta Crystallogr.* **6**, 495–500.
Manenc, J., Vagnard, G., and Benard, J. (1962). *C. R. Acad. Sci.* **254**, 1779.
Manes, L., and Manes-Pozzi, B. (1975). *Int. Conf. Plutonium Other Actinides. 5th, Baden-Baden, Germany, September 1975.*
Markin, T. L., Wheeler, V. J., and Bones, R. J. (1968). *J. Inorg. Nucl. Chem.* **30**, 807–817.
Martin, R. L. (1974). *J. Chem. Soc. Dalton Trans.*, 1335–1350.
Marucco, J. F. (1974). *J. Solid State Chem.* **10**, 211–218.
Marucco, J., Gerdanian, P., and Dodé, M. (1970). *J. Chim. Phys. Phys. Chim. Biol.* **67**, 906–913.
Matsui, T., and Naito, K. (1975). *J. Nucl. Mater.* **56**, 327–335.
Merritt, R. R., Hyde, B. G., Philip, D. K., and Bursill, L. A. (1973). *Philos Trans. R. Soc. London Ser. A* **274** (1245), 627–661.
Mott N. F., and Littleton, M. J. (1938). *Trans. Faraday Soc.* **34**, 485–499.
Naito, K., Ishii, T., Hamaguchi, Y., and Oshima, K. (1967). *Solid State Commun.* **5**, 349–352.
Navrotsky, A. (1974). *In* "Transition Metals" (H. F. Emeleus and D. W. A. Sharp, eds.) Vol. 5, pp. 29–70. (Ser. 2, Vol. 5 of MTP Int. Rev. of Sci, Inorganic Chem.) University Park, Baltimore and Butterworth, London.
Panlener, R. J., Blumenthal, R. N., and Garnier, J. E. (1975). *J. Phys. Chem. Solids* **36**, 1213–1222.
Raccah, P., and Vallet, P. (1965). *Rev. Met. (Mem. Sci.)* **62** (4), 1–2.
Rand, M. H., and Kubaschewski, O. (1963). "The Thermochemical Properties of Uranium Compounds," 93 pp. Oliver & Boyd, Edinburgh.
Rees, A. L. G. (1954). *Trans. Faraday Soc.* **50**, **1**, 335–342.
Roberts, L. E. J., and Markin, T. L. (1967). *Proc. Br. Ceram. Soc.* **8**, 201–217.
Saito, Y. (1974). *J. Nucl. Mater.* **51**, 112–125.
Schmitz, F., and Marajofsky, A. (1974). *Proc. Symp. Thermodynamics Nuclear Materials, IAEA, Vienna, October 1974.*
Scholtz, A. (1964). *Phys. Status Solidi* **7**, 973–982.
Seltzer, M. S., and Hed, A. Z. (1970). *J. Electrochem. Soc.* **117**, 815–818.
Sorensen, O. Toft (1976). *J. Solid State Chem.* **18**, 217–233.
O'Keeffe, M., and Stone, F. S. (1962). *Proc. R. Soc.* London Ser. A **267**, 501–517.
Swalin, R. A. (1972). "Thermodynamics of Solids," 2nd. ed. Wiley, New York.
Tetenbaum, M., and Hunt, P. D. (1968). *J. Chem. Phys.* **49**, 4739–4749.

Thorn, R. J. (1966). *Ann. Rev. Phys. Chem.* **17**, 83–118.
Thorn, R. J., and Winslow, G. H. (1966). *J. Chem. Phys* **44**, 2032–2642.
Thorn, R. J., and Winslow, G. H. (1967). Advances in High Temperature Chemistry" (LeRoy Eyring, ed. Vol. 1, p. 153. Academic Press, New York.
Thornber, M. R., and Bevan, D. J. M. (1970). *J. Solid State Chem.* **1**, 536–544.
Tilley, R. F. D. (1972). "Solid State Chemistry" (H. F. Emeleus and L. E. F. Roberts, eds.), Vol. 10, pp. 279–313 (Ser. 1 of MTP Int. Rev. of Sci., Inorganic Chem.) Butterworth, London and University Park, Baltimore.
Turcotte, R. P., Jenkins, M. S., and Eyring, L. (1973). *J. Solid State Chem.* **7**, 454–460.
Ubbelohde, A. R. (1957). *Q. Rev. Chem. Soc. London* **11**, 246–272.
Ubbelohde, A. R. (1966). *J. Chim. Phys. Phys. Chim. Biol.* **62**, 33–42.
Von Landuyt, J., and Amelinckx, S. (1973). *J. Solid State Chem.* **6**, 222–229.
Wadsley, A. D. (1963). *Adv. Chem. Ser.* **39**, 23–36.
Wagner, C., and Schottky, W. (1930). *Z. Physik. Chem. Abt.* **B11**, 163–210.
Watanabe, D., Cassees, J. R., Jostons, A., and Malin, A. S. (1967). *Acta Crystallogr.* **23**, 307–313.
Watanabe, D., Terasaki, O., Jostsons, A., and Castles, J. R. (1970). *In* "The Chemistry of Extended Defects in Non-Metallic Solids" (L. Eyring and M. O'Keeffe, eds.), p. 238. North-Holland Publ., Amsterdam.
Westman S., and Nordmark, C. (1960). *Acta Chem. Scand.* **14**, 465–470.
Wheeler, V. J., and Jones, I. G. (1972). *J. Nucl. Mater.* **42**, 117–121.
Willis, B. T. M. (1964). *J. Phys.* **25**, 431–439.

Spinodal Decomposition— Phase Diagram Representation and Occurrence*

C. M. F. JANTZEN† *AND H. HERMAN*

DEPARTMENT OF MATERIALS SCIENCE
STATE UNIVERSITY OF NEW YORK
STONY BROOK, NEW YORK 11794

* This research was sponsored by grants from the U.S. Army Research Office, Research Triangle, North Carolina, and the North Atlantic Treaty Organization.

† Present address: Department of Chemistry, University of Aberdeen, Old Aberdeen, Scotland, United Kingdom.

ISBN 0-12 053205-0

I. INTRODUCTION

A. Spinodal Decomposition

When a two-component alloy or ceramic system is brought, with no concurrent unmixing, from the single to the two phase region of the equilibrium phase diagram, it is in a supersaturated condition. On annealing this as-quenched system within the two phase region at a temperature sufficiently high to permit diffusion, phase separation can ensue with the development of a two-phase product. The path (or paths) by which unmixing is accomplished and the form (both structural and morphological) of the product phases have been subject to numerous investigations [reviewed by Kelly and Nicholson (1963), Cahn (1968), and Hilliard (1970)]. In ideally visualized systems, the as-quenched state is initially metastable, and a free energy barrier stabilizes the solution. The system will thus decompose by nucleating the new phase, and the growth of this phase will be limited by diffusion.

Spinodal decomposition represents a departure from the traditional theories of diffusion-controlled precipitation. In the more familiar nucleation and growth processes, where the solid solution is initially metastable, the principal limit to the establishment of the new phase is an interface, either chemical, structural, or both. In spinodal decomposition, on the other hand, a supersaturated solution is formed by rapidly quenching from a single phase region to the region of the miscibility gap where the curvature of the free energy–composition plot is negative, i.e., $\partial^2 f/\partial c^2 < 0$. This is the *spinodal region* and here the smallest fluctuation of composition is stable for growth, a nucleation step not being required for phase decomposition. No discontinuous surface need be established for the introduction of the new phase. Hence, the limitation to the formation of a new phase in the spinodal region is kinetic, since the main barrier to decomposition is diffusion (Cahn, 1961, 1962a, 1965, 1968).

The early theories of spinodal decomposition were concerned with fluids and simple isotropic cubic systems, the first experiments being directed mainly at glasses and cubic alloys. The principal developers of the early theories were Hillert, Cahn, and Hilliard. Though many aspects of these ideas still remain unsettled, the linear theory, due mainly to Cahn, generated considerable interest in the metallurgical, ceramic, and mineralogical communities.

Most of the experimental work has been on metal alloys and oxide glasses, with little study having been directed to nonmetallic crystalline systems, such as oxides and alkali halides. In this paper we shall review those ideas and theories of spinodal decomposition which are relevant to crystalline nonmetals, especially oxides. We shall also present recent results from this laboratory on a number of mixed crystalline oxides.

B. Spinodal Decomposition vis-a-vis Nucleation and Growth

A long history surrounds the theoretical development of nucleation and growth in condensed systems. Modern solid state nucleation concepts really evolved from the ideas embodied in the Volmer–Weber–Becker–Döring formulations, where there is a statistical distribution of embryos, each presumably being able to make an excursion towards a critical size. In the somewhat more advanced theoretical development, both size and composition are variables in the nucleation equation, and two nuclear attributes become associated with the fluctuation which is critical for growth. It is of central importance to note that when a subnuclear embryo increases in size it will initially introduce a positive change in free energy. This free energy barrier comes about from a surface energy term, and it is this term, proportional to the embryo surface area, that constitutes the barrier which must be surmounted before growth can occur. Another characteristic of nucleation and growth processes is the development of distinctly separated particles of the new phase which grow, at first independently, and which finally mature by coarsening processes. Furthermore, the nucleation processes of the classic sort which we are describing occur in the region of the free energy–composition field where the curvature is positive, i.e., $\partial^2 f/\partial c^2 > 0$.

Phase separation by the spinodal mode, on the other hand, occurs by small fluctuations over a large volume. Within the spinodal region, where $\partial^2 f/\partial c^2 < 0$, there is no surface free energy barrier to viable growth, and the solution is unstable to the smallest fluctuation in concentration. The spinodal reaction occurs at a rate controlled mainly by the activation energy for diffusion and no incubation time is observed.

Spinodal decomposition is characterized, in a morphological sense, by mutually connected phases. That is, the major and minor phases are self-connected within the volume of the solid. The interface between the phases is initially very diffuse, a specific spacing, or decomposition wavelength, developing with time. Eventually, the interface sharpens and classic surface energies come into play. It is thus to be expected that coarsening becomes important during some phase of spinodal decomposition.

It is to be emphasized that nucleation and growth *within the metastable region* and spinodal decomposition *within the unstable region* are different processes yielding different decomposition products during the early period of isothermal phase separation. The distinctive features of spinodal decomposition, however, are the kinetics and morphologies by which unstable solutions decompose and the manner in which the resulting morphologies can be determined and distinguished from nucleation and growth.

II. THERMODYNAMIC CONSIDERATIONS

A. Free Energy–Composition–Temperature Relations

Thermodynamically, the spinodal is defined as the locus of $f'' = (\partial^2 f/\partial c^2)_{T,P} = 0$ (Fig. 1). To comprehend the meaning of the "limit of metastability," the spinodal line, consider the following [after Cahn (1965, 1968)]. A binary fluid system of molar free energy F has a free energy curve as given in Fig. 1a. To obtain the free energy of a mole of fluid of average composition c_0, Gibbs' graphical technique is employed to connect by a straight line the points on the free energy curves representing two compositions of the inhomogeneous system. Figure 2 shows two extreme examples of F versus c plots for use in the following argument. F, the molar free energy of the mixture, is given by the intersection of this line at the average composition, c_0. If a homogeneous phase of composition c_0 is in the region of the phase diagram where $f'' < 0$, it is unstable (Fig. 2b) because the very smallest deviation from c_0 lowers the free energy. Therefore, separation proceeds spontaneously as governed by the lowering of the successive lines of Fig. 2b until the line representing the lowest free energy state is attained. This state is defined by the common tangent of the two-phase (binodal) mixture, c_α and $c_{\alpha'}$. The tendency to oppose homogenization really means that a diffusional flux is operating against concentration gradients, yielding a negative diffusion coefficient in the spinodal region. The region of negative curvature in the free energy–composition curves defines the region within which spinodal decomposition is predicted.

Conversely, if $f'' > 0$, then c_0 represents a metastable phase (Fig. 2a). Small departures in composition from c_0 raise the free energy. However, large composition changes toward $c_{\alpha'}$ can lead to a decrease in the free energy. Accompanying these large composition variations is a positive interfacial free energy term which dominates for small regions. This portion of the phase diagram depicts metastable processes, and phase decomposition proceeds by nucleation and growth.

If the free energy curve for a binary system is continuous, with continuous first derivatives (e.g., Fig. 1a), a spinodal region will exist below the critical temperature. In such a system the two phases in equilibrium are related so

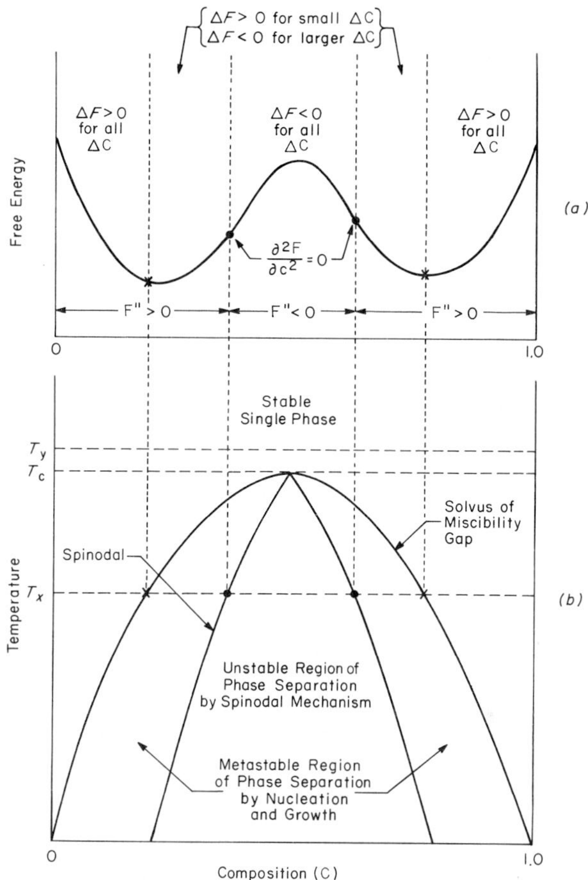

Fig. 1. (a) Schematics of the equilibrium miscibility gap (the locus of the common tangent points, ×) and the spinodal (the locus of the inflection points of the free energy curve, ●), T_c is the critical temperature of the miscibility gap. (b) Free energy curve as a function of composition, at temperature T_x, illustrating the regions of stability (modified from Shewmon, 1969).

that one phase can be transformed continuously into the other by a compositional change. For simplicity, consider an isotropic strain-free binary homogeneous solid solution in which the molar volume is independent of composition and pressure. If this system of equimolar composition is quenched at an infinitely fast rate from a solutionizing temperature T_y to T_x (Fig. 1b), the alloy at T_x will retain a solute distribution which is identical to that state at T_y just prior to quenching. If it is assumed that no diffusion occurs at T_x, then the alloy is in a state of high supersaturation and the thermodynamics requires that phase separation occurs.

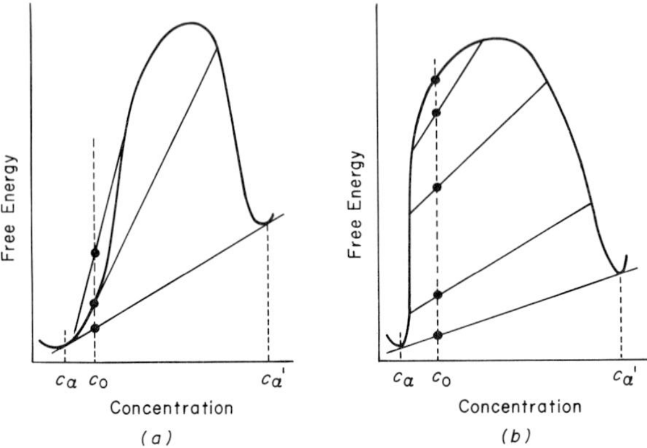

Fig. 2. Enlarged schematic of the free energy–composition curves. The diagram to the left (a) represents the free energy changes during decomposition of a metastable phase of composition c_0, and the diagram to the right (b) represents the free energy change of an unstable phase c_0 (from Cahn, 1968).

Assuming that a solid solution of composition c_0 (Fig. 2b) is uniform, then the free energy F_0 of the homogeneous system is given by,

$$F_0 = \int_v f(c_0)\, dV \tag{1}$$

where $f(c_0)$ is the local free energy and V is the volume.

In a single phase which is nonuniform in composition (containing some quenched-in fluctuations), the local free energy will depend both on the local composition and on the composition gradient. Using the thermodynamics of nonuniform systems and noting that the local free energy, $f(c)$, is a continuous function of both the composition, c, and the derivatives of the composition with respect to distance, Cahn and Hilliard (1958) treated this situation for fluid systems and isotropic solids. The total Helmholtz free energy for a nonuniform system was given as

$$F = \int_v [f(c) + \kappa(\nabla c)^2]\, dV \tag{2}$$

where $f(c)$ is the Helmholtz free energy of a unit volume of homogeneous material of composition c, $\kappa(\nabla c)^2$ the first term of an expansion representing the increase in free energy due to the introduction of a composition gradient, and κ the gradient energy coefficient, a positive proportionality constant (for clustering systems). The second term within the integral of Eq. (2) is an incipient surface tension and represents the contribution of the gradients in composition to the total free energy.

The difference in the free energy per unit volume between the initial homogeneous solution [Eq. (1)] and a solution with composition fluctuations

[Eq. (2)] is

$$\Delta F = (F - F_0) = \int_v [f(c) - f(c_0) + \kappa(\nabla c)^2] \, dV \tag{3}$$

Since the local deviation of uniformity leading to an instability is of primary interest, $f(c)$ can be expanded about the average composition, c_0, and substituted into Eq. (3) yielding, for a conservative system,

$$\Delta F = \int_v [\tfrac{1}{2}(c - c_0)^2 f'' + \kappa(\nabla c)^2] \, dV \tag{4}$$

Equation (4) represents the preliminary thermodynamic part of spinodal theory. Within the spinodal region, f'' is negative, making the first term within the integral negative. For a clustering system, the second term will of course be positive. It is thus possible, within the spinodal region, to always obtain some fluctuation which is stable for growth. The detailed rates and expected morphologies, however, must await kinetic considerations.

In order to examine the thermodynamic effect of a simple composition gradient, prior to kinetic considerations, introduce a one-dimensional composition modulation of amplitude A and wavenumber β ($\beta = 2\pi/\lambda$) (Cahn, 1961),

$$c - c_0 = A \cos(\beta x) \tag{5}$$

where c_0 is the average composition. Inserting this modulation into Eq. (4) yields

$$\Delta F = \tfrac{1}{4} V A^2 [f'' + 2\kappa\beta^2] \tag{6}$$

which is the difference in free energy per unit volume of uniform material and material containing the composition wave described by Eq. (5) (Cahn, 1961, 1968). Within the spinodal region, ΔF can have either a negative or positive value depending upon the magnitudes of the various quantities. For example, for a given value of f'', when β increases from a very low value (long wavelength) to a very high value (short wavelength), ΔF goes from negative (an unstable solution), through zero, to a positive value (a stable solution). There is thus a critical value of β, called β_c, above which the solution is stable to fluctuations in composition. These higher frequency values of $\beta > \beta_c$ contribute a large positive free energy (high incipient surface energy) to ΔF, and, therefore, from a thermodynamic viewpoint, the system would prefer to decompose by the development of low β (long λ) fluctuations. The critical value of β above which $\Delta F > 0$ is

$$\beta_c = (-f''/2\kappa)^{1/2} \tag{7}$$

or in terms of $\lambda_c = 2\pi/\beta_c$,

$$\lambda_c = -(8\pi^2\kappa/f'')^{1/2} \tag{8}$$

It is important to note that the gradient energy term limits decomposition on too fine a scale (Cahn, 1961).

Consideration of the kinetic aspects of spinodal decomposition requires the solution of a general diffusion equation in which the diffusional flux is driven not by a gradient in composition but by a gradient in chemical potential. Cahn (1961, 1968) related the spontaneous diffusional flux to the gradient in chemical potential and showed that for the flux to be spontaneous, it must lead to a decrease in free energy. He thus introduced this constraint into the generalized diffusion equation, and thereby formulated the diffusion equation for the early-stage kinetics of spinodal decomposition as

$$\partial c/\partial t = (M/N_v)[f'' \nabla^2 c - 2M\kappa \nabla^4 c + \text{nonlinear terms}] \tag{9}$$

where N_v is the number of atoms in a given volume and M is a positive atomic mobility given by

$$M = \tilde{D} N_v / f'' \tag{10}$$

The interdiffusion coefficient $\tilde{D}$ takes the sign of f'' and is thus *negative* in the spinodal region. The result of $\tilde{D} < 0$ is that diffusion can occur up gradients in concentration, and, thus, fluctuations can increase in amplitude. This "uphill" diffusion is the essence of spinodal decomposition.

To obtain an analytic solution to Eq. (9), Cahn (1961) used a linear approximation, and the solution is therefore limited only to fluctuations which are small in amplitude. This approximation imposes a severe, but necessary, experimental limitation: only the earliest regime of spinodal decomposition is amenable to study. This limitation has plagued experimental investigations and has generated much debate on experimental demonstrations of spinodal decomposition. This point will be discussed later.

Aside from the linear approximation there are other assumptions inherent in Cahn's solution. For example, f'', κ, and M are considered to be independent of composition. With these "early-stage" approximations, a continuum solution is found to be

$$(c - c_0) = A(\beta, t) \cos(\beta r) \tag{11}$$

where the amplitude A is a function of β and time t. Here β is a vector in wavenumber space and r is the position vector. The general solution to Eq. (11) over all β and r yields

$$c(r, t) - c_0 = (1/2\pi)^3 \int_\beta A(\beta, t) \exp(i\beta r)\, d\beta \tag{12}$$

This general solution represents all possible Fourier components of a general fluctuation. The time-dependent amplitude of a given component is related to the "as-quenched" amplitude at $t = 0$,

$$A(\beta, t) = A(\beta, 0) \exp[R(\beta)t] \tag{13}$$

where $R(\beta)$ is an important kinetic parameter called the "amplification factor," which gives the rates of growth or decay of a Fourier component having wavenumber β, and is given by

$$R(\beta) = -(M/N_v)[f'' + 2\kappa\beta^2]\beta^2 \tag{14}$$

or in terms of the diffusion coefficient,

$$R(\beta) = -\tilde{D}[1 + (2\kappa/f'')\beta^2]\beta^2 \tag{15}$$

Plots of $R(\beta)$ versus wavenumber and wavelength are given by the solid curves in Figs. 3 and 4, respectively.

The amplification factor carries the meaning of a "kinetic distribution function," and conveys within it the central aspect of the spinodal concept: within the spinodal region, where the solution is unstable to fluctuations in compositions, decomposition occurs by the evolution of fluctuations having wavenumbers around a specific growth rate $R(\beta_M)$. For low β's (high λ), the partitioning of the atomic components is slow, and for high β's (short λ), the incipient surface energy contribution is high, forcing these $\beta > \beta_c$ to decay. Thus, around β_M there is a "window," determined by a composite of

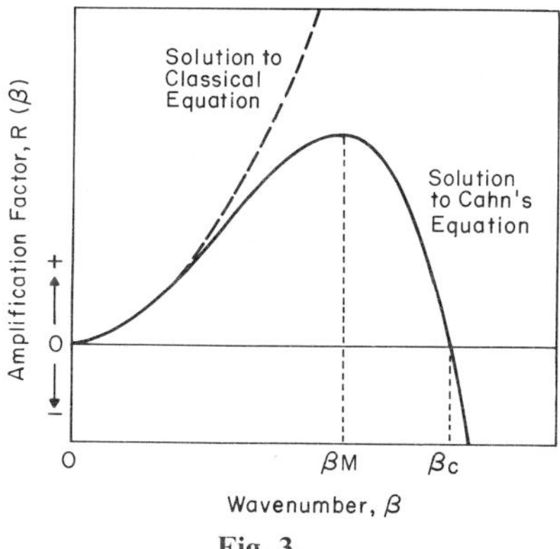

Fig. 3.

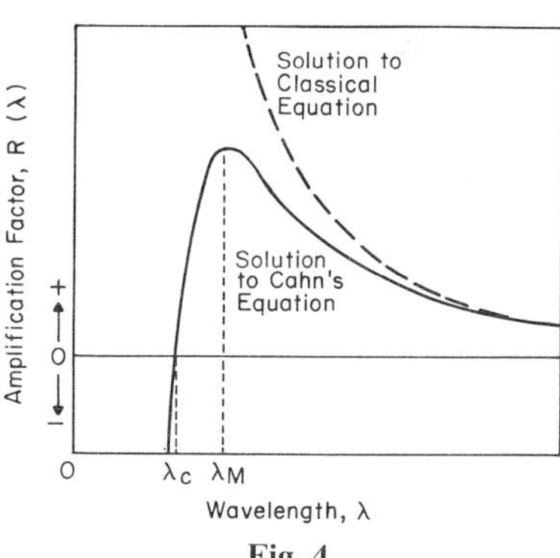

Fig. 4.

Fig. 3. Amplification factor $R(\beta)$ *versus* wavenumber β. The dashed curve represents the solution to the classical diffusion equation [Eq. (9) without the gradient energy term and without the nonlinear terms]. The solid curve is the solution to Eq. (9), neglecting only the higher-order nonlinear β terms. β_m is the wavenumber receiving the maximum amplification and β_c is the critical wavenumber (from Hilliard, 1970).

Fig. 4. The amplification factor $R(\beta)$ plotted as a function of wavelength $\lambda(=2\pi/\beta)$ rather than wavenumber β for the same solutions as given in Fig. 3 (from Hilliard, 1970).

thermodynamics and kinetics, which specifies the range of the fluctuations that will grow at the maximum rate. β_M is thus generally referred to as the "spinodal wavenumber."

Since κ, M, and f'' are here assumed to be essentially independent of composition, it is important to examine these assumptions and their effect on the derivations and experimental demonstrations of spinodal decomposition. It would, of course, be possible to eliminate these assumptions by maintaining the terms with higher derivatives [for a discussion of the higher-order terms of later stage spinodal decomposition and coarsening, see Hopper and Uhlmann (1973a,b,c)]. In fact, there exists experimental evidence (Cahn and Hilliard, 1958; Cook and Hilliard, 1969) that the gradient energy coefficient is not dependent on composition. However, the treatment of M and f'' is not as simple.

As shown in Fig. 5, the assumption of a constant f'' is equivalent to fitting the free energy–composition curve by a parabola (Hilliard, 1970). This approximation is unsatisfactory when the average composition, c_0, approaches the spinodal. Again, because of the linearization of the diffusion equation, the solution is valid only for small fluctuations, i.e., early times of decomposition. However, this limitation is not as severe as it might appear since a most important characteristic of the transformation, the morphology, is established in the early stages and thereafter changes only slowly with time (Hilliard, 1970).

On the matter of the constancy of M, a compositionally dependent diffusion coefficient can be applied at later stages of spinodal decomposition and to the beginning of coarsening (Cahn, 1966).

Cahn (1966) considered the terms which originated from the composition dependence of the diffusion coefficients and solved the nonlinear diffusion equation by successive approximations. Cook (1970) reformulated Cahn's continuum theory to account for the influence of thermal fluctuations. The

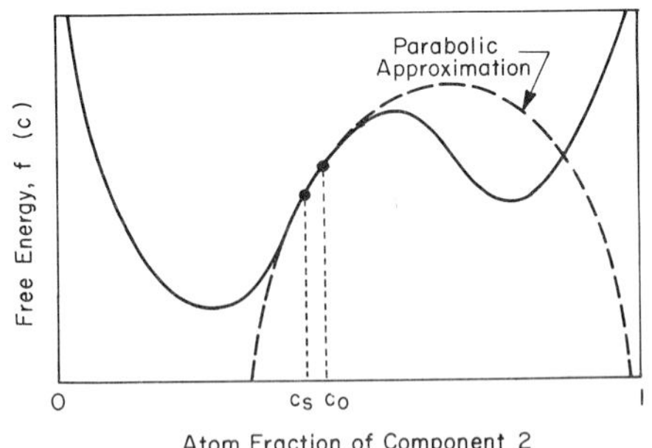

Fig. 5. The derivation of Cahn's diffusion equation assumes that f'' is independent of composition. This is equivalent to fitting the free energy function (solid curve) by a parabola (dashed curve) (from Hilliard, 1970).

use of a discrete model for the diffusion equation in Bravais lattices (Cook, de Fontaine and Hilliard, 1969) led to an equation analogous to Cahn's (1961, 1962a) continuum solution. However, these treatments are phenomenological in that they ignore the atomistic diffusion mechanism and do not consider that the atoms have to overcome potential barriers to change their positions. Recently, Langer (1973), Langer *et al.* (1975), Bortz *et al.* (1974), Marro, *et al.* (1975), and Lebowitz and Kalos (1976) have addressed themselves to this problem and to the manner in which the atomistic interactions explain the nonlinear effects often found during experimentation. The later stages of spinodal decomposition, where the linear approximations are no longer valid, have been examined in detail by both Langer (1973), Langer *et al.* (1975), and Tsakalakos (1977).

B. The Coherent Spinodal

Spinodal decomposition occurs by a continuous process, i.e., a continuous free energy function must exist from one phase to another. When applying the continuum theory to crystalline systems the coexisting phases must be crystallographically similar to one another, or a structurally similar metastable phase must exist. However, in crystalline solutions there is commonly a variation of lattice parameter and, hence, a change of volume with composition. In order for the lattice of such a solution to remain continuous or coherent in the presence of a composition fluctuation, work must be performed in straining the lattice.

Relative to phase separation by the development of coherent fluctuations, Cahn (1962b) introduced a function $\phi(c)$, having all the properties of the Helmholtz free energy $f(c)$, but containing an elastic term. He showed that coherent fluctuations in solids could be treated in a manner analogous to fluids by substitution of $\phi(c)$ for $f(c)$, where

$$\phi(c) = f'(c) + [\eta^2 E/(1 - \nu)]c^2 \qquad (16)$$

where η is the linear change in lattice parameter per unit composition change $(1/a_0)(\partial a/\partial c)$, where a_0 is the lattice parameter at the average composition c_0; E is Young's modulus; and ν is Poisson's ratio. Therefore, the free energy $\phi(c)$ for a coherent system is raised by the reversible elastic work performed to match the lattices. This is shown schematically in Fig. 6, the locus of the coherent spinodal being defined as

$$\partial^2\phi/\partial c^2 \equiv [f'' + 2\eta^2 Y] \equiv 0 \qquad (17)$$

where

$$Y = E/(1 - \nu)$$

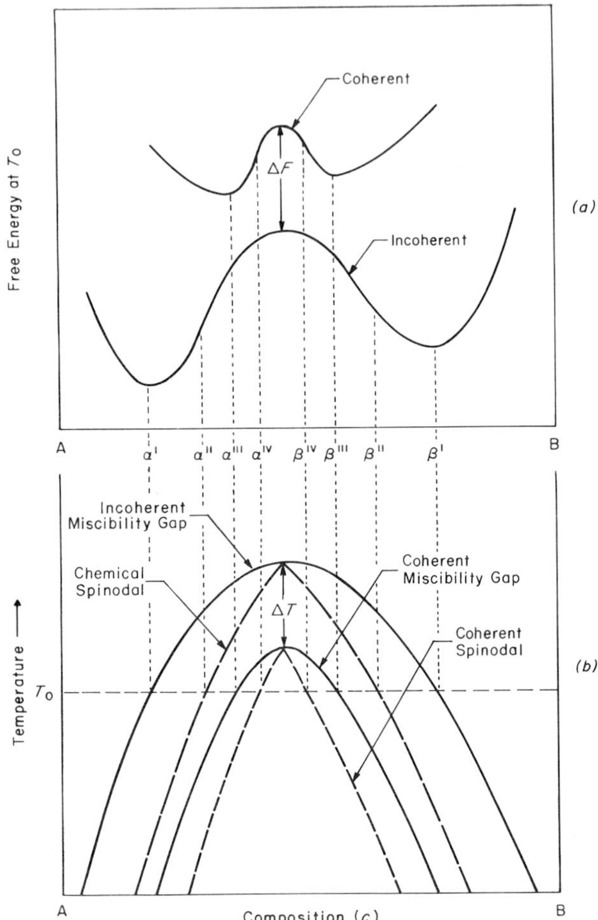

Fig. 6. (a) Corresponding coherent and incoherent free energy curves for a binary solid solution at temperature T_0. The coherent spinodal is given by the inflection points on the free energy curve at α^{IV} and β^{IV}. The coherent free energy curve is higher than the incoherent equilibrium free energy curve (after Hilliard, 1966). (b) Coherent and incoherent miscibility gaps and the respective coherent and chemical spinodals.

The coherent phase diagram is always metastable and lies within the unstressed equilibrium phase diagram. The coherent phase diagram is a real, metastable phase diagram since it involves reversible metastable equilibrium, which is subject only to the constraint that the lattices remain continuous. The spinodal which leads to spontaneous decomposition is thus the spinodal of the coherent free energy curve of the metastable coherent phase diagram (Cahn, 1968).

Carrying the strain term throughout the above derivation, and assuming that this term is also compositionally independent, produces an amplification factor:

$$R(\beta) = -(M/N_v)[f'' + 2\eta^2 Y + 2\kappa\beta^2]\beta^2 \tag{18}$$

and $R(\beta)$ is positive if

$$f'' + 2\eta^2 Y + 2\kappa\beta^2 \leq 0 \tag{19}$$

The strain term is important in crystalline systems where it can give rise to directionality in the developing fluctuations. Since Y is a function of crystalline direction, it follows that the temperature of the coherent spinodal will vary with the direction of the wavevector, and the temperature of the coherent spinodal will be at a maximum for those directions which minimize $2\eta^2 Y$ (Hilliard, 1970). The difference ΔT between the temperature of the chemical or incoherent spinodal T_{chem} and the temperature of the coherent spinodal T_{coh} can be estimated by expanding f'' about T_{chem} (Hilliard, 1970) and substituting into Eq. (17) to obtain

$$\Delta T = T_{\text{coh}} - T_{\text{chem}} \simeq -\frac{2\eta^2 Y}{(\partial f''/\partial T)_{T_{\text{chem}}}} \simeq \frac{2\eta^2 Y}{s''} \tag{20}$$

where $f'' = 0$ at T_{chem}; $s =$ entropy/unit volume, and $s'' = \partial^2 s/\partial c^2$. If the entropy of mixing is ideal, then

$$s'' = \partial^2 s/\partial c^2 = -N_v k/c(1-c) \tag{21}$$

where k is Boltzmann's constant. Combining Eqs. (20) and (21) yields a value for ΔT related to the coherency strains:

$$\Delta T = T_{\text{coh}} - T_{\text{chem}} \simeq -2\eta^2 Yc(1-c)/kN_v \tag{22}$$

where Y is the minimum value of

$$\begin{aligned} Y_{100} &= \frac{(c_{11} + 2c_{12})(c_{11} - c_{12})}{(c_{11})} \\ Y_{111} &= \frac{6(c_{11} + 2c_{12})c_{44}}{(c_{11} + 2c_{12} + 4c_{44})} \end{aligned} \tag{23}$$

Variations in ΔT are extremely sensitive to η^2 since this term can exhibit larger variations than the elastic anisotropy factor Y for most systems. De Fontaine (1969) has derived an analytical expression for predicting the *maximum* average wavelength attainable by a coherent composition modulation in a binary solid solution. In fact, for values of η greater than $\sim 10\%$, loss of coherency may occur prior to spinodal decomposition.

In cubic crystals Cahn (1962b) demonstrated that the elastic coefficient Y reduces to $E/(1 - \nu)$ as the anisotropy $(2c_{44} - c_{11} + c_{12})$ reduces to zero. The elastic coefficient is least for β parallel to $\langle 100 \rangle$ and maximum for β parallel to $\langle 111 \rangle$ when $(2c_{44} - c_{11} + c_{12}) > 0$. When this summation is negative, Y is maximized for β parallel to $\langle 100 \rangle$ and minimized for β parallel to $\langle 111 \rangle$. The composition modulation will thus choose the elastically soft directions of a cubic anisotropic crystal and the coherent limit of metastability can be defined as

$$f'' + 2\eta^2 Y_{\min} \equiv 0 \tag{24}$$

The concept of a coherent spinodal and its calculation as presented in the previous discussion was developed from the simple theory of fluid and isotropic crystalline systems and then applied to cubic metal systems exhibiting anisotropy. In cubic ionic substances which exhibit anisotropy, it must be recalled that nearest neighbor lattice sites are occupied by unlike atoms. A recent treatment by Fancher and Barsch (1971) for the calculation of ΔT for anisotropic ionic systems gives

$$\Delta T = T_{\text{coh}} - T_{\text{chem}} = 2\eta^2 Y \frac{c(1-c)}{kN_0} \times 10^3 \tag{25}$$

where k is the Boltzmann's constant; c the composition in atomic fraction; N_0 the number of molecules per unit volume, $N_0 \equiv \frac{1}{2}\{r_{(1)}^3(1-x) + r_{(2)}^3(x)\}$; and $r_{(i)}$ the nearest neighbor distance of components i, where $i = 1, 2$; η the derivative of the nearest neighbor distance in the solid solution with respect to composition, $\eta \equiv (1/r)(\partial r/\partial x)$. The value of Y, the orientation dependent elastic coefficient, is defined as the minimum value of Y_{100} or Y_{111}, as given by

$$Y_{100} = \frac{\{h_{11}(c) + 2h_{12}(c)\}\{h_{11}(c) - h_{12}(c)\}}{h_{11}(c)} \quad \text{when} \quad \alpha < 0 \tag{26}$$

$$Y_{111} = \frac{6\{h_{11}(c) + 2h_{12}(c)\}\{h_{44}(c)\}}{4h_{44}(c) + h_{11}(c) + 2h_{12}(c)} \quad \text{when} \quad \alpha > 0 \tag{27}$$

where the elastic anisotropy factor, α, is defined as

$$\alpha \equiv \frac{h_{11}(c) - h_{12}(c) - 2h_{44}(c)}{2h_{44}(c)} \tag{28}$$

Here, the use of the at-temperature elastic coefficients h_{uv}, where $uv = 11$, 12, 44 enables a better approximation of the temperature dependency of both the anisotropy and the locus of the coherent spinodal. The temperature dependence of the elastic constants is given as

$$h_{uv}(i) = c_{uv}(i)\{1 + 10^{-3}\tau_{uv}(i)(T - 300^{\circ}\mathrm{K})\} \tag{29}$$

where $h_{uv}(i)$ is the temperature dependent elastic constants for component i, $i = 1, 2$; $c_{uv}(i)$ the room temperature elastic constants; $\tau_{uv}(i)$ the temperature derivatives of the elastic constants; and the composition dependence of the at-temperature elastic constants is considered to obey Vegard's law and is defined as

$$h_{uv}(c) = h_{uv(1)}(1 - c) + h_{uv(2)}(c) \tag{30}$$

In work to be described here, consideration is given to the temperature dependence of the elastic constants as well as to the temperature dependence of the lattice parameters (i.e., the coefficients of linear expansion) and their concomitant effects on the anisotropy and ΔT. Usually, the temperature derivatives of the elastic constants become softer as the temperature increases; however, in several oxide systems the reverse is true, causing the elastically soft directions to change and ΔT to become smaller. Similarly, the η term often becomes smaller at higher temperatures, since varying thermal expansion rates cause the lattice parameters to converge. The calculations and their relation to experimentation in binary metallic, alkali halide, and oxide systems will be discussed later.

In summary, coherency has two effects on spinodal decomposition. First, coherency strains cause the depression of the spinodal curve relative to the equilibrium phase diagram and thereby increase the stability of the solid solution. Second, the work term influences morphology due to the elastic anisotropy which is introduced.

III. THE EXPERIMENTAL SITUATION

The Ideal versus Reality

There have been a number of studies to demonstrate the occurrence of a spinodal mode during the decomposition of quenched systems. The situation is somewhat simplified by the use of "linearity" in the theoretical approach outlined above. However, there remain numerous ambiguities. But in the main, it is clear that spinodal decomposition is an operative and observable phenomenon. We shall here briefly review the experimental situation and how the results in metallic alloys will influence studies carried out on ceramics.

This discussion will be limited to simple binary systems which exhibit stable and metastable miscibility gaps (Fig. 7a). It is, however, important to note that a eutectic system in which the two terminal solid solutions are of

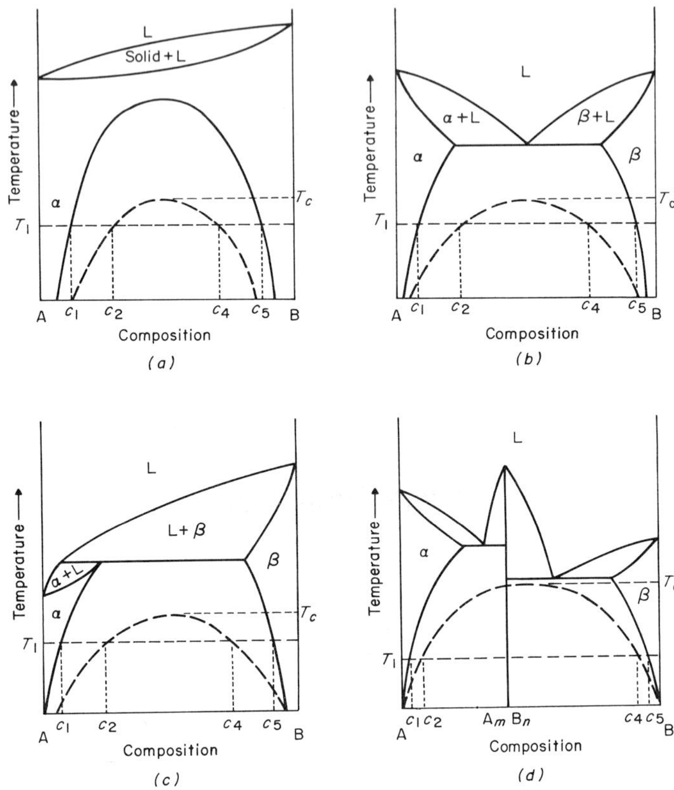

Fig. 7. If the terminal solid solutions, α and β exhibit similar structures, then a metastable miscibility gap can exist in most systems at low temperatures. Metastable solid solubility exists at temperatures above T_c of the metastable miscibility gaps. (a) Schematic of a binary phase diagram exhibiting both a stable miscibility gap (solid curve) with solvus composition c_1 and c_5 at T_1 and a metastable miscibility gap (dashed curve) with solvus compositions c_2 and c_4 at T_1. (b) Schematic binary system with two terminal solid solutions exhibiting an equilibrium eutectic with solvus compositions c_1 and c_5 at T_1 and a metastable miscibility gap (dashed curve) with solvus compositions c_2 and c_4 at T_1. (c) Schematic binary system with two terminal solid solutions exhibiting an equilibrium peritectic with solvus compositions c_1 and c_5 at T_1 and a metastable miscibility gap (dashed curve) with solvus compositions c_2 and c_4 at T_1. (d) Schematic binary phase diagram with two terminal solid solutions and an intermediate congruently melting compound A_mB_n. The equilibrium solvus compositions at T_1 are c_1 and c_5 and the metastable miscibility gap (dashed curve) solvus compositions are c_2 and c_4 at T_1 (after Fine, 1972).

similar structure are merely binary systems with a liquidus minimum that intersects the solid solubility gap because the excess free energy of the solid $\Delta G^{\mathrm{xs(sol)}}$ is greater than the excess free energy of the liquid $\Delta G^{\mathrm{xs(liq)}}$, i.e., the miscibility gap extends up to temperatures where the liquid is still stable.

Therefore, a metastable coherent miscibility gap can be predicted for eutectic systems as well (Fig. 7b). Likewise, the intersection of a simple solid–liquid loop with a solid miscibility gap due to $\Delta G^{xs(sol)} > \Delta G^{xs(liq)}$, causes a peritectic diagram. If the terminal solid solutions are of similar phase structure a metastable coherent miscibility gap can again exist (Fig. 7c). Last, the most complex type of binary system, exhibiting two eutectics and a congruently melting intermediate compound A_mB_n can also exhibit a metastable miscibility gap if the α and β terminal solutions have similar structures (Fig. 7d). In each of the four types of phase diagrams represented in Figs. 7a–d, decomposition at T_1 of a quenched composition, c_0, will result in the coherent end-product phases α of c_2 and β of c_4.

In principle, spinodal decomposition should be expected to occur by quenching from above the critical temperature and aging at a temperature within the spinodal region. It is also possible to study the kinetics of the process by slower cooling, and in fact, c-curve behavior for spinodal decomposition has been predicted by Huston *et al.* (1966). They have formulated various kinetic and morphological dependencies on quench rate.

If diffusion can be limited following the quench to the spinodal region, the specimen can be aged at some higher temperature and, presumably, the kinetics of the decomposition process can be followed. Only in glasses or in solids for which the coherent spinodal is well below the melting point, is there any hope of obtaining a time scale long enough to suppress the reaction during quenching, and then studying it isothermally (Cahn, 1968). For quench rates which are rapid enough to preclude significant decomposition, the composition–amplitude spectrum of spinodal decomposition should be independent of the quench rate. It is of some interest to note that Agarwal and Herman (1973a) have found that very little decomposition occurs both during and after liquid quenching of aluminum alloys, and have suggested that the ultrarapid quench from the melt is the ideal way to obtain a homogeneous solid solution for studies of spinodal decomposition.

B. Morphology

There has been a tendency to employ morphology as a mark of spinodal decomposition. For example, a specific repetitive spacing or mutual connectivity between the conjugate product phases has been used to confirm the operation of the process. Much of this has been obtained from electron microscopy, and a number of the early experiments used spacing exclusively.

It has been pointed out by several authors (Hilliard, 1970; Herman and MacCrone, 1971) that morphology alone cannot be used to unambiguously determine that a process of phase decomposition is spinodal. The situation is clearly rather more involved, requiring a study of the evolution of the new

TABLE I

SOME FACTORS WHICH DIFFERENTIATE BETWEEN NUCLEATION AND GROWTH AND SPINODAL MECHANISMS DURING ISOTHERMAL PHASE SEPARATION[a]

Nucleation and growth	Spinodal decomposition
Invariance of second phase composition to time at constant temperature	Continuous variation of both extremes in composition with time until equilibrium compositions are reached
Interface between phases is always same degree of sharpness during growth	Interface between phases initially is very diffuse, eventually sharpens
Tendency for random distributions of particle sizes and positions in matrix	Regularity of second phase distribution in size and position characterized by a geometric spacing
Tendency for separation of second phase spherical particles with low connectivity	Tendency for separation of second phase, nonspherical particles with high connectivity

[a] From Cahn and Charles (1965).

phases with time. Actually, a number of factors differentiate nucleation and growth from spinodal decomposition. These are listed in Table I [after Cahn and Charles (1965)].

Figure 8 shows schematic representations of the evolution of the concentration profiles as a function of time. Illustrated are composition profiles which developed from nucleation and growth or from spinodal decomposition. The variations depicted are (1) an embryo achieving criticality by growing only in size (Fig. 8a), (2) an embryo increasing its composition (Fig. 8b), and (3) an embryo increasing both its size and composition (Fig. 8c). The concentration gradient shown in Fig. 8d depicts the amplitude evolution of a sine wave shaped composition fluctuation and demonstrates the diffuseness of the early and middle stages of spinodal decomposition. The diffuse boundary of early spinodal decomposition (Fig. 8d) eventually sharpens in the later stages until the morphology resembles that of nucleation and growth. The distinction between the morphologies of nucleation and growth and that of spinodal decomposition is therefore *limited* to early stage morphologies where the amplitude of the modulations increases as a function of time, but the wavelength of the growing fluctuation remains essentially constant. The morphology of either isothermal spinodal decomposition or spinodal decomposition during continuous cooling are the same; a superimposition of sinusoidally varying composition modulations clustered about a wavelength λ_{max} that receives maximum amplification. The wavelength for which the growth rate is a maximum depends on the temperature and on the solution parameters but it is on the order of 100 Å.

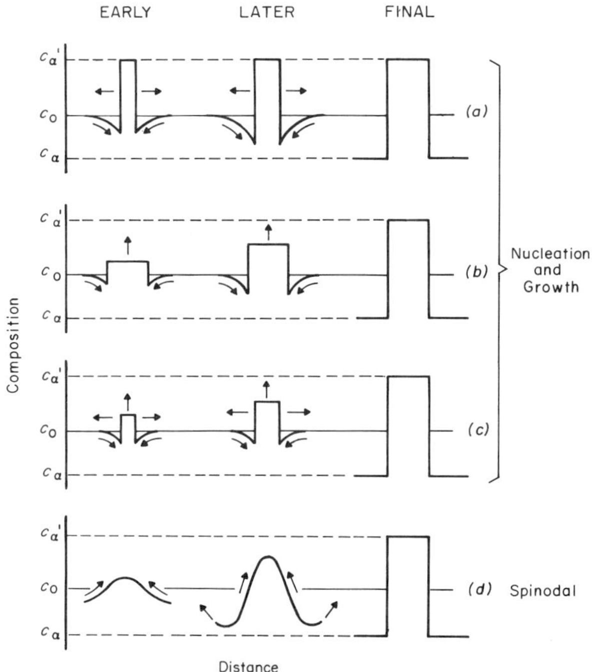

Fig. 8. Schematic concentration profiles illustrating the different development of phase decomposition by nucleation and growth (a–c) and spinodal mechanisms (d). (a) A nucleus growing in size at a fixed composition with the corresponding depleted matrix zones; the arrows indicate diffusion of material down a concentration gradient. (b) A nucleus of constant size, but with changing composition. (c) A nucleus both growing and changing composition simultaneously. (d) A composition fluctuation growing in size and concentration by "uphill" diffusion.

Cahn's (1965) computer simulation of phase connectivity is based on superpositioning of sinusoidal composition fluctuations with a Gaussian distribution of amplitudes of fixed wavelength, but random orientation and phase for isotropic solids (Fig. 9a). He demonstrated that the connectivity of the two conjugate phases is maintained over a very wide range of volume fractions ($\sim$0.15–0.85) and suggested that interconnectivity over such wide composition ranges is the factor which distinguishes the morphology of spinodal decomposition from nucleation and growth in isotropic systems. The discrete particles which form during nucleation and growth (Fig. 9b) become interconnected only when the two conjugate phases have approximately the same volume fractions (Haller, 1965). Goldstein (1968) has shown theoretically that the formation of an interconnected structure from discrete particles is highly unlikely during the early stages of phase separation. However, Seward *et al.* (1968) were able to follow the development of discrete

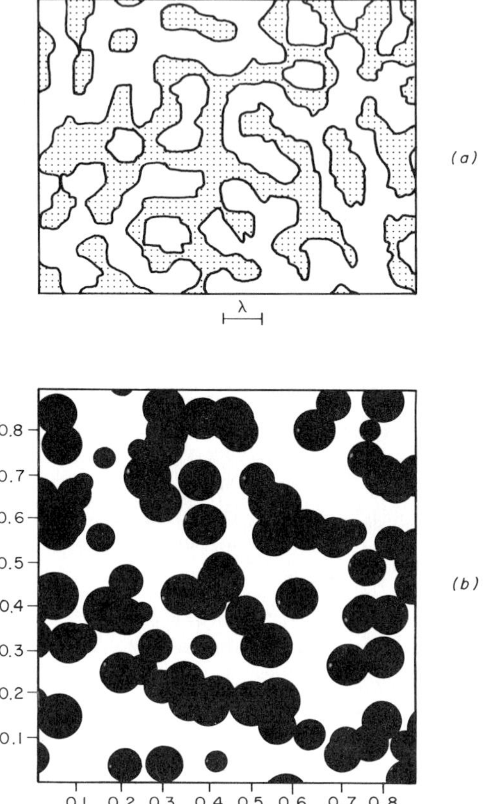

Fig. 9. (a) Cross section of a spinodal structure in an isotropic material computer simulated by superposition of sinusoidal composition fluctuations of fixed wavelength, but random orientation and phase. The points define regions where the concentration is greater than average. Note the high degree of connectivity and the lack of periodicity (from Cahn, 1965). (b) Cross section of a three-dimensional matrix of equal spheres, the positions of which have been randomized by computer. Note the connectivity (from Haller, 1965).

particles in the BaO–SiO_2 system into an interconnected structure. They postulated that this occurred by Brownian motion establishing contact and coalescence of neighboring spherical particles in a viscous matrix. Cahn, however, has suggested that in systems such as BaO–SiO_2, where the mobility is strongly dependent on composition, the formation of early-stage discontinuous structures in spinodal decomposition is not inconsistent with a spinodal mechanism (Hilliard, 1970).

For anisotropic crystalline systems, the specific volume depends on concentration, and the strain energy contributes to the total free energy during the development of composition modulations. This will introduce crystallographic directionality and result in a periodic structure. In anisotropic crystals the coherency strains have both a local and a long range influence on diffusion (Cahn, 1969a) and since the fastest growing Fourier components dominate the process, the resulting structure is composed of only a few components. For example, if the anisotropy factor $(2c_{44} - c_{11} + c_{12})$ is positive, $\langle 100 \rangle$-type waves dominate and the resulting structure is defined by three sinusoidal waves of wavelength λ_{max} in [100], [010], and [001]. These $\langle 100 \rangle$-type modulations produce a characteristic "basket weave" or "tweed" structure representative of spinodal decomposition in anisotropic systems (Cahn, 1962b, 1965, 1969a; de Fontaine, 1970; and Rundman, 1973).

Since phase connectivity and periodicity cannot be considered to be unequivocal evidence of spinodal decomposition, conventional transmission electron microscopy (TEM) cannot alone substantiate spinodal decomposition. In addition, the modulated structures observed by TEM are usually later stages of decomposition, where the simplifying assumptions of linear spinodal theory are not applicable.

A rather more direct and meaningful approach to studies of spinodal decomposition arose from work of Rundman and Hilliard (1967). They employed small-angle x-ray scattering methods to delineate the early stage process in a metal alloy system. In the following, their work and its important implications will be reviewed.

C. Small-Angle X-Ray Scattering (SAXS)

Neither the morphology of decomposition nor the integrated intensity measurements of SAXS experiments* can explicitly identify the occurrence of spinodal decomposition. However, the SAXS analysis given by Rundman and Hilliard (1967) (RH) provides a distinctive method for identifying spinodal decomposition for early stages of decomposition.

Rundman and Hilliard (1967) recognized the correspondence between the Fourier amplitudes of the composition fluctuations and the amplitude of the x rays scattered by these fluctuations. The amplitude of the intensity $A(s)$ scattered by a material of volume V at a point s in reciprocal space is given from scattering theory as

$$A(s, t) = \int_v \rho(r, t) \exp(-2\pi i s \cdot r)\, dr \tag{31}$$

* It has been shown that an integrated intensity function Q can be used to determine the volume of phase-separated material (Porod, 1967). It still remains unclear, however, how to differentiate the spinodal process from nucleation and growth, using this method.

where $\rho(r, t)$ is the electron density at point r at time t in Fourier space. The electron density can be related to both the composition and the structure factors by

$$\rho(r, t) = \rho_0 + (f_A - f_B)[c(r, t) - c_0] \tag{32}$$

where ρ_0 is the average electron density and f_A and f_B are the atomic scattering factors for atoms A and B, respectively.* Assuming that the scattering factors are independent of angle for small angles, the amplitude can be expressed as

$$A(s, t) = (f_A - f_B) \int_v [c(r, t) - c_0] \exp(-2\pi i s \cdot r)\, dr \tag{33}$$

The scattered intensity can be related to the amplification factor by substituting $s = \beta/2\pi$ into Eq. (33), where β is a vector in Fourier space, the magnitude of which is the wavenumber. Noting that the diffracted intensity $I(s)$ is equal to $A(s) \cdot A^*(s)$ yields an expression for the temporal evolution of intensity

$$I(\beta, t) = I(\beta, 0) \exp[2R(\beta)t] \tag{34}$$

where $I(\beta, 0)$ depicts the "as-quenched" intensity profile, i.e., at $t = 0$. According to RH, linear spinodal behavior is indicated by a linear plot of ln $I(\beta, t)$ *versus* time, where the slope $2R(\beta)$ will be positive for $\beta < \beta_c$ and negative for $\beta > \beta_c$ (Fig. 10a). $R(\beta)$ must be positive for growth to occur, while $R(\beta) < 0$ indicates decay. Again, the $R(\beta)$ values are obtained from the ln I *versus* t data plots; and when $R(\beta)$ is plotted *versus* β, a maximum is obtained at β_m for a spinodal process. The amplification factor shows a negative value for $\beta > \beta_c$, as seen in Fig. 10b.

As stated previously, the physical significance of the maximum in the $R(\beta)$ *versus* β plot can be seen by considering two competing factors. For low β values, the diffusion distance is large and consequently the rate of interdiffusion is diminished; for high β values, the "incipient surface energy" $\kappa(\Delta c)^2$ contributes a large positive value, giving an excessive positive free energy. Thus, due to this balance between kinetic and thermodynamic features, the system decomposes by way of a "growth window," where β_m is the component growing at the maximum rate.

Relative to experimental tests for spinodal decomposition it has been shown that the ideal ratio of β_c/β_m (for the linearized theory) should be $\sqrt{2}$.

* X rays are scattered at low-angles by variations in electron density (i.e., composition fluctuations). This diffuse scattering can be analyzed by the use of models to characterize the electron density variation within the solution. The technique employed by RH requires no model and actually depicts the distribution of the composition fluctuations. The atomic species must differ sufficiently in atomic number (atomic scattering factor) for the composition variations to give rise to an effective variation in electron density, and hence to give a measurable scattered intensity.

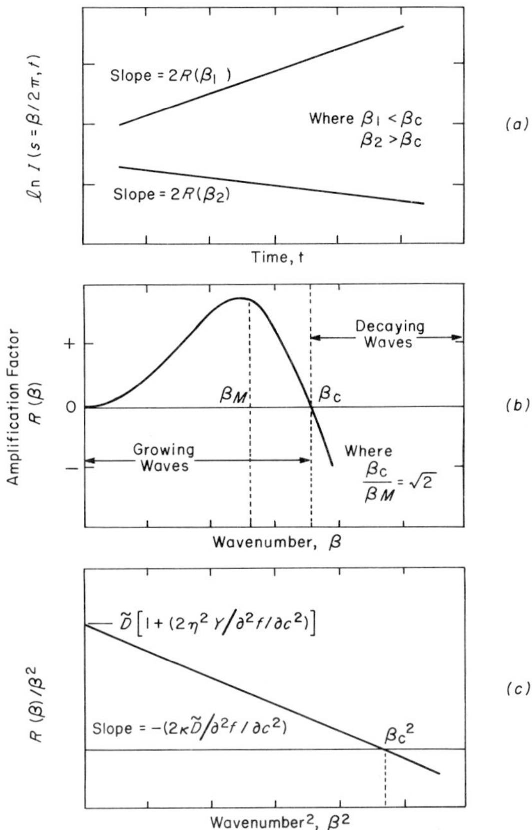

Fig. 10. Schematic representation of the Rundman and Hilliard (1967) analysis. (a) Plot of ln $I(\beta,t)$ *versus* t, for solution of Eq. (34). (b) $R(\beta)$ *versus* β. Plot of Eq. (15). (c) $R(\beta)\cdot\beta^{-2}$ *versus* β^2. Plot of Eq. (35). See Text.

The number, in fact, is usually found experimentally to be larger. Furthermore, upon rearrangement of Eq. (18) and using Eq. (10),

$$R(\beta)\cdot\beta^{-2} = -(\tilde{D}/f'')[f'' + 2\eta^2 Y + 2\kappa\beta^2) \tag{35}$$

It is seen that a plot of $R(\beta)\cdot\beta^{-2}$ *versus* β^2 should yield a straight line with a slope equal to $-(2\kappa\tilde{D}/f'')$ and an intercept on the $\beta^2 = 0$ axis of $-\tilde{D}[1 + 2\eta^2 Y/f'']$; Fig. 10c. Therefore, both the interdiffusion coefficient and the gradient energy coefficient can in priniciple be determined if certain thermodynamic and elastic data are known. In addition, as has been shown by Hilliard (1970)

$$\beta_m{}^2 \simeq [T - T_{\text{chem}})s'' - 2\eta^2 Y]/4\kappa \tag{36}$$

where T_{chem} is the temperature of the chemical spinodal and s'' is again the second derivative of entropy with respect to composition. Therefore, if SAXS measurements are obtained at more than one temperature, a plot of $\beta_m{}^2$ or $(2\pi/\lambda_m)^2$ *versus* temperature should be linear with a slope of $s''/4\kappa$ and, using Eq. (20), it is possible to obtain the temperature T_{coh} corresponding to the coherent spinodal.

An analysis similar to that given here can of course be applied to small-angle neutron scattering (SANS). Because of the long wavelengths that are possible for neutrons ($\sim$4–20 Å) SANS can be used to measure extremely small scattering vectors. Of great interest is the potential to use SANS to study materials for which atomic scattering factors are too similar to be distinguished with SAXS, e.g., for Al–Si, Al–Mg, Au–Pt, Al_2O_3–SiO_2, MgO–Al_2O_3. In addition, for sufficiently long wavelength neutrons (beyond the Bragg cut-off), double-Bragg scattering effects are precluded.

D. X-Ray Diffraction and Other Techniques

Daniel and Lipson (1943, 1944) were the first to recognize sidebands adjacent to the major Bragg reflections in x-ray diffraction. They interpreted the satellites to be representations of a cubic lattice which has undergone deformation by the segregation of the different atoms. Rigorously, it has been shown that a composition modulation in a planar lattice is responsible for the production of satellites about a Bragg peak (de Fontaine, 1966; Guinier, 1963). Therefore, diffraction from a one-dimensional grating with sinusoidal modulation consists of the main reflection of the grating plus satellite reflections. The intensities of the satellites are governed by variations of scattering power and spacing (Daniel and Lipson, 1943, 1944; Hargreaves, 1951). The relationship which describes the position of the sidebands is

$$L = \frac{h \tan \theta}{(h^2 + k^2 + l^2)\,\Delta\theta} \tag{37}$$

where L is the wavelength of the original lattice spacing, θ the Bragg angle, h, k, l the Miller indicies of the Bragg reflection, $\Delta\theta$ the angular displacement of the sideband from the main reflection. The wavelength of the modulation λ is related to the crystal lattice by $\lambda = L/a$, where a is the lattice parameter. These relationships are useful for x-ray diffractometry. In addition, Cook and Hilliard (1969) and Philofsky and Hilliard (1969) have developed expressions for the effective interdiffusivity and gradient energy coefficient from the decay rates of satellite intensities with time.

Flewitt (1974) has modified these relationships for application to microdensitometer traces across given hkl reflections in electron diffraction patterns and has suggested that the greater intensity associated with electron diffraction allows the low-intensity sidebands of early decomposition to be measured.

Since sidebands can be caused by either a periodic variation of the lattice parameter in a certain direction, by periodic variations in atomic scattering power, or by a combination of these effects (Daniel and Lipson, 1943, 1944; Tiedema *et al.*, 1957), the appearance of x-ray diffraction satellites does not preclude decomposition by a spinodal mechanism. Hargreaves (1951) demonstrated that the Daniel and Lipson solution can be applied to a composition modulation of rectangular waveform, yielding satellites, and, decomposing by a nucleation and growth mechanism. In the system studied by Hargreaves (Cu–Ni–Fe), the structure was envisaged to be composed of a regular arrangement of lamellae of intermediate phases which retain their coherence with the matrix crystal and modulate it. Hillert (1961) attributed the modulated structures and the existence of x-ray satellites to "exchange transformations" in which periodic variations in composition occur due to a simple interchange of atoms. Exchange-type transformations as well as lamellar decomposition products are common in oxide and silicate systems. Therefore, the application of x-ray diffraction satellite measurements to determine the operation of spinodal decomposition in ceramic systems is at best ambiguous.

A recent technique related to small angle scattering has been applied to the study of spinodal decomposition in glasses by Zarzycki and Naudin (1968, 1969). They demonstrated that by suitable photographic reduction of a transmission electron micrograph, a diffraction grating can be generated and a record made of the Fraunhofer diffraction spectrum in visible light. A photometric trace across the diffraction ring can then be obtained. This analysis is analogous to SAXS, since the differences in blackening on the photographic plate are related to visible light just as the fluctuations of electron density are related to x rays. A spinodal wavelength, therefore, can be obtained directly from the diffraction ring.

Recently, lattice imaging by TEM has been claimed to enable direct detection at the atomic level of the localized variations in lattice parameter which is characteristic of spinodal decomposition (Sinclair and Thomas, 1974; Gronsky *et al.*, 1975). Measurement of the modulation wavelengths are in apparent agreement with those determined by diffractometer satellites for the alloys studied (Gronsky *et al.*, 1975).

In summary, neither morphology, as determined with conventional TEM nor sideband x-ray diffraction studies, can be considered as unequivocal proof of spinodal decomposition. It is our belief that small-angle scattering, or perhaps some form of lattice imaging technique, can give a clearer indication of the kinetics of spinodal decomposition.

TABLE II ELASTIC CONSTANTS, TEMPERATURE DERIVATIVES OF THE ELASTIC CONSTANTS, COEFFICIENTS OF LINEAR EXPANSION AND LATTICE PARAMETER DATA FOR SEVERAL METALS, ALKALI HALIDES, AND OXIDES

Material	Lattice parameter (Å)	Elastic constants ($\times 10^{11}$ dyn/cm^2)			Temperature derivatives ($\times 10^3$/°K)			Coefficient of linear expansion ($\times 10^{-6}$/°C)
		c_{11}	c_{12}	c_{44}	τ_{11}	τ_{12}	τ_{44}	
Al	4.0522[a]	9.13[b]	5.47[b]	2.386[b]	−0.32[b]	−0.14[b]	−0.44[b]	25.0[c]
Zn	3.9422[a]	13.68[b]	5.03[b]	2.838[b]	−0.48[b]	−0.16[b]	−0.81[b]	35.0[c]
Au	4.0786[d]	19.00[b]	16.10[b]	4.23[b]	−0.18[b]	−0.15[b]	−0.33[b]	17.30[e]
Pt	3.9231[f]	34.70[b]	25.10[b]	7.65[b]	−0.10[b]	−0.03[b]	−0.035[b,g,h]	11.69[e]
Ni	3.5238[i]	24.80[b]	15.30[b]	11.60[b]	−0.27[b]	−0.05[b]	−0.30[b]	17.10[e]
AgCl	5.549[j]	6.01[b]	3.62[b]	6.20[b]	−1.04[b]	−0.29[b]	−0.44[b]	33.10[k]
NaCl	5.6402[l]	4.77[m,b]	1.13[m,b]	1.27[m,b]	−0.80[b,n]	0.17[b,h,n]	−0.27[b,n]	44.00[o]
KCl	6.2931[p]	3.97[m,b]	0.615[m,b]	0.63[m,b]	−0.86[b]	0.56[b,h]	−0.25[b]	36.00[o]
CoO	4.260[q]	26.17[r]	14.50[r,b]	8.23[r,b]	0.89[b,g]	−1.23[b,g]	0.29[b,g]	9.98[s]
MgO	4.213[t]	28.99[u]	8.57[u]	15.49[u]	−0.24[b]	−0.11[b]	−0.12[b]	13.30[v]
CaO	4.8105[w]	22.20[x]	8.20[x]	8.10[x]	−0.353[x]	−0.266[x]	−0.72[x]	15.25[y]
NiO	4.1769[z]	33.40[aa,g]	6.60[aa,g]	9.80[aa,g]	−0.06[aa,g]	−0.40[aa,g]	0.29[aa,g]	9.98[s]
$MgAl_2O_4$	8.0831[bb]	27.90[b]	15.30[b]	15.30[b]	−0.262[cc]	−0.0997[cc]	−0.0878[cc]	9.17[y]
Al_2O_3	7.989[dd]	49.50[b]	16.00[b]	14.60[b]	−0.75[b]	0.40[b,h]	−1.800[b]	8.58[y]

[a] Rundman and Hilliard (1967).
[d] JCPDS card file No. 4–784.
[f] JCPDS card file No. 4-802.
[h] Some nonlinearity.
[i] JCPDS card file No. 4-850.
[l] JCPDS card file No. 5-0628.
[o] Hilton and Jones (1967).
[r] Aleksandrov, Shabanova, and Reshchikova (1968).
[v] Porter (1965).
[y] Krikorian (1960).
[bb] JCPDS card file No. 21-1152.
[dd] Calculated from effective "a_0" of Clarke, Howe, and Badger (1934).
[b] Landolt–Börnstein tables (1966, 1969), Bechmann and Hearmon (1966); Bechmann *et al.* (1969); Landolt–Börnstein Numerical Data.
[j] JCPDS card file No. 6-0480.
[m] Bartels and Schuele (1965).
[p] JCPDS card file No. 4-0587.
[s] Batelle Memorial Institute (1966).
[w] JCPDS card file No. 4-0777.
[z] JCPDS card file No. 4-0835.
[cc] Chang and Barsch (1973).
[c] Weast (1973).
[e] Shaffer (1964).
[g] Calculated from graphical data for temperature range of interest.
[k] Fauchaux and Simmons (1964).
[n] Haussühl (1960).
[q] JCPDS card file No. 9-402.
[t] JCPDS card file No. 4-0829.
[u] Liebfried and Ludwig (1961).
[x] Bartels and Vetter (1972).
[aa] du Plessis, Van Tonder, and Alberts (1971).

IV. THE COHERENT SPINODAL: THEORY AND EXPERIMENT

A comparison of the theoretically derived coherent spinodal and the spinodal determined by various experimental techniques in metallic, alkali halide, and oxide systems will follow.* The coherent spinodal is determined in several different ways following the basic calculation technique of Fancher and Barsch (1971). Calculations will be made of the "at-temperature" and "room temperature" coherent spinodals. The "at-temperature coherent spinodal" is reserved here for the coherent spinodal derived by using the at-temperature constants (i.e., the temperature derivatives of the elastic constants in conjunction with the at-temperature lattice parameters), while the term "room temperature coherent spinodal" refers to calculations based on ambient lattice parameter data (except for Al–Zn) and ambient elastic data. It will be demonstrated that calculations of the coherent spinodal using at-temperature data are in better agreement with experimentation than previous determinations based on ambient data. In addition, it will be shown that temperature of the coherent spinodal can change by more than 5000°C in systems where the temperature derivatives of the elastic constants decrease sharply with increasing temperature. The data necessary for the calculations have been summarized and graphically extrapolated from a number of sources (Table II).

In the following a brief review will be presented of phase decomposition within the spinodal region for some binary metallic systems. A comprehensive review was made by Hilliard (1970), so the examples of alloys contained herein will be brief and will be used mainly for comparison with the results from nonmetallic systems.

A. Metallic Systems

1. Al–Zn

The RH analysis was originally carried out for SAXS measurements on the Al-base Zn system. The calculation of the coherent spinodal, however, is not straightforward because aluminum is fcc and zinc is hcp. Based on the similarity between the two structures, and the lattice parameters for Al–22% Zn at 65°C (Ellwood, 1952), RH calculated the ΔT at this composition to be 26 ± 15°C, placing the T_{coh} at 237 ± 25°C. Cahn (1961) calculated a ΔT of about 40°C for the critical composition and Rundman (unpublished) calculated a ΔT of about 32°C for 40% Zn, using the excess entropy (Hilliard

* All values given in *atomic percent* unless otherwise noted.

et al., 1951; Corsepius and Munster, 1959) and elastic constants adjusted for temperature dependence (Koster, 1948).

The elastic constants (Table II), together with the 65°C lattice parameters (RH), yields a ΔT of 35°C at 22% Zn and a ΔT of 51°C at the critical composition 38% Zn (see Table III). However, it is important to note that the use of the at-temperature data yields a smaller ΔT of ~25°C at 22% Zn and a ΔT of ~35°C for the critical composition. Use of Ellwood's (1952) at-tempera-

TABLE III

INTERRELATIONSHIP OF COHERENT TEMPERATURE DEPRESSION ΔT TO ELASTIC ANISOTROPY FACTORS

System	ΔT	Y_{min} (direction)	(value)	$\eta^2 Y$	α
CoO–MgO	1	100	26.21	1.89×10^{-3}	−0.425
	2[a]	100	32.40	4.66×10^{-3}	−0.340
AgCl–NaCl	20	100	4.49	1.35×10^{-3}	−0.684
	18[a]	100	5.68	1.11×10^{-3}	−0.572
Al–Zn	35	111	8.69	6.20×10^{-3}	0.252
	51[b]	111	10.39	9.25×10^{-3}	0.226
$MgAl_2O_4$–Al_2O_3	66	111	59.56	1.02×10^{-2}	−0.283
	39[a]	111	50.89	0.73×10^{-2}	0.124
CoO–NiO	88	111	32.87	1.30×10^{-2}	0.553
	83[a]	111	31.69	1.20×10^{-2}	0.086
MgO–$MgAl_2O_4$	100	100	11.69	2.34×10^{-2}	−0.767
	168[a]	100	32.08	4.40×10^{-2}	−0.391
CaO–MgO	105	111	13.94	2.78×10^{-1}	−0.988
	4205[a]	100	29.54	5.61×10^{-1}	−0.293
Au–Pt	145	100	11.56	2.31×10^{-2}	−0.621
	148[a]	100	17.39	2.64×10^{-2}	−0.449
KCl–NaCl	604	100	3.06	3.90×10^{-2}	−0.083
	779[a]	111	3.88	5.00×10^{-2}	0.649
Au–Ni	1062	100	10.56	2.64×10^{-1}	−0.715
	1723[a]	100	17.91	4.48×10^{-1}	−0.598
CoO–CaO	2706	111	24.09	2.89×10^{-1}	1.039
	3043[a]	100	24.88	3.54×10^{-1}	−0.211
CaO–NiO	3196	100	16.60	3.49×10^{-1}	−0.409
	4376[a]	111	28.74	5.17×10^{-1}	0.099

[a] Calculated using room temperature lattice parameters and without use of the temperature derivatives of the elastic constants.

[b] Calculated using lattice parameter data at 65°C [after Rundman and Hilliard (1967)].

ture lattice parameter data for the critical temperature, rather than the value obtained from the coefficients of linear expansion, lowers the ΔT to $\sim 20°C$ and elevates the T_{coh} to $\sim 335°C$. As can be seen in Fig. 11a, using the at-temperature lattice parameters and elastic constants decreases the ΔT between the chemical and coherent spinodals and raises the critical temperature T_c of the coherent spinodal. Use of the at-temperature data, as calculated here for Al–Zn, lowers the $\eta^2 Y$ factor but slightly increases the anisotropy factor α (Table III).

From the experimental viewpoint, several Al–Zn alloy compositions have been studied with a number of techniques. Using SAXS, it was found that GP zones form near ambient temperature for: 6.8% Zn (Gerold and Schweizer, 1961); 7% Zn (Agarwal, 1974); 7% Zn (Harkness *et al.*, 1969). Herman *et al.* (1963) studied quench aging in 5.3% Zn using electrical resistance measurements and SAXS, confirming and extending the earlier work of Panseri and Federighi (1960).

Murakami *et al.* (1969), using SAXS, determined the coherent spinodal to be about 120–130°C for 6.8% Zn. However, this temperature and composition lie directly on the α/α' coherent solvus and hence these authors may have actually observed the reversion of GP zones. The compositions and temperatures given above all lie outside of the at-temperature and room temperature coherent spinodals of Fig. 11a.

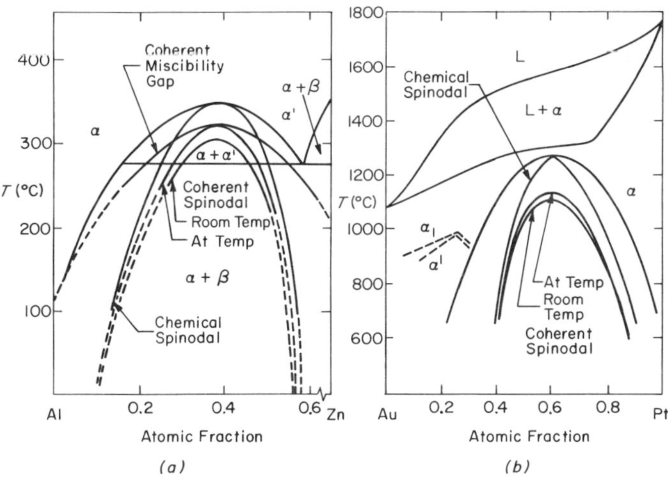

Fig. 11. Equilibrium phase diagram (from Fink and Willey, 1936) for Al–Zn with the at-temperature and room temperature coherent spinodals and the incoherent chemical spinodal. (b) Equilibrium phase diagram of Au–Pt (from Darling *et al.*, 1952; Tiedema *et al.*, 1957) with coherent at-temperature and room temperature spinodals and the incoherent chemical spinodal.

Bonfiglioli and Guinier (1966) studied the aging kinetics of several Al–Zn alloys (20.7, 38.5, 49.1% Zn) at −45°C, with SAXS measurements being made at −150°C. They suggested that a spinodal reaction may precede the zone-formation stage. Ardell *et al.* (1969) studied several concentrated Al–Zn alloys at high temperatures with TEM. For 28.4 and 39.3% Zn, a modulated structure was found below 310 and 335°C, respectively. It was suggested that the critical range for the α/α' coherent solvus was 310–320°C for the 28.4 and 50.3% Zn alloys and 335–340°C for the 39.3% Zn alloy. These values for the α/α' solvus are somewhat higher than the coherent miscibility gap predicted here (Fig. 11a), but the values of Bonfiglioli and Guinier (1966) lie well within the predicted spinodal.

Graf and Lenormand (1964) detected x-ray satellites in 22% Zn aged at 250–275°C. As discussed previously, RH used SAXS to study this alloy during quench aging at 65°C, and concluded that spinodal decomposition is operating. However, an unexplained early transient, among other things, led to some uncertainty as to the exact mode of decomposition (Gerold and Merz, 1967). There was also some question as to whether decomposition occurred during the quench or shortly thereafter at ambient temperature (Gerold and Merz, 1967; Bartel and Rundman, 1975). During even the fastest quench, solute clustering will occur which can confuse the interpretation of SAXS data. Agarwal and Herman (1973a,b) have therefore repeated the RH experiment, using liquid quenching (LQ) techniques in order to capture an earlier stage of the decomposition process and to minimize solute clustering during quenching. Their data on 22% Zn is in good agreement with theory and improves somewhat on the results of RH. In addition, using TEM, Agarwal and Herman (1973a,b) observed a modulated structure with a wavelength of ~40 Å, which is close to that found from the SAXS experiments.

Studies of Al–Zn alloys beyond 22% Zn are rather confused by a discontinuous mode of precipitation which gives rise to SAXS (Agarwal and Herman, 1973a) and other complicating factors (Rao *et al.*, 1966, 1967). Because of these complications, compositions greater than ~22% Zn will not lead to clear experimental evidence for spinodal decomposition. There is evidence, however, that spinodal decomposition appears as a precursor mode for these higher compositions.

The situation relative to Al–Zn is certainly not settled and more experimentation is required. As pointed out in much of the literature cited above, complications are introduced when the alloy is not near the middle of the miscibility gap. In fact, the linear theory as developed by Cahn, Hilliard, de Fontaine, and others is best tested at the center of the gap. The SAXS approach, as introduced by RH, is an extremely powerful approach to studies of early stage phase separation and, in fact, using neutrons, is extendable to systems which cannot be studied with x rays, due to the proximity of the atomic scattering factors of the alloy components. It is to be noted that very recently small-angle neutron scattering (SANS) was employed to study the formation of GP zones in Al–7% Zn (Raynal *et al.*, 1971). Due to high resolution and the ability to study angles near zero, a number of previously unsettled issues were resolved in that study. More recently, D. Schawhn (unpublished) studied at-temperature SANS near the critical point and actually found a depression of 28°C for the coherent gap, which compares favorably with the ΔT of ~35°C as calculated using the at-temperature data (Table III).

2. Au–Pt

A simple miscibility gap exists in the Au–Pt system (Fig. 11b). The pictured chemical spinodal was calculated after the technique of Cook and Hilliard (1965) and lies close to that of Van der Toorn and Tiedema (1960), in which the free energy of mixing was used to locate the inflection points on the free energy curve. Cahn (1961) calculated a ΔT of 200°C for the critical composition in this system (60% Pt), whereas Rundman (unpublished) calculated a ΔT of 90–120°C at 40% Pt from thermodynamic and elastic data. Using Eq. (25) and the data of Table II for the critical composition, the ΔT is found to be 148 and 145°C, for the at-temperature and room temperature data of Table III, respectively.

A number of studies have been carried out of phase decomposition on Au–Pt using x ray, optical microscopy, and electrical resistivity techniques (Tiedema *et al.*, 1957; Van der Toorn, 1960; Kralik *et al.*, 1969). The study by Van der Toorn (1960) of 86.6% Pt single crystals revealed sidebands after 2 and 4 min of aging at 700°C, while sidebands were absent in a composition of 90.1% Pt under the same conditions. Van der Toorn suggested that the lower composition alloy was within the spinodal. In fact, 86.6% Pt is almost coincident with the coherent spinodals calculated here, while the 90.1% Pt composition lies well outside.

Further x-ray experiments by Van der Toorn (1960) on polycrystalline samples of 31.7, 41.8, 81.2, and 89.6% Pt at 600°C revealed that only the 41.8 and 81.2% Pt alloys exhibited sidebands and, hence, were likely to be within the spinodal. These experimental results are in substantial agreement with the boundaries of the coherent spinodal as calculated here and shown in Fig. 11b. Similarly, modulated structures typical of spinodal decomposition, were found in $\langle 100 \rangle$ by x-ray techniques in ~40, ~60, and ~80% Pt solid solutions aged between 500 and 600°C (Carpenter, 1967). These compositions and aging temperatures are well within the coherent spinodals as predicted

TABLE IV

CRITICAL TEMPERATURE OF THE COHERENT SPINODAL FOR SEVERAL OXIDE, ALKALI HALIDE, AND METAL SYSTEMS

System	Critical composition (at. fract.)	ΔT Room temp. data	ΔT At-temp. data	$Y_{\text{min at-temp.}}$	Critical temperature of coherent spinodal using at-temp. data (°C)
CoO–CaO	0.52	3043	2706	111	<RT
CaO–NiO	0.58	4376	3196	100	<RT
CoO–NiO	0.53	83	88	111	682
CaO–MgO	0.63	4205	105	111	2955
CoO–MgO	0.98	2	1	100	879
$MgO–MgAl_2O_4$	0.17	168	100	100	2180
$MgAl_2O_4–Al_2O_3$	0.97	39	66	111	2024
AgCl–NaCl	0.50	18	20	100	170
KCl–NaCl	0.67	779	604	100	<RT
Al–Zn	0.38	51[a]	35	111	320
Au–Pt	0.60	148	145	100	1130
Au–Ni	0.70	1723	1062	100	<RT

[a] Calculated using lattice parameter data at 65°C after Rundman and Hilliard (1967).

here (Fig. 11b) and, indeed, the calculations predict modulations along $Y_{\min} = \langle 100 \rangle$ (Tables III and IV).

The similarity of the atomic scattering factors severely limits studies of the Au–Pt system by SAXS. Hence, small-angle scattering using neutrons has been carried out on this system by Singhal *et al.* (1977). Using SANS, an evolving, early-stage product has been detected for aging Au–60% Pt at 600°C following the quench. The spacing found with both SANS and TEM is about 100 Å. Additional, ongoing experiments by these workers involve aging of foils obtained by ultrarapid quenching from the melt. Liquid quenching yields a much less clustered as-quenched solid solution and enables an examination of a far earlier stage of phase separation (Singhal *et al.*, 1977).

3. Au–Ni

The Au–Ni system exhibits a simple miscibility gap with a critical temperature at $\simeq$820°C for a composition of 70% Ni. The chemical spinodal was calculated by Cook and Hilliard (1965) using the equilibrium miscibility gap determined by Münster and Sagel (1958). Cahn (1961) estimated the ΔT to be 2000°C, placing the coherent spinodal some 1000°C below ambient. The large ΔT term is caused by the greater than 13% difference in the lattice parameters of gold and nickel at room temperature. Rundman's calculations (unpublished) considered the excess entropy and predicted the coherent critical point to be close to absolute zero temperature (−273°C). Golding and Moss (1967) recalculated the coherent spinodal using the measured entropy and accounted for the nonlinearity of Y_{100} with composition. This shifted the critical point to a composition of 40% Ni and a temperature of 0°C. On the basis of experimental evidence they considered that a range of ±100–200°C was a reasonable error in their calculations (Moss and Averbach, 1967; Woodilla and Averbach, 1968).

In the present study the coherent spinodal was calculated for 70% Ni, assuming $Y_{\min}$ to be linear with composition. The data of Table II were used, and gave a ΔT of 1062°C, using the at-temperature data, and a ΔT of 1723°C for room temperature data. The at-temperature data give a coherent critical temperature of −240°C, in good agreement with Rundman (unpublished), and the room temperature result is in close agreement with Cahn's (1961) original ΔT of 2000°C. All of the theoretically determined coherent spinodals, except that of Golding and Moss (1967), preclude spinodal decomposition at or above ambient temperature.

The first experimental evidence of spinodal decomposition in Au–Ni appeared in the electron diffraction studies of Fukano (1961). A periodic modulation of ≃68 Å along ⟨110⟩ and ⟨100⟩ was observed for 20% Ni at 165°C. Moss (1966) detected satellites in x-ray diffraction patterns of quenched 40% Ni single crystals. This was later attributed to clustering (Moss and Averbach, 1967).

More recently diffusion experiments have been carried out by Yang (1971). The dependence of the diffusivity in Au–Ni films, containing composition modulations, on the wavelengths of the fluctuations and annealing temperatures allowed determination of the coherent spinodal. The ⟨111⟩ coherent spinodal was determined to be 284°C for 44% Ni and 306°C for 59% Ni. The ⟨100⟩ coherent spinodal should occur at higher temperatures since ΔT is directly proportional to Y, and here $Y_{100} < Y_{111}$ (Table III). Likewise, recent lattice imaging of the (200) planes for 77% Ni has shown composition modulations of 25 Å, presumably formed by spinodal decomposition at 150°C (Gronsky *et al.*, 1975). Using electron diffraction techniques Woodilla and Averbach (1968), also found modulations in ⟨100⟩ directions at 180°C in 40% Ni and 220°C in 45% Ni.

At 40% Ni our at-temperature calculations predict the coherent spinodal to be near 0°C and $Y_{\min} = \langle 100 \rangle$ (Tables III and IV). The large difference in the lattice parameters (≃13%) and the nonlinearity of Y with composition makes an estimate of the η^2 term critical to approximating the coherent spinodal.

B. Alkali Halide Systems

There are few comprehensive studies of precipitation in alkali halide systems prior to 1960 (Miyake and Suzuki, 1954; Suzuki, 1955, 1958; Dreyfus and Nowick, 1962; Lilley and Newkirk, 1963) and indeed, thermodynamic and diffusion data is lacking in such systems. Therefore, theoretical calculations must rely on techniques mainly involving lattice parameter data since these are well documented for alkali halides. Because the calculation of the coherent spinodal is largely dependent on the lattice parameter term, η, accurate predictions can, in fact, be made of the position of the coherent spinodal in alkali halide systems.

1. AgCl–NaCl

The metastable miscibility gap was determined in 1962 by Stokes and Li and shows a very flat solvus at the critical composition. Theoretical calculations suggest that the miscibility gap is below 100°K (Hendricks, 1964). The chemical spinodal was herein calculated (the technique of Cook and Hilliard, 1965). The room temperature coherent spinodal is depressed only 18°C from the critical temperature, and the at-temperature ΔT is 20°C (Fig. 12a), as calculated here by the methods of Fancher and Barsch (1971) previously described [Eq. (25)]. The small ΔT is due to the similarity of the interatomic distances in NaCl and AgCl which yields the smallest strain term, $\eta^2 Y$, for any system given in Table III. The calculations predict that spinodal decomposition should occur in this system over almost the entire miscibility

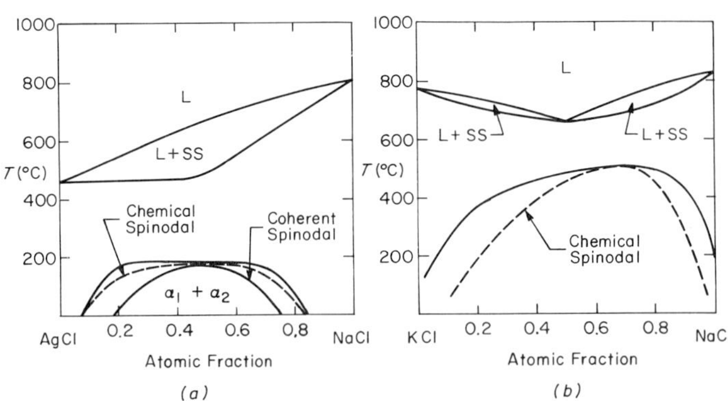

Fig. 12. (a) Equilibrium phase diagram and miscibility gap of AgCl–NaCl (from Stokes and Li, 1962) with the chemical and coherent spinodals, as calculated in the present study. (b) Equilibrium phase diagram of KCl–NaCl (from Scheil and Stadelmaier, 1952; Bunk and Tichelaar, 1953) with the calculated chemical spinodal. The at-temperature coherent spinodal is depressed 604°C from the critical temperature of 490°C for 67% NaCl. Note that SS represents a homogeneous solid solution.

gap (Fig. 12a; 20–75% NaCl). Likewise, the composition modulations should occur along the elastically soft ⟨100⟩ direction (Tables III and IV). The mechanism and kinetics of precipitation in the AgCl–NaCl system was studied by Hendricks (1964), Hendricks *et al.* (1964), and Hendricks and Borie (1967) and was shown to exhibit formation of GP zones. The compositions studied by Hendricks (1964) were near 25 and 35% AgCl at 25, 50, and 100°C. These points are close to the limits of the metastable miscibility gap at ambient temperature and, indeed, lie outside of the coherent spinodal proposed here. A 50:50 at. % alloy was chosen for Wong's (1971) SAXS study. Samples were solutionized at 200 and 372°C in the single phase region, quenched to room temperature and aged at 25, 55, and 80°C. The decomposition was found to be very rapid, even on aging at ambient temperature; SAXS measurements indicated that decomposition was by nucleation, growth and coarsening. This data is in agreement with the data of Hendricks (1964); however, the large mass differences between an Ag and Na restrict good mixing of the initial alloy, making it difficult to obtain a well-homogenized starting ingot.

2. KCl–NaCl

Theoretical and experimental determinations of the coherent spinodal have been documented for KCl–NaCl (Table V). The combined equilibrium diagrams of Bunk and Tichelaar (1953) and Scheil and Stadelmaier (1952) are

TABLE V

INCOHERENT AND COHERENT CRITICAL TEMPERATURES IN KCl–NaCl

Critical temperature of the chemical spinodal (°C)	Depression of the coherent spinodal, ΔT	Critical temperature of the coherent spinodal (°C)	At. % NaCl
490[a]	≃415	<75[b]	~70
610[c]	490[c]	120	~70
490[a]	604[d]	−114[d]	~67

[a] Experimental value of Bunk and Tichelaar (1953), Schiel and Stadelmaier (1952).

[b] Wolfson *et al.* (1966).

[c] Fancher and Barsch (1971).

[d] This study, using the calculation technique of Fancher and Barsch (1971) and at-temperature elastic and lattice parameter data.

given in Fig. 12b, along with the calculated chemical spinodal. The calculations of Fancher and Barsch (1971) suggest a ΔT of 490°C, however, they used a theoretically calculated T_c of 610°C, suggesting that the coherent spinodal at the critical composition is ~120°C. Wolfson *et al.* (1966), using the experimental T_c of 490°C [from Bunk and Tichelaar (1953)], calculated the spinodal temperatures to be <75°C on the basis of Cahn's theory. However, this value still predicts spinodal decomposition to occur above ambient temperature. Calculation of the at-temperature coherent spinodal from the data of Table II yields a ΔT of 604°C at the critical composition of 67% NaCl. If the experimental T_c of 490°C [from Bunk and Tichelaar (1953)] is used, then the coherent spinodal is depressed below ambient temperature. The room temperature coherent spinodal is depressed by $\Delta T \simeq 779°$ (Table III), causing the coherent spinodal to be near absolute zero. Use of the room temperature data also changes the predicted direction of Y_{min} from the at-temperature direction of $\langle 100 \rangle$ to $\langle 111 \rangle$ (Tables III and IV).

The depression of the coherent spinodal at 67% NaCl by 604°C is in agreement with the available experimental data of Wolfson *et al.* (1966), Fancher and Barsch (1971), and Bhardwaj and Roy (1971) (Table V). These authors failed to find experimental evidence for spinodal decomposition in the KCl–NaCl system at or above ambient temperature. Had Fancher and Barsch (1971) used their coherent depression of ~490°C and the experimental T_c of 490°C rather than a theoretical T_c, they would have predicted the spinodal to be below ambient.

C. Crystalline Mixed Oxides

Other than for glasses, there has been limited work on phase decomposition in oxide systems. Due to their inherent isotropy, glass-forming systems were the obvious candidates for early searches of spinodal decomposition. Studies of spinodal decomposition in Na_2O–SiO_2 glasses were carried out by Neilson (1969), Tomozawa (1968), and Tomozawa *et al.* (1969, 1970a,b).

Fine (1970a) reviewed precipitation in mixed crystalline oxides, including some early spinodal decomposition studies. Limited experiments seeking spinodal decomposition have been carried out in the crystalline oxides. It is expected that studies of spinodal decomposition in oxides will develop strongly in the future.

Cahn's 1961 calculation for the determination of the coherent spinodal in cubic systems has been extended to tetragonal and rhombohedral oxide systems (Schultz *et al.*, 1969; Stubican and Schultz, 1970; Stubican *et al.*, 1970; Russo *et al.*, 1972; Wu and Mendelson, 1973; Park *et al.*, 1975). For the tetragonal TiO_2–SnO_2 system, Stubican and Schultz (1968, 1970) and

Schultz (1967, 1970) observed a lamellar microstructure. The periodicity was along [001], the wavelength of which was found to first decrease and then increase with aging time at 1000°C. Park *et al.* (1974), in a study of this system, performed SAXS measurements from samples annealed at 900°C within their calculated coherent spinodal. They observed no crossover in I *versus* β plots. Such a crossover is a basis of the kinetic argument for spinodal decomposition (Hilliard, 1970), and yields a proper $R(\beta)$ *versus* β curve (Fig. 10). More recently, Park *et al.* (1976) have published further x-ray satellite results on TiO_2–SnO_2 and claim more explicit evidence for spinodal decomposition.

In the rhombohedral Al_2O_3–Cr_2O_3 (Stubican *et al.*, 1970) and Al_2O_3–Cr_2O_3–Fe_2O_3 (Schultz and Stubican, 1970; Stubican, 1972) oxide systems, the observation of lamellar structures by dark field TEM and the occurrence of spot doubling in preferred directions led to the conclusion that spinodal decomposition is operating. However, further experimental evidence is needed, since, as stated above, these strictly morphological observations can, in fact, result from processes other than a spinodal mode.

Phase separation was studied in the pseudobinary spinel region of the $CoFe_2O_4$–Co_3O_4 system using TEM (Takahashi *et al.*, 1971; Takahashi and Fine, 1970). Spinodal decomposition was concluded to be the mode of decomposition.

In other oxide systems, spinel-structured precipitates are known to commonly form in NaCl-structured cubic oxide matrices (Fine, 1970a,b). Some examples are $Mg_{1.2}Fe_{1.8}O_{3.9}$ precipitates in Fe^{3+}-doped MgO (Groves and Fine, 1964; Wirtz and Fine, 1967, 1968; Fine, 1968, 1970a,b; Woods and Fine, 1969; Kruse and Fine, 1972), Fe_3O_4 precipitates from $Fe_{1-x}O$ (Koch, 1966; Fine, 1972), $CoFe_2O_4$ precipitates from Fe^{3+}-doped CoO (Fine, 1970a), $MgAl_2O_4$ from Al^{3+}-doped MgO (Stubican and Roy, 1965), $MgMn_2O_4$ from Mn^{3+}-doped MgO and metastable precipitation of $Mg(Al_{1-x}Cr_x)_{26}O_{40}$ in MgO–Al_2O_3–Cr_2O_3 (Greskovich and Stubican, 1968). In oxide solid solutions the rate of precipitation is shown to increase proportionally to the deviation from stoichiometry and concentration of vacancies (Herai and Manenc, 1964), and often grain boundary precipitation prevails (Stubican and Viechnichi, 1966).

The above are the principal nonglass oxide systems in which decomposition by a spinodal mode has been suggested, other than geologic silicate systems, e.g., feldspars and pyroxenes. Caution must be exercised in studies of oxide systems, since lamellar products of exsolution or cellular decomposition, as observed in TEM, as well as Magneli phases due to nonstoichiometric oxides, might imitate a spinodal morphology.

The calculation of the coherent spinodal and observations of spinodal decomposition in some further oxide systems will now be discussed in three parts: Those cubic crystalline oxides not involving MgO (e.g., CoO–CaO, CoO–NiO, CaO–NiO, and $MgAl_2O_4$–Al_2O_3); cubic oxides containing MgO (e.g., CoO–MgO, CaO–MgO, and $MgAl_2O_4$–MgO); and glass-forming systems. Lastly, some of the recent experimental work in geologic systems will be discussed (Section IV.D).

1. CoO–CaO

The coherent spinodal has been calculated for the CoO–CaO system based on the equilibrium diagram determined by Tikkanen (1962). At the critical composition of 52% CaO, a ΔT of 2706°C is calculated (Table III), and since the equilibrium temperature is 1560°C, the coherent spinodal is below ambient. In actual fact, there are no experimental data available for this system, likely due to the prohibitively high temperatures required, the high reactivity of CaO, and the instability of CoO.

Ultrarapidly quenched ($\sim 10^6$ °C/sec) laser-melted 60% CoO was studied in this laboratory and revealed a solid solution and formation of a compound; either $Ca_3Co_2O_6$ or $Ca_9Co_{12}O_{28}$ (Jantzen, 1978).

2. NiO–CaO

The coherent spinodal was calculated from the equilibrium diagram of Smith *et al.* (1969) and using the data of Table II. From this calculation the coherent spinodal is placed well below ambient temperature; Table III. Decomposition by a spinodal mode is thus not expected in this system and, in fact, there is no evidence of this in the available experimental reports (Smith *et al.*, 1973; Dhalenne *et al.*, 1972; Revcolevschi, 1976; Jantzen, 1978). The experimental study of Smith *et al.* (1973) shows that products of exsolution form preferentially as {100} plates. This is actually the elastically soft direction, Y_{min}, predicted by the at-temperature elastic and lattice parameter data (Table IV). Additional studies in this system involve LQ of material of eutectic composition, 58% NiO, the quenched specimen exhibiting lamellar morphology typical of eutectic solidification (Dhalenne *et al.*, 1972; Revcolevschi, 1976). Liquid quenching of laser-melted 56% NiO shows decomposition, attendant to quenching, into CaO and Ni_5Ca_2 (Jantzen, 1978).

3. CoO–NiO

The recent phase diagram of Kinoshita *et al.* (1973) was used for calculation of the coherent spinodal in the CoO–NiO system. The lattice parameters are known to follow Vegard's law (Sakata and Sakata, 1958; Bowen *et al.*, 1972), and, hence, the approximation of linearity with composition, inherently

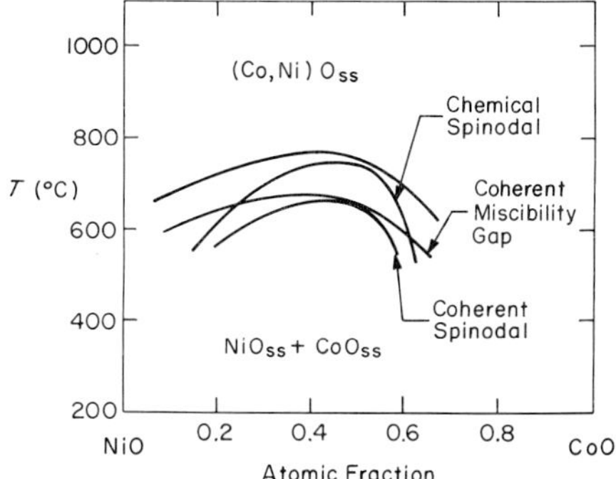

Fig. 13. Equilibrium phase diagram of NiO–CoO (from Kinoshita *et al.*, 1973) with the incoherent chemical spinodal and the at-temperature coherent spinodal as calculated in this study.

assumed in the use of Eq. (25) for the calculation of the coherent spinodal, is valid. Due to the similarity of the lattices and the elastic constants for the two conjugate phases, the coherent critical point is depressed only 88°C using the at-temperature data, and 83°C using room temperature data: Table III. The at-temperature coherent spinodal is shown in Fig. 13, along with the metastable miscibility gap as calculated from the spinodal boundary by the technique of Cook and Hilliard (1965).

Experimentally, this transition metal oxide must be studied in a controlled environment. Under a controlled oxygen atmosphere, Kinoshita *et al.* (1973) found satellite spots and/or streaks on their electron diffraction patterns. However, they did not attribute these satellites to a particular mechanism of phase decomposition. Chemical vapor deposition (CVD) of several compositions in the NiO–CoO system and subsequent TEM and electron diffraction revealed what was thought to be a two-phase interpenetrating structure on the scale of 100 Å for 34% CoO at a temperature estimated to be 550°C (Bowen *et al.*, 1972). Other compositions within the coherent miscibility gap, as proposed here (Fig. 13), were studied by Bowen *et al.* (1972). These compositions (55, 52, 46% CoO) did not exhibit inhomogeneous structures in vacuum at 550°C, but annealing in air at 750°C for 5–96 hr produced microregions of a spinel-structured phase. Annealing of $Ni_{0.50}Co_{0.50}O$ at p_{O_2} of 10^{-8} atm at 700°C gave rise to microseparation of 100–400 Å particles after only 7 days. Transmission electron microscropy studies of 41% CoO solid quenched in air ($p_{O_2} = 0.21$ atm) revealed a modulated structure of 300–400 Å spacing (Jantzen, 1978). The corresponding electron diffraction pattern exhibited streaks on the (111) diffraction spots and spot doubling on the (001) reflections. Although {111} is the soft direction, and hence the direction in which composition modulations are predicted (Tables III and

IV), further aging in the electron beam caused the {111} diffraction spots to split, suggesting the development of a new phase. Liquid quenching of a laser-melted 47% CoO specimen showed only the presence of a solid solution. However, the small size of this specimen made aging studies most difficult to carry out. It can thus only be said that decomposition by a spinodal mode cannot be precluded for CoO–NiO and additional experiments are needed.

4. $MgAl_2O_4$–Al_2O_3

$MgAl_2O_4$–Al_2O_3 is a subsystem of the MgO–Al_2O_3 system. However, it is necessary to replot the phase diagram of Roy *et al.* (1953) in terms of atomic fraction Al_2O_3 in $MgAl_2O_4$–Al_2O_3 (Fig. 14), since the lattice parameter and elastic data necessary for the calculation of the coherent spinodal are those of the end members, spinel ($MgAl_2O_4$) and Al_2O_3. The recent evidence of slight solid solubility on the Al_2O_3-rich portion of the phase diagram

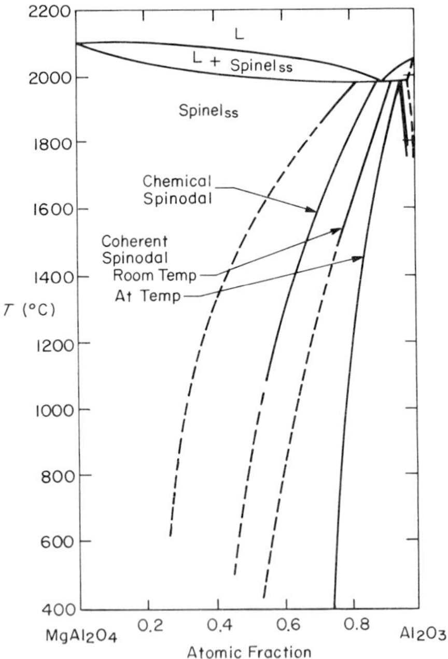

Fig. 14. Equilibrium diagram of the $MgAl_2O_4$–Al_2O_3 subsystem of the MgO–Al_2O_3 system (after Roy *et al.*, 1953) replotted in terms of atomic fraction Al_2O_3 in $MgAl_2O_4$–Al_2O_3. Superimposed are the incoherent chemical spinodal and the coherent spinodals, using both at-temperature and room temperature elastic and lattice parameter data as calculated in the present work. At the critical point, $\Delta T \approx 66°C$, using the at-temperature data; and $\Delta T \approx 39°C$, using room temperature data.

(Viechnicki *et al.*, 1974) allows the extrapolation of the Al_2O_3-rich solvus necessary for a calculation of the coherent spinodal. The effective "cubic a_0" lattice parameter (Table II) was calculated for hexagonal Al_2O_3 from the data of Clark *et al.* (1934). The solid solutions for the entire system MgO–Al_2O_3 are known to obey Vegard's law (Sarjeant, 1967; Sarjeant and Roy, 1967). The coherent spinodals determined here using the at-temperature data yield at ΔT at the critical point of ~66°C and a room temperature ΔT at the critical point of ~39°C. The loci of the at-temperature and room temperature spinodals are given in Fig. 14.

Precipitation strengthening has long been known to occur in spinel with excess Al_2O_3 (Rinne, 1928; Eppler, 1943, 1947; Barnes, 1945; Jagodzinski, 1957; Grabmaier and Falckenberg, 1969; Fine, 1968, 1970a,b). A precipitate phase has been observed optically and by x-ray diffraction techniques at low temperatures (Saalfeld and Jagodzinski, 1957; Jagodzinski and Saalfeld, 1958) in compositions between 50–80 mole % Al_2O_3 in MgO–Al_2O_3 (0–75% Al_2O_3 in $MgAl_2O_4$–Al_2O_3). This preprecipitate was believed to be a precursor of an intermediate monoclinic phase which eventually transforms to the stable α-Al_2O_3 phase. The monoclinic phase forms as lamellae on (311) planes of the spinel matrix, with [110] in the spinel being parallel to [010] in the precipitate. Additional TEM investigations of several compositions (66–80 mole % Al_2O_3 in MgO–Al_2O_3; 50–75% Al_2O_3 in $MgAl_2O_4$–Al_2O_3) at various temperatures between 700 and 1300°C suggested the occurrence of two monoclinic intermediate phases, one of thin planar form, Type I precipitates, where

$$\langle 113 \rangle_{\text{matrix}} // \langle 001 \rangle_{\text{ppt}}$$

and Type II precipitates composed of thin laths, where (Lewis, 1969)

$$\langle 100 \rangle_{\text{matrix}} // \langle 010 \rangle_{\text{ppt}}$$

The laths coalesce upon long aging to form composite plate shapes. Type I precipitates were identified for 78 mole % Al_2O_3 in MgO–Al_2O_3 (71% in $MgAl_2O_4$–Al_2O_3) upon aging at 850°C and Type II precipitates for 71.4% Al_2O_3 in MgO–Al_2O_3 (60% in $MgAl_2O_4$–Al_2O_3) at 800°C (Lewis, 1969). These compositions lie outside of the at-temperature spinodal as predicted here.

The most recent experimental evidence on single crystals—boules of 71% Al_2O_3 in $MgAl_2O_4$–Al_2O_3 at 850°C by TEM (Bansal and Heuer, 1974)—suggests two stages of preprecipitation: preprecipitate I (GP zones) after 10 min of aging and preprecipitate II, "GP precipitates," after 60 min of aging. The preprecipitates lead subsequently to intermediate precipitates and ultimately to stable α–Al_2O_3 (Bansal and Heuer, 1974).

Experiments were undertaken in this laboratory to further document the preprecipitation processes. Using SANS, no evidence of spinodal decomposition was found in single-crystal spinel of 71% Al_2O_3 annealed at 650, 750, 850, or 1050°C. As predicted here, this composition and these temperatures lie outside of the at-temperature coherent spinodal. However, x-ray diffraction analyses (Jantzen, 1978; Jantzen and Herman, unpublished) reveal the presence of the monoclinic intermediate phase in the single crystal boules (as pulled from the melt). The monoclinic phase develops only at early times, becoming more Al_2O_3-rich, with the concurrent depletion of Al_2O_3 in the matrix. As early as 2 min at 850°C (5 min at 750°C and 10 min at 650°C) the α-Al_2O_3 phase grows at the expense of both the monoclinic phase and the spinel matrix. Discrete precipitates of $\sim$25 Å dimension are visible in dark-field TEM in samples aged at 850°C for 10 min (Jantzen, 1978). At later stages, satellites to the (440) and (511) Bragg reflections of the spinel phase are observed in x-ray diffraction. These can be attributed to the formation of the closely related incipient structures. Saalfeld and Jagodzinski (1957) suggested that the precursor monoclinic phase grows in preference to the α-Al_2O_3 phase because of the great difficulty in nucleating α-Al_2O_3 directly.

Laser-melted specimens of 59 and 90% Al_2O_3 in the $MgAl_2O_4$–Al_2O_3 system were quenched by a hammer-and-anvil technique (Krepski, 1975; Krepski *et al.*, 1975). For the 59% Al_2O_3 samples aged for 60 min at 800°C, mottled structures, such as those observed by Bansal and Heuer (1974), were found in TEM (Krepski, 1975). This temperature and composition lies on the tentative extrapolation of the (room temperature) coherent spinodal as calculated here (Fig. 14). However, samples of this composition aged at 1000°C for 120 min yielded an extensively mottled structure and satellites on the corresponding selected area diffraction patterns. In the 90% Al_2O_3 sample, a mottled preprecipitate and needlelike precipitate were observed by Krepski (1975) in the as-LQ samples. The latter precipitates were interpreted as a monoclinic intermediate phase, and aging at 1000°C enhanced its development. Interestingly, as shown in Fig. 14, this composition and temperature lies within the range of the coherent spinodal as calculated here, and it is evident that even cooling on the order of 10^5 °C/sec, as achieved in liquid quenching, is not rapid enough to prevent decomposition. This represents some evidence that spinodal decomposition may occur in this system.

5. MgO–$MgAl_2O_4$

One of the classic MgO-containing binary systems is MgO–$MgAl_2O_4$, the MgO-rich subsystem of MgO–Al_2O_3. As in the previous section, this subsystem will be treated in terms of the two conjugate end members, MgO and spinel. The phase diagram of Alper *et al.* (1962) is thus replotted in terms

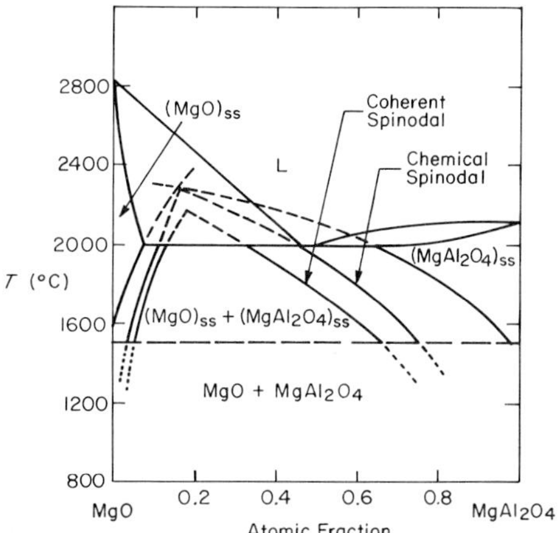

Fig. 15. Equilibrium diagram of the MgO–$MgAl_2O_4$ subsystem of the MgO–Al_2O_3 system (after Alper *et al.*, 1962), replotted in terms of atomic fraction $MgAl_2O_4$ in MgO–$MgAl_2O_4$. Superimposed are the incoherent chemical spinodal and the at-temperature coherent spinodal as calculated in this study.

of atomic fraction of spinel (Fig. 15). To our knowledge, no previous theoretical estimates of the position of the coherent spinodal exist for the MgO–$MgAl_2O_4$ system. At the critical composition 16% $MgAl_2O_4$, the equilibrium critical temperature 2280°C is depressed by 100°C, using the at-temperature data, and by 168°C, using the room temperature data (Table III). Only the at-temperature coherent spinodal is plotted in Fig. 15.

Table VI shows that calculations with and without the temperature derivatives of the elastic constants, using either room temperature or at-temperature lattice parameter data, markedly effect results for MgO-containing systems. For MgO–$MgAl_2O_4$, use of room temperature lattice parameters, without the temperature derivatives of the elastic constants,

TABLE VI

THE EFFECT OF THE TEMPERATURE DERIVATIVES OF THE ELASTIC CONSTANTS τ_{uv} ON SYSTEMS CONTAINING MgO

System	ΔT (°C) without τ_{uv} Room temp. "a_0"	At-temp. "a_0"	ΔT (°C) with τ_{uv} Room temp. "a_0"	At-temp. "a_0"
MgO–$MgAl_2O_4$	168	275	61	100
MgO–CaO	4206	5209	85	105
MgO–CoO	2	1	2	1

yields a ΔT of 168°C. This value decreases to a ΔT of 61°C when the temperature derivatives are included in the calculation. Likewise, use of the temperature derivatives decreases the ΔT from 275 to 100°C when at-temperature lattice parameter data is used (Table VI). Hence, the use of at-temperature lattice parameters causes a slightly greater mismatch in the η term (i.e., using at-temperature elastic constants: for room temperature a_0, $\Delta T = 61$°C, and for at-temperature a_0, $\Delta T = 100$°C).

Experimentally, MgO–$MgAl_2O_4$ represents a challenging system due to the prohibitively high temperatures involved and the vaporization of MgO. Liquid quenching of laser-melted specimens was attempted by Krepski (1975), but vaporization of MgO was severe under the laser beam. Previously, LQ experiments were carried out to determine the range of metastable solid solubility in MgO–Al_2O_3 (Sarjeant and Roy, 1967; Sarjeant, 1967), and eutectic solidification studies in the MgO–$MgAl_2O_4$ portion of the system revealed MgO whiskers in a spinel matrix, where $(hkl)_{MgO}//(hkl)_{MgAl_2O_4}$ (Kennard *et al.*, 1973, 1976). Three compositions within the two-phase region (0.48, 0.40, and 0.24% $MgAl_2O_4$), but close to the equilibrium solvus, were studied at 1350 and 1100°C after solutionizing at 1950°C (Stubican, 1967; Stubican and Roy, 1965). At these compositions and temperatures diffusion-controlled nucleation and growth was found to occur; however, an induction period was absent.

Precipitation hardening was found by Shannon *et al.* (1974) at 1250°C for 97 and 92% $MgAl_2O_4$, which was attributed to a metastable precipitate phase. After long annealing times for 97% $MgAl_2O_4$, a second-stage precipitation of MgO was observed. For 84% $MgAl_2O_4$, preprecipitates of MgO were observed in the as-quenched samples.

In summary, all of the compositions studied in the MgO–$MgAl_2O_4$ system to date have been outside of the limits of the at-temperature determined coherent spinodal as proposed here, and no evidence of decomposition by a spinodal mode has been found.

6. MgO–CaO

The MgO–CaO system is unique due to the large influence that the at-temperature elastic data have on the calculation of the coherent spinodal. Use of the room temperature elastic data gives a depression of the coherent spinodal of 4206–5209°C, whereas consideration of the temperature derivatives of the elastic constants reduces the ΔT to only 85–100°C for the critical point (Tables III and VI). Therefore, spinodal decomposition might be observable in this system. Figure 16 shows the calculated coherent spinodal, using the temperature derivatives of the elastic constants and the at-temperature ($\Delta T = 105$°C) and room temperature ($\Delta T = 85$°C) lattice parameters.

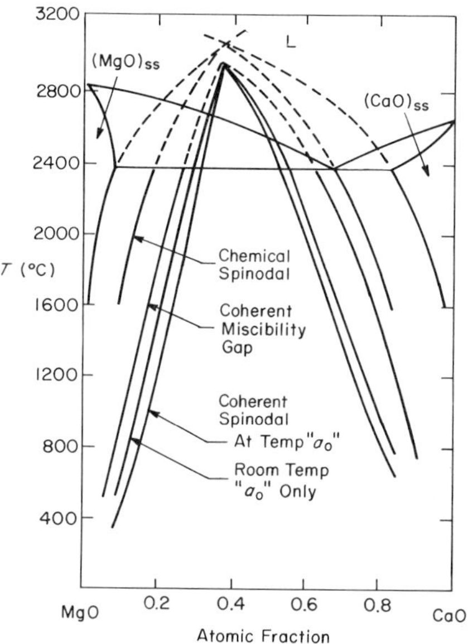

Fig. 16. Equilibrium diagram of MgO–CaO (from Doman *et al.*, 1963) with the chemical and coherent spinodals as calculated in this study. The room temperature coherent spinodal is depressed below ambient temperature and does not appear on this diagram.

The use of the at-temperature lattice parameter data increases the ΔT for the critical point by only 20°C (Table VI). The dramatic effect arising from the temperature derivatives of the elastic constants results from the decrease of the c_{11}, c_{12}, c_{44} of MgO with increasing temperature above 400°C. In addition, the c_{11} and c_{12} of CaO decrease with temperature above 75°C. Therefore, the two conjugate phases become elastically similar at higher temperatures and $Y_{\min}$ shifts from the $\langle 100 \rangle$ as predicted for room temperature, to $\langle 111 \rangle$ (Table III).

Experimentation in this system is difficult due to the high liquidus temperatures, vaporization of MgO and chemical reactivity of CaO. However, several compositions have been laser melted and hammer-and-anvil quenched and the solubility has been extended (Krepski, 1975). Small cubic crystallites have been observed in several as-LQ samples, but annealing experiments have not been attempted.

7. MgO–CoO

The miscibility gap in MgO–CoO is highly asymmetric, with a critical composition at 98% MgO (Robin, 1955). The lattice parameters of the two end-member phases are closely related (Table II), and since η is small, ΔT

is only 1–2°C (Table III). There is little change in ΔT even when the temperature derivatives of the elastic constants are used (Table VI), because the miscibility gap occurs at below 800°C and the main effects of decreasing τ_{uv} in MgO occur for temperatures above 400°C. To our knowledge, there have been no experimental phase decomposition studies in this system, although, as calculated here, spinodal decomposition is clearly possible. However, it is expected that experimental problems will result due to the lack of stability of CoO and the vaporization of MgO.

D. Glasses

Stable and metastable miscibility gaps are common in glass-forming oxide systems. As a homogeneous glass solution is cooled from the single phase region to within the gap in miscibility unmixing is favored.* Spinodal decomposition has been of special interest in glass-forming systems because of their inherent isotropy. However, anisotropy, texture, regularity, or preferred growth can develop due to variations in conditions during melting and quenching (Charles, 1973; Rindone, 1975).

Many binary, ternary, and quaternary systems exhibit liquid immiscibility, leading to metastable miscibility gaps: e.g., $CaO–SiO_2$ (Ol'shanskiĭ, 1951; Burnett and Douglas, 1970); $Na_2O–SiO_2$ (Andreev *et al.*, 1964; Hammell, 1966; Porai-Koshits and Aver'yanov, 1968); $Li_2O–SiO_2$ (James and McMillan, 1970; Tomozawa, 1972); $Al_2O_3–SiO_2$ (MacDowell and Beall, 1969; Risbud and Pask, 1975); $B_2O_3–SiO_2$ (Charles and Wagstaff, 1968); $B_2O_3–Na_2O$ (Shaw and Uhlmann, 1968) and other borate, silicate, phosphate, and germanium-bearing binary systems.

There has been considerable effort in extending the theory and experimentally confirming the mechanism of spinodal decomposition in glass systems (Tomazawa *et al.*, 1969, 1970a,b; Cahn, 1969b; Cahn and Charles, 1965; Zarzycki and Naudin, 1967, 1968, 1969; Neilson, 1969, 1970). In addition, Srinivasan *et al.* (1976) and Sarkar *et al.* (1973) have attempted to estimate the spinodal temperature in glasses using TEM. Calculations of the free energy inflection points from the metastable liquidus in glass systems have been made by Shul'ts (1973). Experimentation has been increasing on glass microstructure by SAXS (Cartz, 1957; Porai-Koshits and Andreev, 1958, 1959; Williams *et al.*, 1965; Goganov and Porai-Koshits, 1965, 1966; Russell and Bergeson, 1965; Brumberger and Debye, 1957; Andreev and Porai-Koshits, 1970; Sakaino *et al.*, 1975) and by light scattering (Goldstein, 1965).

* Reviews of phase separation in glass systems are given by Porai-Koshits and Aver'yanov (1973), Vogel (1966, 1975), Bobkova and Trunets (1973), Rindone (1975), Haller and Mecedo (1968), and Ohlberg *et al.* (1965).

The most detailed experimental work on spinodal decomposition has focused on the Na_2O–SiO_2 system, shown in Fig. 17. Fourier analysis of wide-angle x-ray data suggested that Na_2O–SiO_2 glasses exhibit some type of structure (Warren and Biscoe, 1938). Evidence of a typical modulated structure by TEM was the first suggestion that the microperiodicity is due to spinodal decomposition (Hammell, 1966). Subsequent SAXS experiments were carried out by Tomazawa (1968), Tomozawa *et al.* (1969, 1970a,b), and Neilson (1969, 1972) and SANS experiments by Roth and Zarzycki (1974).

Small-angle x-ray scattering in conjunction with TEM was carried out for 12.6 and 13.2 mole % Na_2O poured glasses in the temperature range 450–650°C (Neilson, 1969). Although these parameters are well within the spinodal as calculated by Tomazawa (1968), shown in Fig. 17, the scattering results were ambiguous. Neilson (1969) suggested that if spinodal decomposition does occur in this system, competing mechanisms are also involved. A further study of the 12.6 mole % Na_2O glass at 770 and 790°C by SAXS

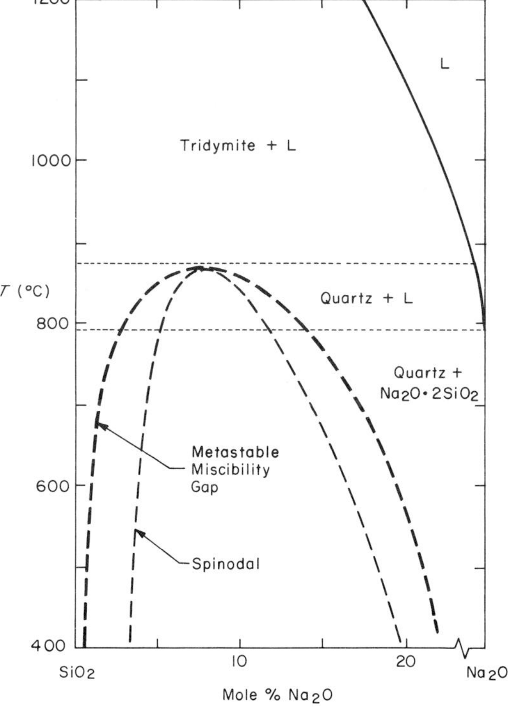

Fig. 17. Equilibrium phase diagram of the SiO_2-rich portion of the Na_2O–SiO_2 system (from Krackek, 1930) with the metastable miscibility gap (from Andreev *et al.*, 1964; Hammell, 1966; and Porai-Koshits and Aver'yanov, 1968) and the chemical spinodal (from Tomozawa, 1968; and Tomozawa *et al.*, 1970b).

indicated that decomposition occurs in this region by nucleation, growth, and coarsening (Neilson, 1972). This is within the metastable nucleation and growth region as proposed by Tomozawa (1968).

A comprehensive study of phase decomposition in the Na_2O–SiO_2 system by Tomozawa (1968) provided kinetic evidence for spinodal decomposition. SAXS was used to follow the time development of the Fourier amplitudes of the composition fluctuations as unmixing occurred in 5–18 mole % Na_2O (Tomozawa *et al.*, 1969, 1970a,b; Tomozawa, 1968). It was noted by these authors that the early stage of phase separation which is obtained by ordinary quenching techniques deviates from Cahn's early-stage solution of spinodal decomposition. However, as decomposition proceeds, the nonlinear terms are seen to progressively influence the amplification factor. Thus, as predicted by de Fontaine (1967), a shifting maximum in the scattering profile and an attendant displacement of the crossover point (corresponding to β_c) are to be expected for off-centered compositions. Therefore, Tomozawa *et al.* calculated the progressive theoretical amplifications factors for Na_2O–SiO_2 glasses following de Fontaine's (1967) solution of Cahn's diffusion equation. With reference to the experimental SAXS analysis, Tomozawa (1968) discussed the kinetic development of later-stage decomposition by a spinodal mode. This deviation from linearlike behavior was also noted in SANS experiments on 12.0 mole % Na_2O glasses at 480 and 560°C, but these effects were attributed to coarsening (Roth and Zarzycki, 1974). However, these workers did not study the time development of the scattering profiles during unmixing.

Al_2O_3–SiO_2 is a glass-forming system in which even equilibrium experimentation is difficult. Inherent problems with melting and crystallization of the intermediate compound, mullite, have precluded a firm determination of the equilibrium phase diagram and have seriously complicated the kinetics of decomposition. Figure 18 gives a portion of the equilibrium diagram of Aramaki and Roy (1962), with the delineation of the region of very rapid crystallization superimposed (Takamori and Roy, 1973). The flattened liquidus (Bowen and Greig, 1924; Aramaki and Roy 1959; Galakhov, 1964) near mullite implies that a region of liquid immiscibility exists in the Al_2O_3–SiO_2 system (Davis and Pask, 1972). The corresponding structures observed in light microscopy and electron microscopy represent additional evidence for liquation or liquid–liquid immiscibility (Galakhov and Konovalova, 1964; Ganguli and Saha, 1967). Haller and Macedo (1968) attributed the interconnectivity in this system to coalescence of spherical particles of large volume fraction.

A detailed TEM study of rapidly quenched Al_2O_3–SiO_2 glass led MacDowell and Beall (1969) to propose the high temperature metastable miscibility gap shown in Fig. 18. The composition limits of this "region of coconnectivity" they defined as the spinodal by the observation of inter-

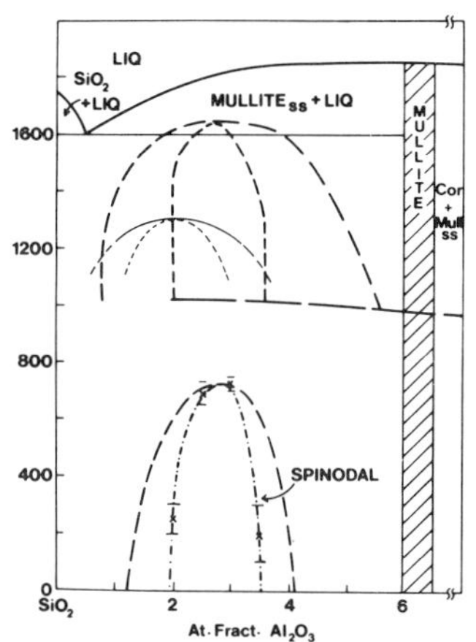

Fig. 18. Equilibrium phase diagram of the SiO_2-rich portion of the Al_2O_3–SiO_2 system (from Aramaki and Roy, 1962) with the metastable miscibility gaps ("limit of coconnectivity") suggested by MacDowell and Beall (1969) and Galakhov *et al.* (1976). Also superimposed is the limit above which rapid crystallization occurs as defined by DTA (Takamori and Roy, 1973). The lower spinodal is suggested on the basis of recent small-angle neutron scattering data (Jantzen, 1978; Jantzen *et al.*, (1978), ------, connectivity limit on rapid quenching (McDowell and Beall, 1969); ——, rapid crystallization (Takamori and Roy, 1973); -·-·-·-·, spinodal (this study); ——-, extrapolated miscibility gap (Galakhov *et al.*, 1976).

connected structures. Recently, Galakhov *et al.* (1976) have also defined the existence of a high temperature metastable miscibility gap by extension from the ternary Ga_2O_3–Al_2O_3–SiO_2 system. However, Takamori and Roy (1973) demonstrated that crystallization is almost complete for the high temperatures at which the metastable miscibility gap is proposed.

In order to clarify these discrepancies in Al_2O_3–SiO_2, rapidly quenched samples prepared by MacDowell (and kindly made available to us) were analyzed by SANS (Jantzen *et al.*, 1978; Jantzen, 1978)*. On the basis of a preliminary RH-type analysis of the neutron scattering data, a new spinodal is proposed for this system (Fig. 18). The metastable miscibility gap was calculated by the technique of Cook and Hilliard (1965) from the delineation of the new spinodal. Further experimentation is currently in progress to more accurately determine this boundary.

E. Geologic Systems

A great deal of mineralogic research is concerned with the study of phase changes in subsolidus systems. Of the many complex geologic silicate systems, the pyroxenes and feldspars have been studied the most closely for evidence of spinodal decomposition. However, these complex monoclinic and triclinic structures, for simplicity, are treated theoretically as being elastically isotropic and invariant in elastic properties with respect to composition (Christie, 1968, 1969; Yund and McCallister, 1970; Fletcher and McCallister,

* In a recent SAXS study of glass Al_2O_3–SiO_2, scattering intensity was in fact detected, which has been intepreted as arising from density changes occuring during unmixing. It is of interest to note that these changes directly reflect the evolving composition modulations (Jantzen *et al.* (1978).

1974; Petrović, 1973). Only Ohashi (1974) and Robin (1974a,b) have considered the effects of triclinic–monoclinic anisotropy on the calculation of the coherent spinodal in the alkali feldspar systems as well as the need for better documentation of the effects of temperature on the lattice parameter and elastic data.

Experimentally, studies of spinodal decomposition in silicate systems have been primarily by morphologic examination with TEM (McConnell, 1971; McLaren, 1974). Fletcher and McCallister (1974) have suggested that spinodal decomposition may be the operative mechanism in natural exsolution.

In the clinopyroxene join, $CaMgSi_2O_6$–$Mg_2Si_2O_6$, a ΔT of 110°C was estimated by Fletcher and McCallister (1974) at the critical composition, and the coherent spinodal was delineated on the Ca-rich side by TEM (McCallister and Yund, 1975). McConnell (1975) has related the microstructure of decomposition in minerals to their cooling history with special reference to spinodal decomposition in alkali feldspars (McConnell, 1969) and plagioclase feldspars (McConnell, 1974; McLaren, 1974; Nord *et al.*, 1974): in these systems, microstructures develop on a scale of ~100 and ~200 Å, respectively.

Christie (1968) originally suggested that alkali feldspars may exsolve by a spinodal mechanism. Later Yund and McCallister (1970) calculated the coherent spinodal at the critical composition to be ~600°C ($\Delta T \simeq 80$°C) for the *bc* plane. McConnell (1971) suggested that the ΔT may be somewhat smaller, and Yund (1974) has determined the T_c of the coherent $KAlSi_3O_8$–$NaAlSi_3O_8$ solvus to be 710°C at 33 mole % $KAlSi_3O_8$. However at 37 mole % $KAlSi_3O_8$, for all samples annealed below 600°C, spinodal textures were observed in TEM (Owen and McConnell, 1971, 1974). In fact, Yund *et al.* (1974) found similar microstructures at 33 mole % $KAlSi_3O_8$ up to 700°C.

More recently, spinodal decomposition has been invoked for the sodic-rich regions of the $NaAlSi_3O_8$–$CaAl_2Si_2O_8$ plagioclase feldspars (Nord *et al.*, 1976). In this laboratory, a $40KAlSi_3O_8$:$60NaAlSi_3O_8$ mole % mixture of natural orthoclase albite was quenched from the melt into water, forming a glass. Annealing for 1 min at 250°C produced a modulated structure ($\lambda \sim 25$ Å) in TEM, and after prolonged heating, lath-shaped crystals appeared (Jantzen, 1978).

V. CONCLUSIONS

Strengthening of materials by controlled phase decomposition has widespread technological applications in metallurgy and ceramics. It appears possible that the spinodal concept can be useful in solving problems of practical alloy and heat treatment design (Charles, 1973). Spinodal alloys and ceramics exhibit several distinct and technologically desirable features due to their uniformity and their insensitivity to structural defects [see Cahn (1968) for a summary]. In addition, a spinodal alloy which exhibits anisotropy, can, to a limited extent, be endowed with anisotropic physical properties. Hence, experimental determination or the accurate theoretical prediction of the coherent spinodal can be technologically significant. Once the pertinent parameters (e.g., gradient energy coefficient) have been determined, predictions can be made pertaining to the structure, and the optimum composition and/or heat treatment can be selected for a given material. Alternatively, addition of alloying elements can selectively alter the coherency stresses, and, hence, the position of the spinodal can be adjusted once these effects can be accurately predicted. Thus, manipulation of mismatch can be used to vary both the structure and the resulting properties.

It is important to be able to predict accurately the systems that will decompose by a spinodal mode and, likewise, to predict the temperature

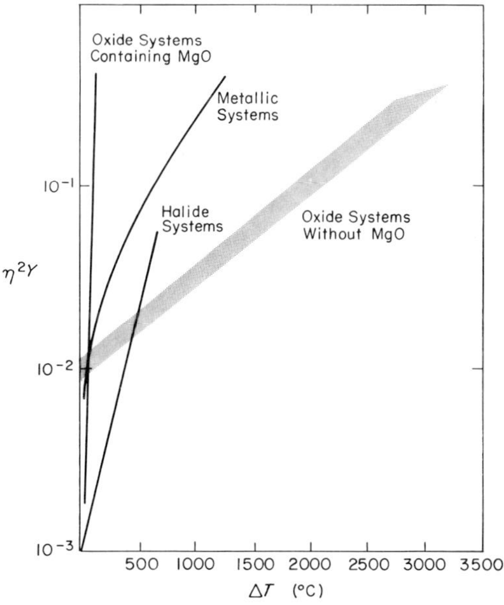

Fig. 19. Variation of the strain energy term $\eta^2 T$ with the depression of the coherent spinodal ΔT for calculations using at-temperature elastic and lattice parameter data. The metallic and alkali halide systems follow similar trends, while the oxide systems show two distinct linear trends, depending on whether or not the elastically peculiar MgO is involved. Systems containing MgO show a large variation in $\eta^2 Y$ over a small range of ΔT while the remaining oxide systems exhibit a wide variation of the strain energy term over a large range of ΔT.

and composition range over which decomposition will occur. It is to be emphasized that use of the at-temperature elastic data and lattice parameters is a realistic approach to determining the coherent spinodals.

The systems listed in Table III are given in order of increasing ΔT, as calculated here from the at-temperature data. Empirically, ΔT is proportional to the strain energy term $\eta^2 Y$. This relationship is clearly observed in Fig. 19, where the elastic energy term $\eta^2 Y$, calculated from the at-temperature data, is plotted against ΔT at the critical composition for various metal, alkali halide and oxide systems. The metallic and alkali halide systems are found to be closely linear. However, the oxide systems follow two distinct linear trends, depending on whether or not the binary system involves MgO, which is elastically peculiar. These relationships are *not* observed between the elastic energy term calculated from room temperature data. Hence, it is suggested that at-temperature data be used for characterization of technologically useful spinodal systems.

References

Agarwal, S. C. (1974). Ph.D. Thesis, S.U.N.Y. at Stony Brook, New York.

Agarwal, S. C., and Herman, H. (1973a). *In* "Phase Transitions and Their Application in Materials Science" (H. K. Henisch and R. Roy, eds.), pp. 207–222. Pergamon, New York.

Agarwal, S. C., and Herman, H. (1973b). *Scripta Metall.* **7**, 503–508.

Aleksandrov, K. S., Shabanova, L. A., and Reshchikova, L. M. (1968). *Soviet Phys.-Solid State* **10**, 1316–1321.

Alper, A. M., McNally, R. N., Ribbe, P. H., and Doman, R. C. (1962). *J. Am. Ceram. Soc.* **45**, 263–268.

Andreev, N. S., and Porai-Koshits, E. A. (1970). *Discuss. Faraday Soc.* **50**, 135–144.

Andreev, N. S., Goganov, D. A., Porai-Koshits, E. A., and Sokolov, Yu. G. (1964). *Catalyzed Crystallization of Glass* **3**, 47–52.

Aramaki, S., and Roy, R. (1959). *Nature* **184**, 631–632.

Aramaki, S., and Roy, R. (1962). *J. Am. Ceram. Soc.* **45**, 229–242.

Ardell, A. J., Nuttal, K., and Nicholson, R. B. (1969). *In* "The Mechanisms of Phase Transformations in Crystalline Solids," Mono. 33, pp. 22–26. Inst. of Metals, London.

Bansal, G. K., and Heuer, A. H. (1974). *Philos. Mag.* **29**, 709–722.

Barnes, M. H. (1945). Tech. Rep. F.I.A.T. **655**.

Bartel, T. L., and Rundman, K. B. (1975). *Metall. Trans.* **6A**, 1887–1893.

Bartels, R. A., and Schuele, D. E. (1965). *J. Phys. Chem. Solids* **26**, 537–549.

Bartels, R. A., and Vetter, V. H. (1972). *J. Phys. Chem. Solids* **33**, 1991–1992.

Batelle Memorial Inst. (1966). "Engineering Properties of Selected Ceramic Materials" (J. F. Lynch, C. G. Ruderer, and W. H. Duckworth, eds.). Am. Ceram. Soc., Columbus, Ohio.

Bechmann, R., and Hearmon, R. F. S. (1966). *In* "Landolt–Börnstein Numerical Data" (K.-H. Hellwege and A. M. Hellwege, eds.), Vol. I. Springer-Verlag, Berlin.

Bechmann, R., Hearmon, R. F. S., and Kurtz, A. M. (1969). *In* "Landolt-Börnstein Numerical Data" (K.-H. Hellwege and A. M. Hellwege, eds.), Vol. II. Springer-Verlag, Berlin.

Bhardwaj, M. C., and Roy, R. (1971). *J. Phys. Chem. Solids* **32**, 1603–1607.

Bobkova, N. M., and Trunets, I. A. (1973). *In* "The Structure of Glass" (E. A. Porai-Koshits, ed.), Vol. 8, pp. 43–47. Consultants Bureau, New York.
Bonfiglioli, A. F., and Guinier, A. (1966). *Acta Metall.* **14**, 1213–1224.
Bortz, A. B., Kalos, M. H., Lebowitz, J. L., and Zendejas, M. A., (1974). *Phys. Rev.* **B10**, 535–541.
Bowen, H. K., Kingery, W. D., Kinoshita, M., and Goodwin, C. A. (1972). *J. Cryst. Growth* **13–14**, 402–405.
Bowen, N. L., and Grieg, J. W., (1924). *J. Am. Ceram. Soc.* **7**, 238–254.
Brumberger, H., and Debye, P. (1957). *J. Phys. Chem.* **61**, 1623–1624.
Bunk, A. J. H., and Tichelaar, G. W. (1953). *Proc. K. Ned. Akad. Wet.* **56B**, 375–384.
Burnett, D. G., and Douglas, R. W. (1970). *Physical Chemistry of Glasses* **11**, 125–135.
Cahn, J. W. (1961). *Acta Metall.* **9**, 795–801.
Cahn, J. W. (1962a). *Acta Metall.* **10**, 179–183.
Cahn, J. W. (1962b). *Acta Metall.* **10**, 907–913.
Cahn, J. W. (1965). *J. Chem. Phys.* **42**, 93–99.
Cahn, J. W. (1966). *Acta Metall.* **14**, 1685–1692.
Cahn, J. W. (1968). *Trans. Metall. Soc. AIME* **242**, 166–180.
Cahn, J. W. (1969a). *In* "The Mechanisms of Phase Transformations in Crystalline Solids," Mono. 33, pp. 1–5. Inst. of Metals, London.
Cahn, J. W. (1969b). *J. Am. Ceram. Soc.* **52**, 118–121.
Cahn, J. W., and Charles, R. J. (1965). *Physics and Chemistry of Glasses* **6**, 181–191.
Cahn, J. W., and Hilliard, J. E. (1958). *J. Chem. Phys.* **28**, 258–267.
Carpenter, R. W. (1967). *Acta Metall.* **15**, 1567–1572.
Cartz, L. (1957). *Nature* **180**, 1115–1116.
Chang, Z. P., and Barsch, G. R. (1973). *J. Geophys. Res.* **78**, 2418–2433.
Charles, R. J. (1973). *Am. Ceram. Soc. Bull.* **52**, 673–686.
Charles, R. J., and Wagstaff, F. E. (1968). *J. Am. Ceram. Soc.* **51**, 16–20.
Christie, O. H. J. (1968). *Lithos* **1**, 187–192.
Christie, O. H. J. (1969). *Lithos* **2**, 285–291.
Clark, G. L., Howe, E. E., and Badger, A. E. (1934). *J. Am. Ceram. Soc.* **17**, 7–8.
Cook, H. E. (1970). *Acta Met.* **18**, 297–306.
Cook, H. E., and Hilliard, J. E. (1965). *Trans. Am. Inst. Min. Metall. Pet. Eng.* **233**, 142–146.
Cook, H. E., and Hilliard, J. E. (1969). *J. Appl. Phys.* **40**, 2191–2196.
Cook, H. E., de Fontaine, D., and Hilliard, J. E., (1969). *Acta Met.* **17**, 765–773.
Corsepius, H., and Münster, A. (1959). *Z. Phys. Chem.* (Frankfurt am Main) **22**, 1–19.
Daniel, V. and Lipson, H. (1943). *Proc. R. Soc.* London Ser. **A181A**, 368–378.
Daniel, V. and Lipson, H. (1944). *Proc. R. Soc.* London Ser. A **182A**, 378–387.
Darling, A. S., Mintern, R. H., and Chaston, J. C. (1952). *J. Inst. Metals* **81**, 125–132.
Davis, R. F. and Pask, J. A. (1972). *J. Am. Ceram. Soc.* **55**, 525–531.
de Fontaine, D. (1966). *In* "Local Atomic Arrangements Studied by X-Ray Diffraction" (J. B. Cohen and J. C. Hillard, eds.), pp. 51–94. Gordon & Breach, New York.
de Fontaine, D. (1967). Ph.D. Thesis, Northwestern Univ., Evanston, Illinois.
de Fontaine, D. (1969). *Acta Metall.* **17**, 477–482.
de Fontaine, D. (1970). *In* "Ultrafine Grain Metals" (J. Burke and V. Weiss, eds.), pp. 93–131. Syracuse Univ. Press, Syracuse, New York.
Dhalenne, G., Revcolevschi, A., and Collongues, R. (1972). *Mater. Res. Bull.* **7**, 1385–1392.
Doman, R. C., Barr, J. B., McNally, R. N., and Alper, A. M. (1963). *J. Am. Ceram. Soc.* **46**, 313–316.
Dreyfus, R. W., and Nowick, A. S. (1962). *Phys. Rev.* **126**, 1367–1377.
du Plessis, P. DeV., Van Tonder, S. J., and Alberts, L. (1971). *J. Phys. C* **4**, 1983–1987.
Ellwood. E. C. (1952). *J. Inst. Met.* **80**, 217–224.

Eppler, W. F. (1943). *Z. Angew. Mineral* **4**, 345–362.
Eppler, W. F. (1947). Tech. Rep. F.I.A.T. No. 1038.
Fancher, D. L., and Barsch, G. R. (1971). *J. Phys. Chem. Solids* **32**, 1303–1313.
Fauchaux, R. D., and Simmons, R. O. (1964). *Phys. Rev. A*. **136A**, 1664–1674.
Fine, M. E. (1968). *Tr. Jpn. Inst. Met. (Suppl.)* **9**, 527–532.
Fine, M. E. (1970a). *In* "Advances in Materials Research " (H. Herman, ed.). Vol. 4, pp. 1–53. Wiley, New York.
Fine, M. E. (1970b). *In* "The Chemistry of Extended Defects in Non-Metallic Solids" (L. Eyring and M. O'Keefe, eds.), pp. 575–608. North-Holland Publ., Amsterdam.
Fine, M. E. (1972). *Am. Ceram. Soc. Bull.* **51**, 510–515.
Fink, W. L., and Willey, L. A. (1936). *Trans. Metall. Soc. AIME* **122**, 244–265.
Fletcher, R. C., and McCallister, R. H. (1974). *Carnegie Inst. Washington Yearb.* **73**, 396–399.
Flewitt, P. E. J. (1974). *Acta Metall.* **22**, 47–63.
Fukano, Y. (1961). *J. Phys. Soc. Jpn.* **16**, 1195–1204.
Galakhov, F. Ya. (1964). *Izv. Akad. Nauk. SSSR, Ser. Khim.* **8**, 1292–1296.
Galakhov, F. Ya., and Konovalova, S. F. (1964). *Izv. Akad. Nauk. SSSR, Ser. Khim.* **8**, 1287–1291.
Galakhov, F., Ya, (1976). Aver'yanov, V. I., Vavilonova, V. T., and Areshev, *Fiz. Khim. Stekla*, **2**, 127–132.
Ganguli, D., and Saha, P. (1967). *Mater. Res. Bull.* **2**, 25–36.
Gerold, V., and Merz, W. (1967). *Scripta Metall.* **1**, 33–35.
Gerold, V., and Schweizer, W. (1961). *Z. Metallkd.* **52**, 76–85.
Gibbs, J. W. (1906). "The Scientific Papers," Vol. II Longmans, Green and Co., London.
Goganov, D. A., and Porai-Koshits, E. A. (1965). *In* "The Structure of Glass" (E. A. Porai-Koshits, ed.), Vol. 5, pp. 82–89. Consultants Bureau, New York.
Goganov, D. A., and Porai-Koshits, E. A. (1966). *In* "The Structure of Glass" (E. A. Porai-Koshits, ed.), Vol. 6, pp. 107–112. Consultants Bureau, New York.
Golding, B., and Moss, S. C. (1967). *Acta. Metall.* **15**, 1239–1241.
Goldstein, M. (1965). *J. Am. Ceram. Soc.* **48**, 126–130.
Goldstein, M. (1968). *J. Cryst. Growth* **3**, 594–599.
Grabmaier, J. G., and Falckenberg, H. R. (1969). *J. Am. Ceram. Soc.* **52**, 648–650.
Graf, R., and Lenormand, M. (1964). *C. R. Acad. Sci. Ser. B* **259**, 3494–3497.
Greskovich, C., and Stubican, V. S. (1968). *J. Am. Ceram. Soc.* **51**, 42–46.
Gronsky, R., Okada, M., Sinclair, R., and Thomas, G. (1975). *Annu. Proc. E.M.S.A., 33rd*, p. 22.
Groves, G. W., and Fine, M. E. (1964). *J. Appl. Phys.* **35**, 3587–3593.
Guinier, A. (1963). "X-Ray Diffraction in Crystals, Imperfect Crystals and Amorphous Bodies." Freeman, San Francisco.
Haller, W. (1965). *J. Chem. Phys.* **42**, 686–693.
Haller, W., and Macedo, P. B. (1968). *Physics and Chemistry of Glasses* **9**, 153–155.
Hammell, J. J. (1966). *Proc. Int. Congr. Glass, VII, Brussels, 1965*, pp. 36.1–36.5.
Hargreaves, M. E. (1951). *Acta Crystallogr.* **4**, 301–309.
Harkness, S. D., Gould, R. W., and Hern, J. J. (1969). *Philos. Mag.* **19**, 115–128.
Haussühl, S. (1960). *Z. Phys.* **159**, 223–229.
Hendricks, R. W. (1964). Ph.D. Thesis, Cornell Univ., Ithaca, New York.
Hendricks, R. W., and Borie, B. S. (1967). *In* "Small-Angle X-ray Scattering" (H. Brumberger, ed.), pp. 319–334. Gordon & Breach, New York.
Hendricks, R. W., Baro, R., and Newkirk, J. B. (1964). *Trans. Metall. Soc. AIME* **230**, 930–931.
Herai, T., and Manenc, J. (1964). *Mem. Sci. Rev. Metall.* **61**, 677.
Herman, H., and MacCrone, R. K. (1971). *J. Am. Ceram. Soc.* **55**, 50.

Herman, H., Cohen, J. B., and Fine, M. E. (1963). *Acta Metall.* **11**, 43–56.
Hillert, M. (1961). *Acta Metall.* **9**, 525–535.
Hilliard, J. E. (1970). *In* "Phase Transformations", pp. 497–560. Am. Soc. Metals, Metals Park, Ohio.
Hilliard, J. E., Averbach, B. L., and Cohen, M. (1951). *Acta Metall.* **2**, 621–631.
Hilton, A. R., and Jones, C. E. (1967). *Appl. Opt.* **6**, 1513–1517.
Hopper, R. W., and Uhlmann, D. R. (1973a). *Acta Metall.* **21**, 35–42.
Hopper, R. W., and Uhlmann, D. R. (1973b). *Acta Metall.* **21**, 267–272.
Hopper, R. W., and Uhlmann, D. R. (1973c). *Acta Metall.* **21**, 377–384.
Huston, E. L., Cahn, J. W., and Hilliard, J. E. (1966). *Acta Metall.* **14**, 1053–1062.
Jagodzinski, H. (1957). *Z. Kristallogr.* **109**, 388–409.
Jagodzinski, H., and Saalfeld, H. (1958). *Z. Kristallogr. Mineral.* **110**, 197–218.
James, P. F., and McMillan, P. W. (1970). *Physics and Chemistry of Glasses* **11**, 59–63.
Jantzen, C. M. F. (1978). Ph.D. Thesis, S.U.N.Y. at Stony Brook, New York.
Jantzen, C. M. F., Schwahn, D., Schelten, J., and Herman, H. (1978). *Int. Conf. Small-Angle Scattering. 4th, Gatlinburg, Tennessee, 3–7 October 1977.*
Jantzen, C. M. F., and Herman, H. (in prep.).
Jones, R. D., and Thomas, K. G. (1970). *Philos. Mag.* **22**, 427–430.
Kelly, A., and Nicholson, R. B. (1963). *Prog. Mater. Sci.* **10**, 151–391.
Kennard, F. L., Bradt, R. C., and Stubican, V. S. (1973). *J. Am. Ceram. Soc.* **56**, 566–569.
Kennard, F. L., Bradt, R. C., and Stubican, V. S. (1976). *J. Am. Ceram. Soc.* **59**, 160–163.
Kinoshita, M., Kingery, W. D., and Bowen, H. K. (1973). *J. Am. Ceram. Soc.* **56**, 398–399.
Koch, F. B. (1966). Ph.D. Thesis, Northwestern Univ., Evanston, Illinois.
Koster, W. (1948). *Z. Metallkd.* **39**, 1–9.
Krackek, F. C. (1930). *J. Phys. Chem.* **34**, 1583–1598.
Kralik, G., Weise, J., and Gerold, V. (1969). *In* "The Mechamisms of Phase Transformations in Crystalline Solids," Mono. **33**, pp. 27–28. Inst. of Metals, London.
Krepski, R. (1975). M.S. Thesis, S.U.N.Y. at Stony Brook, New York.
Krepski, R., Swyler, K., Carleton, H. R., and Herman, H. (1975). *J. Mater. Sci.* **10**, 1452–1454.
Krikorian, O. H. (1960). U.C.R.L. Rept. No. 6132, Univ. of California Research Laboratory, Berkley, California.
Kruse, E. W., III, and Fine, M. E. (1972). *J. Am. Ceram. Soc.* **55**, 32–37.
Langer, J. S. (1973). *Acta Met.* **21**, 1649–1658.
Langer, J. S., Bar-on, M., and Miller, H. D., (1975). *Phys. Rev.* **A11**, 1417–1429.
Lebowitz, J. L., and Kalos, M. H., (1976). *Scipta Met.* **10**, 9–12.
Leibfried, G., and Ludwig, W. (1961). *Solid State Phys.* **12**, 275–444.
Lewis, M. H. (1969). *Philos. Mag.* **20**, 985–998.
Lilley, E., and Newkirk, J. B. (1963). *Adv. X-Ray Anal.* **7**, 195–202.
Marro, J., Bortz, A. B., Kalos, M. H., and Lebowitz, J. L., (1975). *Phys. Rev.* **B12**, 2000–2011.
McCallister, R. H., and Yund, R. A. (1975). *Carnegie Inst. Washington Yearb.* **74**, 433–436.
McConnell, J. D. C. (1969). *Philos. Mag.* **19**, 221–229.
McConnell, J. D. C. (1971). *Min. Mag.* **38**, 1–20.
McConnell, J. D. C. (1974). *In* "The Feldspars" (W. S. Mackenzie and J. Zussman, eds.), pp. 460–477. Manchester Univ. Press, Manchester, England.
McConnell, J. D. C. (1975). *Annu. Rev. Earth Planet. Sci. Lett.* **3**, 129–155.
MacDowell, J. F., and Beall, G. H. (1969). *J. Am. Ceram. Soc.* **52**, 17–25.
McLaren, H. C. (1974). *In* "The Feldspars" (W. S. Mackenzie and J. Zussman, eds.), pp. 378–423. Manchester Univ. Press, Manchester, England.
Miyake, S., and Suzuki, K. (1954). *J. Phys. Soc. Jpn.* **9**, 702–712.
Moss, S. C. (1966). *In* "Local Atomic Arrangements Studied by X-Ray Diffraction" (J. B. Cohen and J. E. Hilliard, eds.), pp. 95–122. Gordon & Breach, New York.

Moss, S. C., and Averbach, B. L. (1967). *In* "Small-Angle X-Ray Scattering" (H. Brumberger, ed.), pp. 335–350. Gordon & Breach, New York.
Münster, A., and Sagel, K. (1958). *Z. Phys. Chem.* (*Frankfurt am Main*) **14**, 296–305.
Murakami, M., Kawano, O., Murakami, Y., and Morinaga, M. (1969). *Acta Metall.* **17**, 1517–1521.
Neilson, G. F. (1969). *Physics and Chemistry of Glasses* **10**, 54–62.
Neilson, G. F. (1970). *Discuss. Faraday Soc.* **50**, 145–154.
Neilson, G. F. (1972). *Physics and Chemistry of Glasses* **13**, 70–76.
Nord, G. L., Heuer, A. H., and Lally, J. S. (1974). *In* "The Feldspars" (W. S. Mackenzie and J. Zussman, eds.), pp. 522–535. Manchester Univ. Press, Manchester, England.
Nord, G. L., Zen, E., and Hammarstrom, J. (1976). *Geol. Soc. Am. Abstr.* **8**, 1030.
Ohashi, Y. (1974). *Carnegie Inst. Washington Yearb.* **73**, 399–403.
Ohlberg, S. M., Hammell, J. J., and Golab, H. R. (1965). *J. Am. Ceram. Soc.* **48**, 178–180.
Ol'shanskiĭ, Ya. (1951). *Dokl. Akad. Nauk SSSR* **75**, 93–96.
Owen, D. C., and McConnell, J. D. C. (1971). *Nature* (*London*) *Phys. Sci.* **230**, 118–119.
Owen, D. C., and McConnell, J. D. C. (1974). *In* "The Feldspars" (W. S. Mackenzie and J. Zussman, eds.), pp. 424–439. Manchester Univ. Press, Manchester, England.
Panseri, C., and Federighi, T. (1960). *Acta Metall.* **8**, 217.
Park, M., Mitchell, T. E., and Heuer, A. H. (1974). *Am. Ceram. Soc. Bull.* **53**, 322.
Park, M., Mitchell, T. E., and Heuer, A. H. (1975). *J. Am. Ceram. Soc.* **58**, 43–47.
Park, M. W., Mitchell, T. E., and Heuer, A. H. (1976). *J. Mater. Sci.* **11**, 1227–1238.
Petrovic, R. (1973). *Contr. Mineral. Petrol.* **41**, 151–170.
Philofsky, E. M., and Hilliard, J. E. (1969). *J. Appl. Phys.* **40**, 2196–2205.
Porai-Koshits, E. A., and Andreev, N. S. (1958). *Nature* **182**, 335–336.
Porai-Koshits, E. A., and Andreev, N. S. (1959). *J. Soc. Glass Technol.* **43**, 235–261.
Porai-Koshits, E. A., and Aver'yanov, V. I. (1968). *J. Non-Cryst. Solids* **1**, 29–38.
Porai-Koshits, E. A., and Aver'yanov, V. I. (1973). *In* "The Structure of Glass" (E. A. Porai-Koshits, ed.), Vol. 8, pp. 28–36. Consultants Bureau, New York.
Porod, G. (1967). *In* "Small-Angle X-Ray Scattering" (H. Brumberger, ed.), pp. 1–15. Gordon & Breach, New York.
Porter, H. B. (1955). N.A.V.O.R.D. Rep. No. 4893.
Rao, K. K., Katz, L. E., and Herman, H. (1966/1967). *Mater. Sci. Eng.* **1**, 263–280.
Raynal, J. N., Schelten, J., and Schmatz, W. (1971). *J. Appl. Crystallogr.* **4**, 511.
Revcolevschi, A. (1976). *J. Mater. Sci.* **11**, 563–565.
Rindone, G. E. (1975). *Bull. Cent. Glass Ceram. Res. Inst. Calcutta* **22**, 119–128.
Rinne, F. (1928). *Neues Jahrb. Mineral. Monatsh* **58A**, 43–108.
Risbud, S., and Pask, J. A. (1975). *Am. Ceram. Soc. Bull.* **54**, 818.
Robin, J. (1955). *Ann. Chim.* (*Paris*) **10**, 389–412.
Robin, P. Y. F. (1974a). *Amer. Mineral.* **59**, 1286–1298.
Robin, P. Y. F. (1974b). *Amer. Mineral.* **59**, 1299–1318.
Roth, M., and Zarzycki, J. (1974). *J. Non-Cryst. Solids* **16**, 93–100.
Roy, D. M., Roy, R., and Osborn, E. F. (1953). *Am. J. Sci.* **251**, 337–361.
Rundman, K. B. (1962). Unpublished.
Rundman, K. B. (1973). *In* "Metals Handbook" (T. Lyman, ed.), Vol. 8, pp. 184–185. Am. Soc. Metals, Metals Park, Ohio.
Rundman, K. B., and Hilliard, J. E. (1967). *Acta Metall.* **15**, 1025–1033.
Russell, C. K., and Bergeson, C. G. (1965). *J. Am. Ceram. Soc.* **48**, 268–271.
Russo, V. F., Bitler, W. R., and Stubican, V. S. (1972). *Philos. Mag.* **25**, 1513–1515.
Saalfeld, H., and Jagodzinski, H. (1957). *Z. Kristallogr. Mineral.* **109**, 87–109.
Sakaino, T., Yamane, M., and Makishima, A. (1975). *Bull. Cent. Glass Ceram. Res. Inst. Calcutta* **22**, 108–115.

Sakata, T., and Sakata, K. (1958). *J. Phys. Soc. Jpn.* **13**, 675–683.
Sarjeant, P. T. (1967). Ph.D. Thesis, The Pennsylvania State Univ., University Park, Pennsylvannia.
Sarjeant, P. T., and Roy, R. (1967). *J. Appl. Phys.* **38**, 4540–4542.
Sarkar, A., Gupta, P. K., Srinivasan, G. R., Volterra, V., and Macedo, P. B. (1973). *J. Chem. Phys.* **59**, 4246–4249.
Schwahn, D. (1976) Unpublished.
Scheil, F., and Stadelmaier, H. (1952). *Z. Metallkd.* **43**, 227–236.
Schultz, A. H. (1967). M.S. Thesis, The Pennsylvania State Univ., University Park, Pennsylvania.
Schultz, A. H. (1970). Ph.D. Thesis, The Pennsylvania State Univ., University Park, Pennsylvania.
Schultz, A. H., and Stubican, V. S. (1970). *J. Am. Ceram. Soc.* **53**, 613–616.
Schultz, A. H., Bitler, W. R., and Stubican, V. S. (1969). *Phys. Status Solidi B* **32**, 117–119.
Seward, T. P., III, Uhlmann, D. R., and Turnbull, D. (1968). *J. Am. Ceram. Soc.* **51**, 634–643.
Shaffer, P. T. B. (1964). "Materials Index of High Temperature Materials: No. 1." Plenum, New York.
Shannon, R. E., Johnson, D. L., and Fine, M. E. (1974). *J. Am. Ceram. Soc.* **57**, 269.
Shaw, R. R., and Uhlmann, D. R. (1968). *J. Am. Ceram. Soc.* **51**, 377–382.
Shewmon, P. G. (1969). "Transformations in Metals." McGraw-Hill, New York.
Shul'ts, M. M. (1973). *In* "The Structure of Glass" (E. A. Porai-Koshits, ed.), Vol. 8, pp. 23–27. Consultants Bureau, New York.
Sinclair, R., and Thomas, G. C. (1974). *Ann. Proc. E.M.S.A., 32nd*, p. 500.
Singhal, S. P., Herman, H., and Kostorz, G. (1977). *AIME Annu. Meet. Atlanta.*
Smith, D. E., Tien, T. Y., and Van Vlack, L. H. (1969). *J. Am. Ceram. Soc.* **52**, 459–460.
Smith, D. E., Tien, T. Y., and Van Vlack, L. H. (1973). *Am. Ceram. Soc. Bull.* **52**, 348.
Srinivasan, G. R., Sarkar, A., Gupta, P. K., and Macedo, P. B. (1976). *J. Non-Cryst. Solids* **20**, 141–148.
Stokes, R. J., and Li, C. H. (1962). *Acta Metall.* **10**, 535–542.
Stubican, V. S. (1967). *In* "The Science of Ceramics" (G. H. Stewart, ed.), Vol. 3, pp. 285–297. Academic Press, London and New York.
Stubican, V. S. (1972). *In* "Reactivity of Solids—7th International Symposium" (J. S. Anderson, M. W. Roberts, and F. S. Stone, eds.), pp. 273–282. Chapman & Hall, London.
Stubican, V. S., and Roy, R. (1965). *J. Phys. Chem. Solids* **26**, 1293–1297.
Stubican, V. S., and Schultz, A. H. (1968). *J. Am. Ceram. Soc.* **51**, 290–291.
Stubican, V. S., and Schultz, A. H. (1970). *J. Am. Ceram. Soc.* **53**, 211–214.
Stubican, V. S., and Viechnicki, D. (1966). *J. Appl. Phys.* **37**, 2751–2754.
Stubican, V. S., Schultz, A. H., and Bitler, W. R. (1970). *Philos. Mag.* **22**, 993–1001.
Suzuki, K. (1955). *J. Phys. Soc. Jpn.* **10**, 794–804.
Suzuki, K. (1958). *J. Phys. Soc. Jpn.* **13**, 179–186.
Takahashi, M., and Fine, M. E. (1970). *J. Am. Ceram. Soc.* **53**, 633–634.
Takahashi, M., Guimarães, J. R. C., and Fine, M. E. (1971). *J. Am. Ceram. Soc.* **54**, 291–295.
Takamori, T., and Roy, R. (1973). *J. Am. Ceram. Soc.* **56**, 639–644.
Tiedema, T. J., Bouman, J. and Burgers, W. G. (1957). *Acta Metall.* **5**, 310–321.
Tikkanen, M. H. (1962). *In* "Phase Diagrams for Ceramists" (E. M. Levin, C. R. Robbins, and H. F. McMurdie, eds.), Vol. I, p. 101. Am. Ceram. Soc., Columbus, Ohio.
Tomozawa, M. (1968). Ph.D. Thesis, Univ. of Pennsylvania, University Park, Pennsylvania.
Tomozawa, M. (1972). *Physics and Chemistry of Glasses* **13**, 161–166.
Tomozawa, M., Herman, H., and MacCrone, R. K. (1969). *In* "Phase Transformations in Crystalline Solids," Mono. **33**, pp. 6–10. Inst. of Metals, London.
Tomozawa, M., MacCrone, R. K., and Herman, H. (1970a). *J. Am. Ceram. Soc.* **53**, 62–63.

Tomozawa, M., MacCrone, R. K., and Herman, H. (1970b). *Physics and Chemistry of Glasses* **11**, 136–150.
Tsakalokos, T. (1977). Ph.D. Thesis, Northwestern University, Evanston, Illinois.
Uhlmann, D. R., and Chalmers, B. (1965). *Ind. Eng. Chem.* **57**, 19–31.
Van der Toorn, L. J. (1960). *Acta Metall.* **8**, 715–727.
Viechnicki, D., Schmid, F., and McCauley, J. W. (1974). *J. Am. Ceram. Soc.* **57**, 47.
Vogel, W. (1966). *In* "The Structure of Glass" (E. A. Porai-Koshits, ed.), Vol. 6, pp. 114–120. Consultants Bureau, New York.
Vogel, W. (1975). *Bull. Cent. Glass Ceram. Res. Inst. Calcutta* **22**, 95–107.
Warren, B. E., and Biscoe, J. (1938). *J. Am. Ceram. Soc.* **21**, 259–265.
Weast, R. C. (1973). "Handbook of Chemistry and Physics," 54th ed. C.R.C. Press, Cleveland, Ohio.
Williams, J. A., Phillips, B., Rindone, G. E., and McKinstry, H. A. (1965). *Ad. X-Ray Anal.* **8**, 58–77.
Wirtz, G. P., and Fine, M. E. (1967). *J. Appl. Phys.* **38**, 3729–3737.
Wirtz, G. P., and Fine, M. E. (1968). *J. Am. Ceram. Soc.* **51**, 402–406.
Wolfson, R. G., Kobes, W., and Fine, M. E. (1966). *J. Appl. Phys.* **37**, 704–712.
Wong, C. K. (1971). M.S. Thesis, S.U.N.Y. at Stony Brook, New York.
Woodilla, J. E., Jr., and Averbach, B. L. (1968). *Acta Metall.* **16**, 255–263.
Woods, K. N., and Fine, M. E. (1969). *J. Am. Ceram. Soc.* **52**, 186–188.
Wu, K. T., and Mendelson, K. S. (1973). *J. Chem. Phys.* **58**, 2929–2933.
Yang, W. M. C. (1971). Ph.D. Thesis, Northwestern Univ., Evanston, Illinois.
Yund, R. A. (1974). *In* "Geochemical Transport and Kinetics" (A. W. Hofmann, B. J. Giletti, H. S. Yoder, Jr., and R. A. Yund, eds.), p. 173–183. Academic Press, New York.
Yund, R. A., and McCallister, R. H. (1970). *Chem. Geol.* **6**, 5–30.
Yund, R. A., McLaren, A. C., and Hobbs, B. E. (1974). *Contrib. Mineral Petrol.* **48**, 45–55.
Zarzycki, J., and Naudin, F. (1967). *C. R. Acad. Sci. Ser. B* **265**, 1456–1459.
Zarzycki, J., and Naudin, F. (1968). *C. R. Acad. Sci. Ser. B* **266**, 145–148.
Zarzycki, J., and Naudin, F. (1969). *J. Non-Cryst. Solids* **1**, 215–234.

IV

Crystal Chemistry, Crystal Growth, and Phase Equilibria of Apatites*

DELLA M. ROY, LARRY E. DRAFALL,[†]
AND RUSTUM ROY

MATERIALS RESEARCH LABORATORY
AND
DEPARTMENT OF MATERIALS SCIENCE AND ENGINEERING
THE PENNSYLVANIA STATE UNIVERSITY,
UNIVERSITY PARK, PENNSYLVANIA

* Part of this work was carried out with support of the ACS-PRF, Grant PRF-7623-AC2,3.
† Present address: Lambda/Airtron, 200 East Hanover Avenue, Morris Plains, New Jersey 07950.

ISBN 0-12 0532-05-0

I. INTRODUCTION

Apatite, named from the Greek words "I deceive" remains to day a deceiver or teaser, for it describes a crystalline mineral family of great complexity. Important in at least three diverse fields, it represents the most abundant of the naturally occurring phosphatic minerals (McConnell, 1973), serving as the source for development of many phosphorus containing compounds and the production of phosphatic fertilizers. During the past three decades, the use of apatite crystalline phases for fluorescent lamp phosphors (Johnson, 1963) has been a major development, and other solid state applications have been explored, such as use for single-crystal laser host materials. The third and perhaps most important field for human kind are the biological apatites which in polycrystalline or semicrystalline form constitute the major inorganic matter of bones and teeth of vertebrates (Jensen and Rowles, 1957a,b; Posner, 1969; Brown, 1973).

The above define principally the family of phosphatic minerals and related synthetic compounds. In addition, apatite also is used to describe a structure type (Naray-Szabo, 1930), the specific crystal structure having the space group symmetry $P6_3/m$ possessed by the compound $Ca_{10}(PO_4)_6F_2$, fluorapatite, or slight distortions thereof. It appears that nearly half the periodic table may be substituted into the position of one of the four ions Ca^{2+}, P^{5+}, O^{2-}, or F^- (the calcium actually exists in two distinct subtypes of site), singly or in various combinations to yield modified compounds or entirely different isostructural chemical species.

The present chapter does not pretend to treat phase equilibrium relations for the whole apatite family, but centers around the phosphatic apatites, those including fluor-, chlor-, and hydroxyanion containing compounds, and is mostly restricted to oxides. For the reasons given above, however, it was felt that the complexity of this structural family should be elucidated first before presenting the phase data. Thus the chapter is divided into three segments: The background crystal chemistry of apatite phases (even this part cannot claim to be totally comprehensive); the phase equilibrium section; and the applications of relevant phase data in crystal growth studies, a selected application for which such phase data have been extensively utilized, with specific examples from each of the three areas.

An understanding of phase equilibrium relations is essential in all the above processes: fertilizer, phosphor, or various industrial single-crystal or polycrystalline material preparations, as well as processes taking place in biological systems. In addition, it has potential application in an area of growing importance, stable ceramic waste forms for radioactive waste isolation (Roy, 1975; McCarthy and Davidson, 1975).

II. CRYSTAL CHEMISTRY*

A. Introduction

The usage of the word apatite in this section refers to the "normal" apatites with $P6_3/m$ space group symmetry and the "apatite-type" structural compounds such as the analogues of vanadinite, mimetite, pyromorphite, and others. McConnell's (1973) book should be consulted for a detailed treatment of the natural apatites and many synthetic phases. Table I is

TABLE I

PRINCIPAL CATIONS AND ANIONS IN APATITE-TYPE MINERALS

		Anions					
Mineral	Cation	F	Cl	OH(O)	CO_3	OH, F	OH, F, Cl
Abukumalite	Ca, Y, etc.					Si, P, Al	
Beckelite	Ca, Ce, etc.			Si			
Britholite	Ca, Ce, etc.			Si, P			
Caracolite	Na, Pb		S				
Chlorapatite	Ca		P				
Dahllite	Ca			P	P		
Dehrnite	Ca, Na, K			P			
Ellestadite[a]	Ca						Si, S
Fermorite	Ca, Sr					P, As	
Fluorapatite	Ca	P					
Francolite	Ca	P			P		
Hedyphane	Ca, Pb		As				
Hydroxyapatite	Ca			P			
Lessingite	Ca, Ce, etc.			Si			
Lewistonite	Ca, K, Na			P			
Mimetite	Pb		As				
Pyromorphite	Pb		P				
Vanadinite	Pb		V				
Voelckerite	Ca			P			
Saamite	Ca, Sr, RE			P			
Svabite	Ca					As	
Strontiapatite	Ca, Sr, etc.					P	
Wilkeite	Ca						P, Si, S

[a] Includes hydroxylellestadite.

* The authors are concerned with the crystal chemistry of most synthetic apatite phases which have been reported in the literature and the present is not intended to be an exhaustive literature survey of all synthetic materials.

derived from this work which summarizes many of the principal cations and anions present in apatite-type minerals.

Apatites have been synthesized by many different methods as follows: chemical precipitation from aqueous solutions at low temperatures and pressures, solid state reactions, hydrothermal techniques (both low and high temperatures and pressures), and crystallization from a melt. A significant number of phases synthesized may not have a natural analog because during synthesis at relatively high growth rates which do not correspond to equilibrium conditions, the resulting crystals may consist of metastable phases. Furthermore, certain of the elements are insignificant in occurrence naturally and therefore not sufficiently concentrated to form major phases, or the different constituents necessary for a particular composition rarely occur together. The data and interpretations of the data available in the literature must be carefully evaluated. In many cases chemical analyses are needed to completely verify assumptions concerning the composition of synthesized phases. One cannot reliably assume that the product obtained contains exactly the same proportions as the initial ingredients without at least utilizing x-ray diffraction or polarized microscopy to determine the presence of an additional phase.

Some of the various families of substitutions which have been experimentally established in the apatites are summarized in Table II. The table shows the structural formulas in the form conveying the maximum structural in-

TABLE II

STRUCTURAL FORMULAS FOR APATITIES[a] $(A_4B_6)(MO_4)_6X_2$

A_{10}	M_6	O_{24}	X_2	Designation
A_4B_6		$O_{12}O_6O_6$		
Ca_{10}	P_6	O_{24}	$(OH)_2$	OHA
Ca_4Ln_6	Si_6	O_{24}	$(OH)_2$	
Sr_{10}	S_3Si_3	O_{24}	$(F)_2$	FAp
Ca_2Ln_8	Si_6	O_{24}	O_2	Oxyapatite
Sr_1Ca_9	P_6	O_{24}	$[\ ]_1$	X-site vacancies
Nd_4Ca_6	Ge_6	O_{24}	$[\ ]_2$	X-site vacancies
Sr_{10}	P_4Si_2	O_{24}	$[\ ]_2$	X-site vacancies
Na_2Ca_8	P_6	O_{24}	$[\ ]_2$	X-site vacancies
$[\ ]_2La_2\ Ca_4La_2$	P_6	O_{24}	$[\ ]_2$	Cation + anion vacancies
$[\ ]_2La_2\ La_2$	Ge_6	O_{24}	$(OH)_2$	Cation vacancies (?)

[a] A summary of structural formulas of various types of apatites, largely from extensive data on high temperature synthetic phases, showing that anion-vacant and cation + anion-vacant apatites are known but that the evidence for cation-vacant high temperature phases is absent. (Brackets indicate vacancies.)

formation, e.g., the known differences in the calcium sites, the oxygen positions, and the anion sites where in a particular substitution, certain site distinction orderings are especially significant.

B. Structural Description of Apatite

A structural description of apatite is helpful in comprehending the many ionic substitutions that occur. The crystal structure of fluorapatite is shown in Fig. 1. The fluorapatite has hexagonal symmetry with the space group $P6_3/m$ where the unit cell contains two formula units. The structure consists of a column of fluorine ions on the hexad (6_3) axis parallel to the crystallographic c axis. The ten calcium ions are distributed among two nonequivalent crystallographic sites. Six Ca_{II}(6h) ions are symmetrically located about the sixfold screw axes and the other four Ca_I(4f) ions are positioned along the threefold axes. Surrounding Ca_I are three oxygen triangles, two of which are located one above and one below Ca_I and the third larger triangle almost at the same height along c as Ca_I. Surrounding Ca_{II} are six oxygens and one fluorine ion.

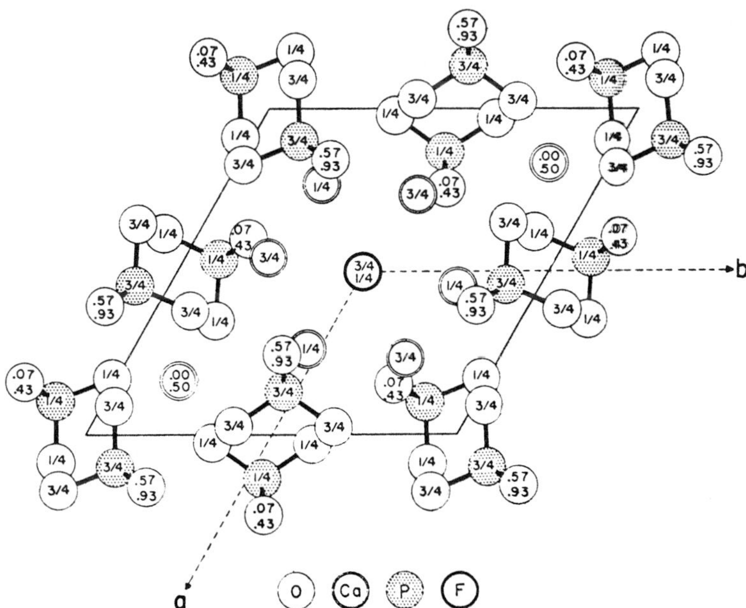

Fig. 1. View of FAp structure projected on ab plane. Numbers shown are z coordinates (heights above ab plane) of atoms. Ca (1) has z coordinates 0.00 and 0.50; Ca (2) has z coordinates $\frac{1}{4}$ and $\frac{3}{4}$ (reproduced from Mackie and Young (1973) by permission of the publisher).

Hydroxy- and chlorapatites crystallize with the same space group ($P6_3/m$) but probably are stabilized by impurities or vacancies (Young, 1974). When vacancies and impurities are sufficiently reduced so that 85–90% or more of the stoichiometric amount of Cl is present, the long range ordering in the Cl columns occurs and the space group becomes $P2_1/b$; the mirror planes changes into glide planes and double the unit cell dimension along one hexagonal axis *b* (Young, 1974). Hydroxy- and chlorapatites with monoclinic cells are a distorted version of the hexagonal cells. Fluorine occurs at the center of and coplanar with a triangle of Ca_{II} ions. In monoclinic hydroxyapatite the oxygens are displaced along the *z* direction from the plane of triangles formed by the Ca_{II} ions. In chlorapatite the larger Cl ions are located at approximate *z* levels of zero and $\frac{1}{2}$, between planes of Ca_{II} triangles; and not at approximate *z* values of $\frac{1}{4}$ and $\frac{3}{4}$ as are OH^- and F^-.

C. Ionic Substitutions

The basic formula $A_{10}(MO_4)_6X_2$ of apatite allows a wide variety of substitution, including both cation and anion vacancies, to yield a vast number of phases with essentially identical structures. Ions "fitting" in the "A" site include divalent Ca, Mg, Ni, Mn, Fe, Sn, Cd, Rb, Ba, Eu; trivalent rare earths, Sc, Y, Tl, Bi, V; monovalent Na, K, Tl, Rb, Cs; and tetravalent Ce. The "M" site may be occupied by P^{5+}, Si^{4+}, S^{6+}, As^{5+}, Ge^{4+}, Cr^{5+}, Mn^{6+}, V^{5+}, Al^{3+}, B^{3+}, and OH (actually substituting for MO_4 as H_4O_4). In some instances CO_3 substitutes partially for the MO_4 group. The "X" site is occupied by F, Cl, OH, O and to a lesser extent Br, I, CO_3, S, N, and possibly H_2O. The number of ions accommodated may extend to half the periodic table if one includes those ions that may be incorporated in small amounts. Substitutions vary from a complete replacement of one kind of ion by another, or by a combination of others. The requirement for overall charge balance is sometimes satisfied by a mixed substitution at other sites when ions of valence different from that of the replaced ion are substituted.

Cockbain (1968) proposed a classification of members of the apatite group $A_{10}(MO_4)_6X_2$ based on the ratios of the mean sizes of the "A" ions to those of the "M" ions.* For over 80 synthetic phases the ratios ranged between 1.89 and 4.43 with discontinuities between the ratios of 2.50 and 2.60 and probably another between 3.25 and 3.35 which provided a structural basis for the division into the following three groups: vanadinite–svabites with A:M ratio less than 2.5, the apatite–mimetites (2.60–3.25), and the pyromorphites (greater than 3.25). He used the known ionic radii in eightfold coordination (Ahrens, 1952) and calculated an "effective" radius for ions which normally did not exhibit this coordination. Cockbain (1968)

* Cockbain used the terminology "X" for "M" and "Z" for "X".

however, did not consider the effect of the size of the "X" ion and that the coordination numbers of the A ions vary depending upon the position in the structure, i.e., whether the ion is located on the Ca_I site (4f) or the Ca_{II} site (6h).

This discussion on the crystal chemistry of apatite is subdivided according to substitutions occurring in various structural positions of a calcium phosphate hydroxyapatite.

1. Cation Substitutions

In general the ions which substitute for Ca in the A position have valences from 1 to 3 and coordination numbers of VII for the Ca_{II}(6h) site and IX for the Ca_I(4f) position. Table III lists the ionic radii for different coordination numbers for most of the substituents in the A position of apatite. Many

TABLE III

Cation Radii (Å) of Possible Apatite Substituents in the A_{10} Site of $A_{10}(MO_4)_6X_2$

	Coordination number[a]				
	Ahrens (1952)	Shannon and Prewitt (1969)			
	VI	VI	VII	VIII	IX
Divalent ions					
Ba	1.34	1.36	1.39		1.47
Pb	1.20	1.18		1.29	1.33
Eu		1.17		1.25	
Sn	0.93			1.22	
Sr	1.12	1.16	1.21	1.25	
Ca	0.99	1.00	1.07	1.12	1.18
Cd	0.97	0.95	1.00	1.07	
Mn		0.83		0.93	
Zn	0.74	0.745			0.90E[b]
Co	0.72	0.735			
Cu	0.72	0.73			
Mg	0.66	0.720		0.89	
Ni	0.69	0.690			
Monovalent ions					
K	1.33	1.38	1.46?	1.51?	1.55?[b]
Na	0.97	1.02	1.13?	1.16	1.32?[b]

[a] The ionic radii for VIII coordination are listed where the value in nine coordination is not known. The newer values of Shannon and Prewitt (1969) in VI coordination are listed for comparison with Ahrens (1952) who determined only one number.

[b] E = interpolated; ? = doubtful.

of the earlier papers used ionic radii reported by Ahrens (1952) and these are listed along with the newer values of Shannon and Prewitt (1969).

a. Complete Substitution for Calcium

Fairly complete substitutions for calcium have been shown for Sr, Ba, Pb, Cd, Eu^{2+}, and Sn^{2+} for the halide and/or hydrophosphates. Based on data from Mooney and Aia (1960) on calcium, strontium, and barium apatites, Simpson (1968) noted that substitution of an increasingly large cation in the calcium site results in about an equal increase in *a* and *c* lattice parameters of the apatite. Balz (1961) synthesized $Sn_{10}(PO_4)_6O\square$ and confirmed the work of Bauer (1959) who formed $Ba_{10}(PO_4)_6O\square$. Fowler (1974) prepared stoichiometric Ca, Sr, and Ba hydroxyapatites by direct thermal combination of the appropriate reactants in the solid state. Akhavan-Niaki (1961) synthesized Sr chlor- and fluorapatite and Ba chlor- and fluorapatite. The Cd apatites, $Cd_{10}(MO_4)_6X_2$ where M = P, As, or V and X = Cl or Br were reported by Engel (1968). Merker and Wondratschek (1959) prepared $Pb_{10}(PO_4)_6X_2$ where X = F or Cl. Wondratschek (1963) synthesized $Pb_{10}(PO_4)_6O\square$ and reported the *c* axis was doubled in such a way as to indicate the oxygens and vacancies were ordered in the regularly alternating fashion described. Europium hydroxyapatite $[Eu_{10}(PO_4)_6(OH)_2]$ was reported (Mayer and Makogon-Loewry, 1969) but the unit cell parameters indicate Eu^{2+} has a radius slightly smaller than Sr or that a small number of Na ions may have occupied sites. The preparation utilized solutions of $EuSO_4$ and Na_3PO_4 and both Na and SO_4 ions can enter the apatite structure without great difficulty (McConnell, 1973). Klement and Haselbeck (1965) and Kreidler and Hummel (1970) reported complete substitution of Mn for Ca, but only in the chlorapatites. Grisafe and Hummel (1970b) and Mayer *et al.* (1974) report complete calcium substitution by two or three different cations.

b. Partial Calcium Substitutions

The calcium ion sites in the apatite structure are not all equivalent which is reflected in some of the substitutions for the four calcium ions in the Ca_I site and/or six calciums in the Ca_{II} site. The reader should consult Naray-Szabo (1930), Mehmel (1930), Posner *et al.* (1958), Mackie *et al.* (1972), and McConnell (1938, 1952, 1973) for a detailed study of a portion of the literature on the apatite structure.

A series of oxyapatites including $Ca_8Y_2(PO_4)_6O_2$ and $Ca_8La_2(PO_4)_6O_2$ have been described by Ito (1968) where all ion sites are inferentially filled with oxygen and the charge is balanced by substituting two Y or La ions for two Ca ions. A calcium phosphate oxyapatite however, has not been synthesized successfully, and that reported by Bredig *et al.* (1932) was actually shown to be hydroxyapatite by Trömel (1932). Alkali ions can replace cations as in $Pb_9Na(PO_4)_6OH$ and $Pb_8Na_2(PO_4)_6\square$ (Merker and Wondratschek, 1957) where vacant X ion sites are present.

Mayer *et al.* (1974) reported compositions with the general formula $Ln_xM_{10-2x}Na_x(PO_4)_6F_2$ where Ln = trivalent La, Pr, Nd, Sm, Eu, Dy, Er, and Lu and M = Ca, Sr, and Ba. All the Ca and Sr compounds had the hexagonal apatite structure ($P6_3/m$) while the Ba compounds possessed two different but apatite-related space groups, $P\overline{6}$ and $P3$. They also reported that the substitution of the rare earth and sodium ions in the Ca or Sr sites proceeds in a disordered way and the ions are statistically distributed among the cation positions. In the Ba system an ordered substitution occurs and is most probably a consequence of the relatively large size difference between Ba and the Ln and Na ions (Mayer *et al.*, 1974).

Precipitates from room temperature aqueous solutions (Trautz *et al.*, 1964) showed that magnesium substituted for calcium ions at most, 0.4 mole %. The magnesium ion content in apatites was believed by McClellan and Lehr (1969) to be related to the amount of phosphate groups replaced by carbonate ions. The larger the number of phosphates replaced, the larger were the amounts of magnesium ions present. Duff (1972) concluded that it was unlikely that magnesium analogs of fluorapatite and hydroxyapatite

TABLE IV

CATION RADII (Å) OF POSSIBLE APATITE SUBSTITUENTS IN THE (MO_4) SITE OF $A_{10}(MO_4)_6X_2$

Coordination number (Shannon and Prewitt, 1969)			
Ions	IV	Ions	IV
U^{6+} [a]	0.48	As^{5+}	0.335
W^{6+}	0.42	Mn^{6+} [b]	0.27
Ge^{4+}	0.40	Si^{4+}	0.26
Al^{3+}	0.39	P^{5+}	0.17
V^{5+}	0.355	S^{6+}	0.12
Cr^{5+}	0.350	B^{3+}	0.12

[a] McConnell (1973) stated that it is not unreasonable to assume U^{6+} substitutes for P and an extremely small number of UO_4^{2-} might be tetrahedrally bonded in the phosphate site.

[b] Pentavalent Mn was first prepared and recognized in a compound by Lux (1946). Apatites containing Mn^{5+} have been reported by Grisafe and Hummel (1970).

TABLE V

UNIT CELL DIMENSIONS OF SYNTHETIC APATITES[a]

Formula	This study				Former studies		
	a(Å)	c(Å)	V(Å^3)	c/a	a(Å)	c(Å)	References
$Mn_{10}(PO_4)_6Cl_2$	9.54	6.20	489	0.650	(9.30)	(6.20)	Klement & Haselbeck (1965)
$Cd_{10}(PO_4)_6F_2$	9.30	6.63	497	0.713	9.32	6.61	Engel (1970)
$Cd_{10}(PO_4)_6Cl_2$	9.67	6.50	526	0.672	9.67	6.46	Engel (1970)
$Cd_{10}(AsO_4)_6Cl_2$	10.03	6.53	569	0.651	10.04	6.51	Engel (1970)
$Cd_{10}(VO_4)_6Cl_2$	10.13	6.55	582	0.646	10.11	6.52	Engel (1970)
$Ca_{10}(PO_4)_6F_2$	9.36	6.88	522	0.735	9.352	6.871	Wallaeys (1952)
$Ca_{10}(AsO_4)_6F_2$	9.63	6.99	561	0.726			
$Ca_{10}(VO_4)_6F_2$	9.68	7.01	569	0.724	6.67	7.01	Aia & Lublin (1966)
$Ca_{10}(PO_4)_6Cl_2$	9.63	6.78	544	0.704	9.610	6.763	Wallaeys (1952)
$Ca_{10}(AsO_4)_6Cl_2$	10.04	6.83	596	0.680			
$Ca_{10}(VO_4)_6Cl_2$	10.18	6.77	608	0.665	10.16	6.79	Aia & Lublin (1966)
$Ca_{10}(SiO_4)_3(SO_4)_3F_2$	9.43	6.93	533	0.735	(9.54)	(6.99)	Klement & Dihn (1941)
$Ca_4Na_6(SO_4)_6F_2$	9.39	6.89	525	0.734	(9.49)	(6.87)	Klement & Dihn (1941)
$Ca_9Mg(PO_4)_6F_2$	9.355	6.867	520	0.734			
$Ca_9Ni(PO_4)_6F_2$	9.364	6.870	522	0.734			

$Sr_{10}(PO_4)_6F_2$	9.71	7.28	594	0.750	9.719	7.276	Akhavan-Niaki (1961)
$Sr_{10}(AsO_4)_6F_2$	9.99	7.40	640	0.741			
$Sr_{10}(VO_4)_6F_2$	10.01	7.43	645	0.742			
$Sr_{10}(PO_4)_6Cl_2$	9.87	7.19	607	0.728	9.874	7.184	Akhavan-Niaki (1961)
$Sr_{10}(AsO_4)_6Cl_2$	10.18	7.28	653	0.715	(10.12)	(7.50)	Klement & Harth (1961)
$Sr_{10}(VO_4)_6Cl_2$	10.21	7.30	659	0.715			
$Pb_{10}(PO_4)_6F_2$	9.75	7.30	603	0.749	9.76	7.29	Merker & Wondratschek (1959)
$Pb_{10}(AsO_4)_6F_2$	10.08	7.42	653	0.736	10.07	7.42	Merker & Wondratschek (1959)
$Pb_{10}(VO_4)_6F_2$	10.11	7.34	650	0.726	10.10	7.34	Merker & Wondratschek (1959)
$Pb_{10}(PO_4)_6Cl_2$	9.99	7.34	634	0.735	9.97	7.32	Merker & Wondratschek (1959)
$Pb_{10}(AsO_4)_6Cl_2$	10.24	7.44	676	0.726	10.25	7.46	Merker & Wondratschek (1959)
$Pb_{10}(VO_4)_6Cl_2$	10.32	7.36	679	0.713	10.32	7.33	Merker & Wondratschek (1959)
$Pb_{10}(SiO_4)_3(SO_4)_3F_2$	9.88	7.41	627	7.750			
$Pb_4Na_6(SO_4)_6F_2$	9.63	7.11	571	0.738			
$Ba_{10}(PO_4)_6F_2$	10.16	7.69	687	0.757	(10.220)	(7.665)	Akhavan-Niaki (1961)
$Ba_{10}(AsO_4)_6F_2$	10.41	7.83	735	0.752			
$Ba_{10}(VO_4)_6F_2$	10.44	7.86	742	0.753			
$Ba_{10}(PO_4)_6Cl_2$	10.26	7.65	697	0.746	10.275	7.647	Akhavan-Niaki (1961)
$Ba_{10}(AsO_4)_6Cl_2$	10.54	7.73	744	0.733	(10.44)	(7.59)	Klement & Harth (1961)
$Ba_{10}(VO_4)_6Cl_2$	10.55	7.75	747	0.734			

[a] Kreidler and Hummel (1970).

could be prepared by precipitation from aqueous solution. High temperature hydrothermal synthesis (850°C, 1 kbar pressure) of magnesium substituted hydroxyapatite by Drafall and Roy (unpublished) showed at least 20 mole % Mg [e.g., $(Ca_8Mg_2)(PO_4)_6(OH)_2$]. Chlorapatites have been synthesized from molten NaCl where partial substituents for Ca were limited to less than 5 Cu, 5 Zn, 4 Mg, 4 Co, 4 Ni, atoms out of the 10 Ca sites (Klement and Haselbeck, 1965). The fluorapatites $Ca_9Mg(PO_4)_6F_2$ and $Ca_9Ni(PO_4)_6F_2$ were reported by Kreidler and Hummel (1970). Substantial $Mg^{2+} + 2F^-$ substitution in OH-apatite was found by Dinger (1974) and Dinger and Roy (1977).

c. Substitutions for Phosphorus

The ionic radii of cations in four coordination which substituted for phosphorus are listed in Table IV. Many of the elements are significantly larger than phosphorus which on the basis of the pyromorphite series (Baker, 1966) will cause an increase in both *a* and *c* lattice parameters. Cation substitution for the phosphorus can be complete by one element or a mixture of two or more elements. Coupled replacement for P and Ca frequently occur together. Table V (Kreidler and Hummel, 1970) lists many of the cation substitutions of the chlor- and fluorapatites of the general formula $A_{10}(MO_4)_6X_2$ where A = Ca, Sr, Ba, Pb, Cd, Mn; M = P, Si, S, As, V; and X = Cl, F. They also plotted structure-field maps (Figs. 2 and 3) which indicate the ionic radii of the various elements and the type of cation substitutions found in hexagonal apatites. Bauer (1968) also reported synthesis of Ca and Sn apatites with AsO_4 and VO_4 groups replacing PO_4. Grisafe and Hummel (1970a,b) reported the complete replacement of phosphorus by pentavalent arsenic, vanadium and chromium in the calcium, strontium and barium fluor-, and chlorapatites. Distortion of the structure relative to the normal hexagonal apatites was noted in calcium fluorvanadate, -arsenate, and -chromate. They reported that chromium and manganese could not be incorporated into lead apatites, and that complete replacement of phosphorus by manganese occurred only in barium apatites.

Machatschki (1939) first noted the isostructural relation of minerals with the general formula $(Na, Ca, Th, Ln)_{10}(Si, P)_6O_{24}(OH)_2$ to calcium phosphatic apatite which was later confirmed by many others. Crystals of $Ca_4Y_6Si_6O_{24}(OH)_2$ (abukumalite) and La- and Nd-britholite $(Na, Ca, La, Nd)_{10}Si_6O_{24}(OH)_2$ were reported by Trömel and Eitel (1957) which were grown by solid state reaction or flame fusion in air at high temperatures. The hydroxyl ions were supposedly introduced from atmospheric moisture. Cockbain and Smith (1967) also synthesized silicate hydroxyapatites as shown in Table VI where the hydroxyl compounds were formed by solid–vapor reaction with steam passing through the reaction tube at high temperatures and atmospheric pressure. They stated that for the composition

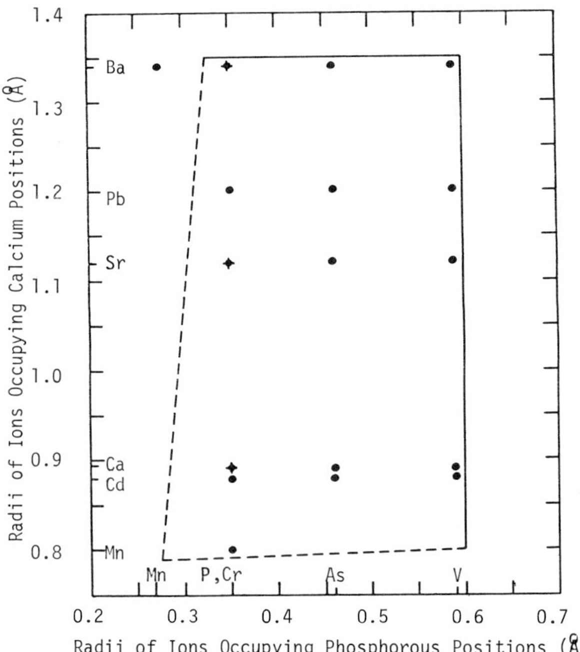

Fig. 2. The structure field of chlorapatite modified from Kreidler and Hummel (1970) with Ahrens radii.

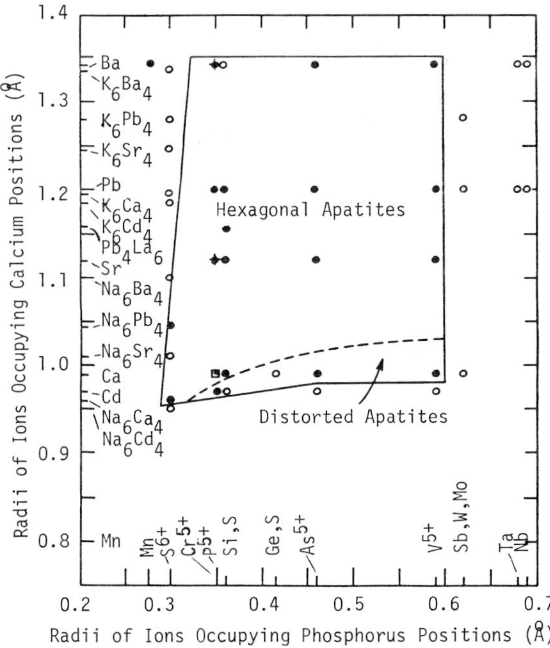

Fig. 3. The structure field of fluorapatite modified from Kreidler and Hummel (1970) with Ahrens radii. (Various combinations of three cations occupying the calcium position or combinations of two cations in the phosphorus site are not included. ●, apatite; ○, no apatite; +, chromium compounds; □, distorted chromium apatites.)

TABLE VI

CELL DIMENSIONS OF SYNTHETIC APATITE-TYPE COMPOUNDS AND RELATED MINERAL MEMBERS OF THE GROUP[a,b]

	Name and formula	*a*	*c*	*c*/*a*	Firing temperature	Atmosphere
Abukumalite	$Ca_4Yt_6(SiO_4)_6(F,OH)_2$[c]	9.31 Å	6.58 Å	0.707	—	
	$Ca_4Yn_6(SiO_4)_6(OH)_2$[d]	9.33	6.78	0.727	1350°C	$N_2 + H_2O + H_2$
	$Ca_{10}La_5Ce_5(SiO_4)_{12}(OH)_2$	9.51	7.00	0.736	1350	Air
	$Ca_6Nd_4(SiO_4)_6$	9.52	7.00	0.735	1350	Air
	$Ca_4Nd_6(SiO_4)_6(OH)_2$	9.52	7.00	0.735	1350	Air
	$Ca_6La_2Ce_2(SiO_4)_6$	9.52	7.00	0.735	1300	Air
	$Ca_4La_3Ce_3(SiO_4)_6(OH)_2$	9.52	7.01	0.736	1350	Air
	$Ca_4Nd_6(SiO_4)_6(OH)_2$	9.52	7.02	0.737	1400[e]	$N_2 + H_2O + H_2$
Ellestadite	$(Ca,C)_4\{(S,Si,P,C)O_4\}_6(Cl,F,O,OH)_2$[f]	9.53	6.91	0.725	—	
	$Sr_4Dd_6(SiO_4)_6(OH)_2$[g]	9.56	7.06	0.738	1370–1400	$N_2 + H_2O + H_2$
Britholite	$(Ca,Ln)_{1.94}(Si,P)_{1.13}(O,OH,F)_{5.12}$[h]	9.61	7.03	0.731	—	—
	$Ca_5La_5(SiO_4)_6OH$	9.62	7.12	0.740	1300	$N_2 + H_2O + H_2$
	$Ca_5La_5(SiO_4)_6OH$	9.63	7.12	0.739	1300	Air
	$Ca_6La_4(SiO_4)_6$	9.63	7.12	0.739	1300	Air
	$Ca_4La_6(SiO_4)_6(OH)_2$	9.63	7.124	0.740	1300	Air
	$Ca_4La_6(SiO_4)_6(OH)_2$	9.63	7.124	0.740	1350	$N_2 + H_2O + H_2$
	$Ca_4La_6(SiO_4)_6F_2$	9.64	7.13	0.740	1375	Cylinder air

Beckelite	$(Ca,Ln)_{1.96}(Si,Al,Zr)_{0.89}(O,OH)_{4.37}$[h]	9.66	7.07	0.732	—	—
Lessingite	$(Ca,Ln)_{2.06}(Si,Al,P)_{1.14}(O,OH,F)_{5.37}$[h]	9.67	7.08	0.732	—	—
	$Sr_4La_6(SiO_4)_6(OH)_2$(?)	9.68	7.21	0.745	1300	$N_2 + H_2O + H_2$
	$Sr_4La_6(SiO_4)_6(OH)_2$(?)	9.69	7.23	0.746	1350	Air
	$Tl_2La_8(SiO_4)_6(OH)_2$(?)	9.69	7.18	0.741	1350	Air
	$La_8(SiO_4)_6$	9.69	7.17	0.740	1300	Air
	$Ba_2La_8(SiO_4)_6O_2$	9.76	7.30	0.748	1300	Air
	$Ca_6La_4(GeO_4)_6$	9.84	7.21	0.733	1350	Air
	$Ca_4La_6(GeO_4)_6(OH)_2$	9.85	7.20	0.731	1350	Air
	$Sr_4La_6(GeO_4)_6(OH)_2$	9.88	7.31	0.740	1350	Air
	$La_8(GeO_4)_6$	9.89	7.26	0.734	1300	Air
	$Ba_2La_8(GeO_4)_6O_2$	9.93	7.37	0.742	1300	Air
	$Ba_3La_7(GeO_4)_6O_{1.5}$	9.99	7.39	0.740	1300	Air

[a] All mixtures for synthetic compounds were held at the stated temperatures for 1 hour unless otherwise specified.

[b] Cockbain and Smith, 1967.

[c] Trömel and Eitel (1957).

[d] "Yttrium oxide" used for this preparation contained only 56% Y_2O_3 the balance being Gd_2O_3 (14%), Dy_2O_3 (13%), and other rare earths (17%).

[e] Fired for $1\frac{1}{2}$ hours.

[f] McConnell (1938).

[g] Dd_2O_3 = "Didymium oxide" mixed rare earths.

[h] Gay, 1957.

$Ca_4La_6(SiO_4)_6(OH)_2$, the 4 calcium and 6 lanthanum ions had a random distribution among the (4f) and (6h) sites and did not show ordering of the 4 Ca ions on the (4f) sites and 6 La on (6h) sites or ordering of 4 La on (4f) sites and 4 Ca and 2 La randomly distributed over the (6h) sites. Ito (1968) hydrothermally synthesized the silicate hydroxyapatites shown in Table VII which are of the type $A_4B_6(SiO_4)_6(OH)_2$ where A = Ca, Sr, Ba, Pb, Mn and Cd; B = La and Y. He also synthesized the closely related compounds $(Na, Li)_2(La, Y)_8SiO_6O_{24}(OH)_2$ where monovalent ions such as Na and Li can replace Ca but K can only partially replace Ca. The introduction of the rare earth, yttrium or lanthanum, completes the valence balancing. Ito (1968) demonstrated the presence of OH in the silicate hydroxyapatites and the absence in oxyapatite by infrared absorption spectrometry, water determination by two different methods, and differential thermal analysis. He concluded that only hydrothermal synthesis can produce hydroxylated end members.

TABLE VII

UNIT CELL DIMENSIONS OF THE SILICATE APATITES (HYDROXYL) SYNTHESIZED HYDROTHERMALLY AT 650°C AND 2 kbar[a,b]

Formula	a(Å)	c(Å)	V(Å^3)	Formula	a(Å)	c(Å)	V(Å^3)
$Mn_4La_6Si_6O_{24}(OH)_2$	9.66	7.05	569	$Sr_4Gd_6Si_6O_{24}(OH)_2$	9.53	7.00	550
$Mn_4Nd_6Si_6O_{24}(OH)_2$	9.50	6.90	539	$Sr_4Dy_6Si_6O_{24}(OH)_2$	9.43	6.92	532
				$Sr_4Y_6Si_6O_{24}(OH)_2$	9.52	6.91	543
$Mn_4Sm_6Si_6O_{24}(OH)_2$	9.43	6.82	525	$Sr_4Ho_6Si_6O_{24}(OH)_2$	9.42	6.91	531
$Mn_4Gd_6Si_6O_{24}(OH)_2$	9.38	6.78	521	$Sr_4Er_6Si_6O_{24}(OH)_2$	9.42	6.83	525
$Mn_4Dy_6Si_6O_{24}(OH)_2$	9.33	6.68	503	$Sr_4Lu_6Si_6O_{24}(OH)_2$	9.50	6.79	532
$Mn_4Y_6Si_6O_{24}(OH)_2$	9.31	6.65	499	$Pb_4La_6Si_6O_{24}(OH)_2$	9.80	7.26	604
$Mn_4Er_6Si_6O_{24}(OH)_2$	9.28	6.63	494	$Pb_4Ce_6Si_6O_{24}(OH)_2$	9.77	7.19	594
				$Pb_4Nd_6Si_6O_{24}(OH)_2$	9.76	7.13	588
$Ca_4La_6Si_6O_{24}(OH)_2$	9.66	7.12	575	$Pb_4Sm_6Si_6O_{24}(OH)_2$	9.74	7.05	579
$Ca_4Ce_6Si_6O_{24}(OH)_2$	9.61	7.09	567	$Pb_4Gd_6Si_6O_{24}(OH)_2$	9.72	6.99	572
$Ca_4Nd_6Si_6O_{24}(OH)_2$	9.56	7.00	554	$Pb_4Dy_6Si_6O_{24}(OH)_2$	9.70	6.90	562
$Ca_4Sm_6Si_6O_{24}(OH)_2$	9.50	6.94	542	$Pb_4Y_6Si_6O_{24}(OH)_2$	9.68	6.86	567
$Ca_4Gd_6Si_6O_{24}(OH)_2$	9.43	6.89	531	$Pb_4Er_6Si_6O_{24}(OH)_2$	9.68	6.84	551
$Ca_4Dy_6Si_6O_{24}(OH)_2$	9.40	6.83	523	$Pb_4Lu_6Si_6O_{24}(OH)_2$	9.64	6.75	543
$Ca_4Y_6Si_6O_{24}(OH)_2$	9.40	6.81	519				
$Ca_4Er_6Si_6O_{24}(OH)_2$	9.38	6.78	517	$Ba_4Nd_6Si_6O_{24}(OH)_2$	9.76	7.18	593
$Ca_4Lu_6Si_6O_{24}(OH)_2$	9.35	6.74	510	$Ba_4Sm_6Si_6O_{24}(OH)_2$	9.73	7.10	582
$Sr_4La_6Si_6O_{24}(OH)_2$	9.71	7.23	591	$Ba_4Gd_6Si_6O_{24}(OH)_2$	9.72	7.03	575
$Sr_4Ce_6Si_6O_{24}(OH)_2$	9.67	7.14	578	$Ba_4Dy_6Si_6O_{24}(OH)_2$	9.66	6.95	561
$Sr_4Nd_6Si_6O_{24}(OH)_2$	9.62	7.10	569	$Na_2Y_8Si_6O_{24}(OH)_2$	9.34	6.78	512
$Sr_4Sm_6Si_6O_{24}(OH)_2$	9.60	7.05	563	$Na_2La_8Si_6O_{24}(OH)_2$	9.74	7.17	589

[a] 450°C for Pb analogs.
[b] Ito, 1968.

The silicate oxyapatites synthesized by Ito (1968) are listed in Table VIII and are of the general formula $A_2B_8(SiO_4)_6O_2$ where A = Ca, Ba, Sr, Pb, Mn, Mg and B = Ln, La, Y. Small groups also exist of the type (Na, Li)$Y_9(SiO_4)_6O_2$, $NaLa_9(SiO_4)_6O_2$, and (Na, Tb, La)$_{10}(SiO_4)_6O_2$. Ito (1968) states that the synthetic compounds of silicate hydroxyapatite prepared by Cockbain and Smith (1967) are probably anhydrous and the compounds of Trömel and Eitel (1957) are also probably oxyapatites because silicate compounds (as distinct from phosphates) do not retain water when synthesized in air either by flame fusion or solid state reaction at high temperature. Ito (1968) shows a distinction between his hydroxyl- and oxyapatites by unit cell volumes versus rare earth ionic radii, chemical analysis, and infrared spectroscopy, whereas, Cockbain and Smith (1967) did not observe differences in their compounds. Felsche (1972) synthesized by solid state reaction a complete series of apatitelike compounds: $RE_{9.33}\square_{0.67}(SiO_4)_6O_2$, $LiRE_9(SiO_4)_6O_2$, and $NaRE_9(SiO_4)_6O_2$ where the RE is La → Lu.

TABLE VIII

UNIT CELL DIMENSIONS OF THE SILICATE APATITES (OXY) SYNTHESIZED AT 1200°C IN AIR[a,b]

Formula	a(Å)	c(Å)	V(Å^3)	Formula	a(Å)	c(Å)	V(Å^3)
$Ba_2La_8Si_6O_{26}$	9.77	7.32	605	$Pb_2Gd_8Si_6O_{26}$	9.54	6.95	548
$Ba_2Nd_8Si_6O_{26}$	9.66	7.16	579	$Pb_2Dy_8Si_6O_{26}$	9.47	6.87	534
$Ba_2Sm_8Si_6O_{26}$	9.62	7.06	566	$Pb_2Y_8Si_6O_{26}$	9.42	6.80	522
$Sr_2La_8Si_6O_{26}$	9.69	7.22	586				
$Sr_2Nd_8Si_6O_{26}$	9.57	7.09	562	$Mn_2La_8Si_6O_{26}$	9.63	7.08	568
$Sr_2Sm_8Si_6O_{26}$	9.51	7.02	550	$Mn_2Nd_8Si_6O_{26}$	9.47	6.91	537
$Sr_2Gd_8Si_6O_{26}$	9.47	6.97	541	$Mn_2Sm_8Si_6O_{26}$	9.42	6.85	526
$Sr_2Dy_8Si_6O_{26}$	9.42	6.90	530	$Mn_2Gd_8Si_6O_{26}$	9.34	6.79	518
$Sr_2Y_8Si_6O_{26}$	9.38	6.86	523	$Mn_2Dy_8Si_6O_{26}$	9.33	6.71	506
$Sr_2Er_8Si_6O_{26}$	9.36	6.81	517	$Mn_2Y_8Si_6O_{26}$	9.32	6.69	503
$Ca_2La_8Si_6O_{26}$	9.63	7.12	571	$Mn_2Er_8Si_6O_{26}$	9.30	6.65	498
$Ca_2Nd_8Si_6O_{26}$	9.52	7.00	549	$Mg_2La_8Si_6O_{26}$	9.59	7.05	561
$Ca_2Sm_8Si_6O_{26}$	9.44	6.93	535	$Mg_2Nd_8Si_6O_{26}$	9.45	6.86	530
$Ca_2Gd_8Si_6O_{26}$	9.39	6.87	525	$Mg_2Sm_8Si_6O_{26}$	9.38	6.80	518
$Ca_2Dy_8Si_6O_{26}$	9.37	6.81	518	$Mg_2Gd_8Si_6O_{26}$	9.33	6.75	509
$Ca_2Y_8Si_6O_{26}$	9.34	6.77	511	$Mg_2Dy_8Si_6O_{26}$	9.31	6.69	504
$Ca_2Er_8Si_6O_{26}$	9.33	6.75	509	$Mg_2Y_8Si_6O_{26}$	9.31	6.64	498
$Ca_2Lu_8Si_6O_{26}$	9.28	6.68	498	$Mg_2Er_8Si_6O_{26}$	9.28	6.58	491
$Cd_2La_8Si_6O_{26}$	9.64	7.09	571	$NaLa_9Si_6O_{26}$	9.69	7.18	583
$Pb_2La_8Si_6O_{26}$	9.71	7.20	588	$NaY_9Si_6O_{26}$	9.33	6.75	509
$Pb_2Nd_8Si_6O_{26}$	9.65	7.12	574	$LiY_9Si_6O_{26}$	9.34	6.72	508
$Pb_2Sm_8Si_6O_{26}$	9.58	7.06	561	(Na,Th,La)$_{10}Si_6O_{26}$	9.66	7.13	576

[a] 900°C for Pb analogs and 1050°C for Mg and alkali analogs.
[b] Ito, 1968.

Another example of complete substitution for phosphorus is the germanium ion, for which the lattice parameters of the germanate compounds increased due to the larger ionic radius of Ge. Strunz *et al.* (1960) reported a germanium apatite $Ca_4Ce_6(GeO_4)_6Cl_2$ and Cockbain and Smith (1967) synthesized other germanates listed in Table VI.

The sulfate apatite $Na_6Ca_4(SO_4)_6F_2$ (Klement and Dihn, 1941) involves substitution of the divalent sulfate ion for the trivalent phosphate group coupled with calcium replacement by sodium to maintain electrostatic neutrality.

*d. Partial Phosphorus Substitution**

Partial substitution for phosphorus has been reported for the compositions† given in the accompanying tabulation. Simpson (1968) reported that four OH ions substituted as $H_4O_4^{4-}$ for PO_4^{3-} should cause enlargements in unit cell dimensions but the extent of this replacement is not certain.

Composition	Reference
$Pb_{10}(GeO_4)_2\square_2$	Cockbain (1968)
$Pb_{10}(SiO_4)(GeO_4)(PO_4)_4\square_2$	
$Ca_9Ba(CO_4)(PO_4)_5OH_2$	
$Sr_{10}(PO_4)_4(SiO_4)_2\square_2$	Schwarz (1968)
$Pb_6K_4(PO_4)_4(SO_4)_2\square_2$	Schwarz (1967a)
$Sr_4La_5(PO_4)(SiO_4)_5\square_2$	Schwarz (1967c)
$Pb_2^{4+}Pb_8^{2+}(PO_4)_4(SiO_4)_2(O)_2$	Ito (1968)
$Ca_2Na_2La_6(SiO_4)_4(PO_4)_2(OH)_2$	
$Ca_4Na_2La_4(SiO_4)_2(PO_4)_4(OH)_2$	
$Ba_{10}(PO_4)_5(BO_4)\square_2$	Bauer (1959)
$Ca_3RE_7(SiO_4)_5(PO_4)O_2$[a]	McCarthy and Davidson (1975)

[a] RE = rare earth.

Aluminum has been found to substitute for both Ca and P (Fisher and McConnell, 1969). The composition with respect to the Al content was intermediate between $Ca_6Al_4(PO_4)_4(AlO_4)_2F_2$ and $Ca_8Al_2(PO_4)_5(AlO_4)F_2$. They reported no distinguishable difference from common apatites regarding the powder diffraction pattern which suggested no change of space group or significant change of *a* or *c* periodicities. The aluminum is present as two coordination numbers IV, VI as shown by infrared absorption spectra, chemical analysis, and *x*-ray powder diffraction where more than twice as much aluminum substitutes for calcium as for phosphorus. Manganese also

* Partial substitutions for phosphorus which form solid solutions are discussed in Subsection 2.

† See Subsection 3 for a discussion of carbonate substitution.

can substitute for either Ca or P but not simultaneously as does aluminum. It seems to occur as either Mn^{2+} or MnO_4^{3-}.

Recently Mehta and Simpson (1975) reported the composition $Ca_6Na_4(PO_3F)_6O_2$ which has three times the fluoride of fluorapatite and has a different structural site for the halogen. They stated the ratio of sodium to calcium requires that the six Ca ions occupy the Ca_{II} (6h) sites and the four Na ions occupy the Ca_I (4f) site. In comparison with $Ca_4Na_6(SO_4)_6F_2$ the location of the Ca and Na is reversed but the sodium–fluorine coordination and the calcium–oxygen bond is preserved by this site switching. They concluded that the sodium–fluoride coordination is of primary importance in these apatites and that a cation can completely replace the calcium in the Ca_I or the Ca_{II} site, but a limited replacement of both sites seemed unlikely. [They could not synthesize $Na_6Ca_4(PO_3F)_6F_2$ or $Na_4Ca_6(SO_4)_6O_2$.]

Urusov and Khudolozhkin (1974) did an energy analysis of cation ordering in the apatite structure and presented experimental data on the distribution of cations isomorphously substituting for one another in two nonequivalent structural positions. They reported that cations larger than Ca (i.e., Sr, Ba) predominantly occupy a smaller Ca_{II} (seven-cornered polyhedron) position and cations smaller than Ca (i.e., Be, Mg, ferric Fe, Mn) occupy a larger Ca_I position. They concluded that cations more electronegative than Ca occupy the more covalent Ca_I position and cations more electropositive than Ca occupy the more ionic Ca_{II} position. Engel *et al.* (1975) also reported in the composition $Pb_6Ca_4(PO_4)_6(OH)_2$ an ordering of Pb on the Ca_{II} (6h) site and of Ca on the Ca_I (4f) positions due to the tendency of lead to form partially covalent bonds. They also observed the same cation ordering in two other apatite crystalline solutions $(Pb, Ca)_{10}(PO_4)_4(SiO_4)_2$ and $(Pb, Ca)_8Na_2(PO_4)_6$.

Blasse (1975) showed that the principle of local charge compensation could predict the site occupation for the two different large cation positions (6h and 4f). Experimental data agreed satisfactorily with his predictions.

There has been no evidence of ferric atoms tetrahedrally bonded in apatites but it is possible that an extremely small number of UO_4^{2-} groups might occur since significant amounts of AlO_4^{5-} have been reported (Fisher and McConnell, 1969).

e. Complete Substitution for Phosphorus by Two Cations

The compound $Sr_{10}(SO_4)_3(SiO_4)_3F_2$ (Schwarz, 1967b) demonstrates complete substitution of SO_4 and SiO_4 for PO_4. Klement (1939) reported the charge compensated substitution $Ca_{10}(SiO_4)_3(SO_4)_3F_2$ (also Sr and Pb analogs where Si^{4+} and S^{6+} replace 2 P^{5+}). Other example of this type are described in the next section where complete or partial crystalline solutions are formed between two end members.

2. Apatite Crystalline Solutions

Crystalline solutions in apatites are frequently encountered where the possibility and type depends on the conditions of formation or preparation, the thermal history after formation, and the end members of the series. Ito (1968) reported numerous crystalline solution series in the oxy- and hydroxy-apatites and constructed plots of unit cell dimensions versus composition for the following:

$Ca_2Y_8(SiO_4)_6O_2$–$Ca_8Y_2(PO_4)_6O_2$
$Ca_2La_8(SiO_4)_6O_2$–$Ca_8La_2(PO_4)_6O_2$
$Ca_2Y_8(SiO_4)_6O_2$–$Y_{10}(SiO_4)_4(BO_4)_2O_2$
$Mg_2Y_8(SiO_4)_6O_2$–$Y_{10}(SiO_4)_4(BO_4)_2O_2$
$Pb_3^{4+}Pb_5^{2+}Y_2(SiO_4)_6O_2$–$Pb_2^{2+}Y_8(SiO_4)_6O_2$
$Ca_{10}(PO_4)_6(OH)_2$–$Ca_4Y_6(SiO_4)_6(OH)_2$
(under hydrothermal conditions)

He also proposes a complete series between $Sr_{10}(PO_4)_6(OH)_2$ and $Sr_4La_6(SiO_4)_6(OH)_2$ because of the similar ionic sizes of Sr and La. A continuous crystalline solution series exists between the end members $Ca_{10}(PO_4)_6F_2$ and $Ca_{10}(VO_4)_6F_2$, which is a "distorted" apatite (Kreidler and Hummel, 1970). (See Fig. 2 which shows a structure-field map for the normal hexagonal apatites and the distorted apatites.) They reported that the distortion of pure calcium fluorovanadate was not observed until the composition $Ca_{10}(PO_4)(VO_4)_5F_2$ was reached and that the absence of a miscibility gap in the system shows the structural similarity of the vanadate compound to $Ca_{10}(PO_4)_6F_2$. The similar chloride compounds $Pb_{10}(PO_4)_6Cl_2$, $Pb_{10}(AsO_4)_6Cl_2$ and $Pb_{10}(VO_4)_6Cl_2$ also show a continuous crystalline solution series (Baker, 1966). Grisafe and Hummel (1970) reported continuous crystalline solutions for the following compounds:

$A_{10}(PO_4, VO_4)_6F_2$	A = Sr, Ba, Pb
$A_{10}(PO_4, CrO_4)_6F_2$	A = Sr, Ba, Pb
$B_{10}(PO_4, MnO_4)_6X_2$	X = F, Cl
$B_{10}(VO_4, MnO_4)_6X_2$	X = F, Cl

Other complete series of crystalline solutions were reported by Engel *et al.* (1975) in the system $Pb_{10}(PO_4)_6(OH)_2$–$Ca_{10}(PO_4)_6(OH)_2$.

Whenever the two end members in a crystalline solution series differ in the type or amount of vacancies, a difference in the occupation of sites occurs in the structure. Wondratschek (1963) reported in crystalline solutions a continuous increase in X-ion vacancies as $Pb_{10}(PO_4)_6F_2$ is gradually replaced by $Pb_8Na_2(PO_4)_6\square_2$.

3. ANION SUBSTITUTIONS

Anions substituting in the X position in the apatite formula $A_{10}(MO_4)_6X_2$ are F, Cl, OH, O and to a smaller extent Br, I, CO_3, S, N, and possibly H_2O. In the structure of apatite the anions Cl, F, or OH occur in columns on the hexad (6_3) axis extending along the c axis of the structure. The symmetry is determined by the position of these anions in the columns. In hydroxy- and fluorapatite the OH or F ions are located at (0, 0, $\frac{1}{4}$) and (0, 0, $\frac{3}{4}$). Chlorapatite originally was presumed to be hexagonal with the chloride ion in the special position $z = \frac{1}{2}$ (Hendricks *et al.*, 1932). However, Young and Elliot (1966) reported that nearly stoichiometric synthetic chlorapatite exhibits the monoclinic space group $P2_1/b$. The Cl atoms are ordered above and below $z = \frac{1}{2}$ (displacement of $\delta = 0.056$) on the pseudohexagonal axis and also are ordered between columns. Hounslow and Chao (1970) reported a monoclinic chlorapatite mineral of nearly stoichiometric chlorine content and observed the dependence of the monoclinic character on this Cl content. [Mackie *et al.* (1972) present a detailed study of the monoclinic structure of synthetic chlorapatite.] When 15% or more of the Cl ions are vacant, the monoclinic form is lost and the hexagonal form results in both synthetic and mineral compounds (Prener, 1967, 1971). The crystal structure determination of cadmium chlorapatite by Sudarsanan *et al.* (1973) indicated hexagonal symmetry ($P6_3/m$) where the chlorine has the same coordinate in the structure as fluorine (0, 0, $\frac{1}{4}$) in calcium fluorapatite, in contrast to the position of chlorine in calcium chlorapatite. Young (1974) stated that the hexagonal crystal structure is probably stabilized in hydroxyapatite by impurities or vacancies. Monoclinic hydroxyapatite has been reported by Elliott (1971) who showed that the hydroxyl ions are ordered similarly to chloride ions in chlorapatite (Elliott *et al.*, 1973). Prener (1967) reported that stoichiometric monoclinic chlorapatite undergoes a reversible transition to hexagonal symmetry at temperatures near 200°C and this transition temperature is lowered by heating the crystals in vacuum which causes nonstoichiometry by loss of $CaCl_2$. A similar reversible transition occurs in monoclinic hydrothermally grown hydroxyapatite single crystals which transform to hexagonal symmetry at 211°C (Van Rees *et al.*, 1973).

The apatites involving substitution of F, Cl, O, or OH were referenced earlier in this section on crystal chemistry. Other anion substitutions were reported by Wondratschek (1963) and Brenner *et al.* (1970), who synthesized compounds with the formula $Pb_{10}(MO_4)_6X_2$ where M = P, As, or V and X = F, Cl, Br, I (except that no iodine phosphate is formed). X-ray single crystal photographs showed hexagonal symmetry and the arsenate showed several kinds of superstructure reflections.

Wallaeys (1952) reported bromapatite $Ca_{10}(PO_4)_6Br_2$ and later Dykes (1974) grew stoichiometric single crystals with the usual apatite hexagonal symmetry ($P6_3/m$) and not the monoclinic ($P2_1/b$) symmetry of stoichiometric chlorapatite. Trombe and Montel (1975) reported the composition $Ca_{10}(PO_4)_6S\square$ which could be synthesized by the following methods: reaction of CaS with oxynitrides $Ln_{10}Si_6O_{24}N_2$ (Ln = La, Nd, Sm, Gd). They also reported the substitution of the lanthanide by Mn or Li in the oxyfluoride apatite series: $Ln_6Mn_4Si_6O_{24}F_2$ (Ln = La, Nd, Sm, Gd, Yb, and Y) and $Ln_8Li_2Si_6O_{24}F_2$ (Ln = La, Nd, Sm, Gd, Ho, Er, and Y).

At first the substitution of the carbonate ion in apatites was not universally accepted. Admixtures of calcium phosphate and calcium carbonate were thought to account for detection of carbonate in materials. Silverman *et al.* (1952) concluded that carbonate ions were present in the apatite structure and Ames (1959) stated that up to 10 wt % carbonate ions are present in the apatite lattice. Wallaeyes (1954) reported a partial replacement of OH ions by carbonate ions by flowing dry CO_2 over hydroxyapatite at 900°C. Simpson (1964, 1965), Le Geros *et al.* (1968) and others studied carbonate apatites precipitated from aqueous solutions and concluded that the carbonate ions substituted for phosphate groups. In papers by Elliot (1969) and Montel (1971), both authors stated that carbonate ions did actually substitute for both hydroxyl and phosphate groups in the apatite structure. Hydrothermal synthesis of various carbonate containing calcium hydroxyapatites were reported by Roy *et al.* (1974) where on the basis of IR, *x*-ray, and chemical investigations, carbonate ions can replace both the OH group and PO_4 tetrahedron earlier in only one of these positions or also combined. The position of CO_3 in the structure is best indicated by IR spectra and Montel (1971) and Nadal *et al.* (1970) have shown that the carbonate bands occupy characteristic sites when CO_3 replaces OH (type A) at 884, 1465, and 1534 cm^{-1} and PO_4 (type B) at 864, 1430, and 1455 cm^{-1}. Substitution at the B site produces contraction of the *a* lattice parameter while CO_3 substitution at the A site produces expansion of the *a* parameter. [McConnell (1973) presents a rather comprehensive discussion of the structural aspects of carbonate substitution in apatites.]

A mixing of the anions in the X site of apatites has been reported in several minerals where F, Cl, and OH occurred together (Cooray, 1970; Harada *et al.*, 1971; McConnell, 1937). A continuous range of compositions between two anions end members such as F and Cl though, is unexpected due to the large differences in ionic radii and differences in coordinate positions. Sudarsanan and Young (1968) showed that there is significant interaction between Cl and F where both ions are slightly displaced from the positions they occupy in pure chlorapatite and fluorapatite, respectively. They also showed by *x*-ray diffraction that O^{2-} can be situated at F^- sites. Piper *et al.* (1965) also found evidence for F^- vacancies and various combinations of F^- vacancies with O^{2-}.

III. PHASE EQUILIBRIA

A. Introduction

While an abundance of work is available on the crystal chemistry and synthesis of apatite structure phases, systematic phase equilibrium studies are relatively sparse. Experimental phase equilibrium studies involving apatite phases have been broadly related to four major fields: phosphor and solid state materials synthesis, fertilizers and slags, mineralogical and geological problems, and applications to biological hard tissue growth processes. The most well defined studies are those of high temperature (and high temperature and pressure) reactions in which the approach to attainment of equilibrium is relatively rapid, where the difficulties encountered are those common to other high temperature/pressure studies, plus the problems of defining stoichiometry in the phases formed. This latter problem has been discussed in some detail in the crystal chemistry section. The low temperature atmospheric pressure solution equilibria having especial relevance to biological processes have been studied extensively, and because the reactions are sluggish and much of the resultant information is subject to questions concerning the degree of attainment of equilibrium conditions, experimental studies have been supplemented by a number of investigations of calculated equilibria. For more extensive treatment of the low temperature processes than given here the reader is referred to review articles such as Brown (1973) and Posner (1969).

B. High Temperature Equilibria in F^-, OH^-, and Cl^- Apatite Systems

1. $CaO–P_2O_5–Ca(OH)_2$ AND $CaO–P_2O_5–H_2O$

a. $CaO–P_2O_5$ *Binary*

The system $CaO–P_2O_5$ has been extensively studied because of its applications to fertilizer and iron and steel slags, as well as its importance for phosphors, mineralogical occurrences, and biological hard tissue. Six authenticated binary compounds were recognized by the time the combined phase diagrams of Hill *et al.* (1944) for the system $P_2O_5–2CaO \cdot P_2O_5$ and Trömel (1943) for the system $2CaO \cdot P_2O_5–CaO$ were published. The latter portion is the more relevant with respect to apatites, since defect apatites are formed from $3CaO \cdot P_2O_5$; and the adjacent compounds on either side are important as they may coexist in equilibrium. A number of the calcium phosphates in this system undergo polymorphism, and a revised phase diagram was published by Trömel *et al.* (1948), involving new polymorphic data for the $\alpha–\beta$ $3CaO \cdot P_2O_5$ inversion.

Nurse *et al.* (1958) reported the formation of a new high temperature polymorph of $3CaO \cdot P_2O_5$ stable above 1430°C, the $\bar{\alpha}$ form, in addition to the previously described α and β forms. In addition Welch and Gutt (1961) noted that crystalline solubility of excess P_2O_5 in $3CaO \cdot P_2O_5$ was present in all three modifications; and they showed a maximum melting temperature (1777°C) on the liquidus curve, with excess P_2O_5, at 53 CaO and 47 P_2O_5 (wt %), higher than the melting temperature of the $3CaO \cdot P_2O_5$ composition (1756°C) (Fig. 4).

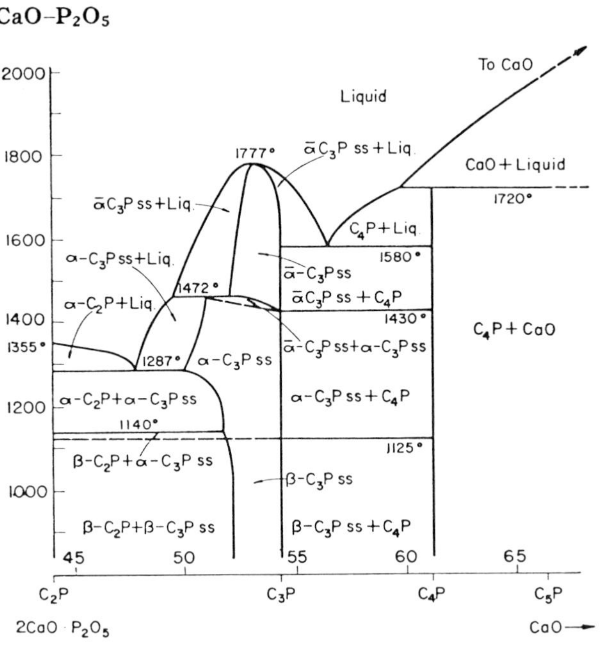

Fig. 4. System CaO–2CaO · P_2O_5. C = CaO, P = P_2O_5 (from Welch and Gutt, 1961).

Welch and Gutt recognized some uncertainty in the effect of solid solution on the temperature of the α–β $3CaO \cdot P_2O_5$ inversion; also, there is some ambiguity in the representation of this inversion in the phase diagram of Welch and Gutt's study. Trömel and Fix (1961) presented a revised diagram for the $2CaO \cdot P_2O_5$–CaO system, also including the presence of $\bar{\alpha}$-$3CaO \cdot P_2O_5$ but a slightly higher temperature congruent melting behavior for the compound.

Kreidler and Hummel (1967) finally presented a tentative revised phase diagram in which they suggested the presence of a seventh compound $7CaO \cdot 5P_2O_5$ that melts incongruently at 970° to β-$2CaO \cdot P_2O_5$ + liquid. Their diagram is given in Fig. 5.

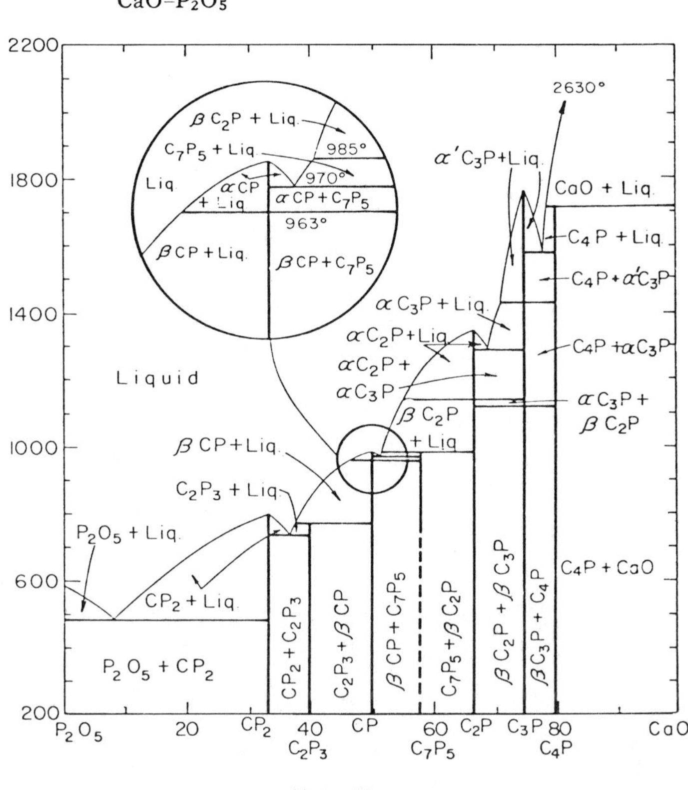

Fig. 5. Phase diagram for the system $CaO–P_2O_5$ (from Kreidler and Hummel, 1967).

b. $CaO–P_2O_5–Ca(OH)_2$

The join $Ca_3(PO_4)_2–Ca(OH)_2$ was investigated by Biggar (1966) as part of an experimental study of systems of petrologic importance. His diagram is given in Fig. 6, which represents equilibria at 1000 bars pressure. Hydroxyapatite forms a eutectic with $Ca(OH)_2$ at 765°C and 96% $Ca(OH)_2$, 4% $Ca_3(PO_4)_2$. In the system with approximately 50% H_2O added the equilibria (including a vapor phase) have a lowering of melting to 735°C, at 1000 bar pressure. Hydroxyapatite crystallized from different compositions along the join was reported to have variable stoichiometry.

c. $CaO–P_2O_5–H_2O$

(*1*) *High temperature equilibria.* The system $Ca(OH)_2–Ca_3(PO_4)_2–H_2O$ was investigated by Biggar (1966) at 795°C and 1000 bar. It is represented as the isothermal section given in Fig. 7, drawn with distorted scale. A two-phase

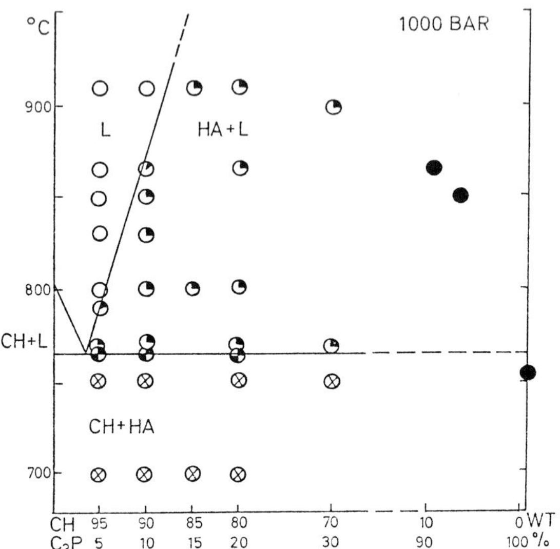

Fig. 6. The join CH–C_3P at 1000 bar; ○, L; ◔, HA + L, ◒, CH + L, ●, charge reacted with capsule; ⊗ subsolidus region. Phase boundaries between C_3P and HA are not shown in view of the reaction with capsule (from Biggar, 1966).

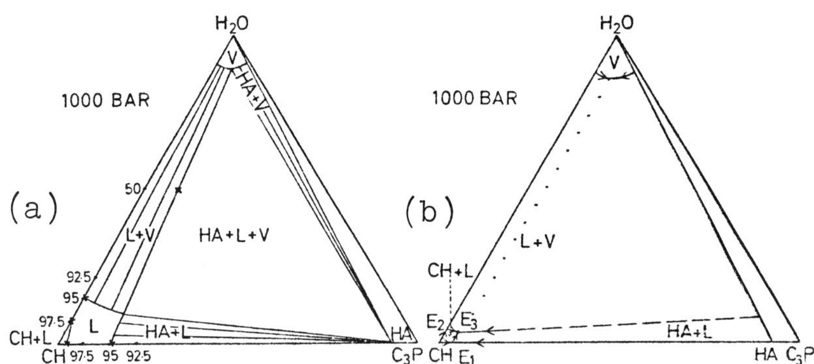

Fig. 7. (a) Isothermal section (795°C) for the region CH–HA–H_2O at 1000 bar with distorted scale. The apatites are not stoichiometric HA. (b) Liquidus (true scale) and vaporous (distorted scale) field boundaries for the join CH–HA–H_2O at 1000 bar, projected on to CH–C_3P–H_2O. The invariant temperatures and liquid compositions are E_1, 765°C, CH 96% C_3P 4%; E_2, 785°C, CH 95% H_2O 5%; E_3, 735°C, CH 92%, C_3P 4%, H_2O 4% (from Biggar, 1966).

region of hydroxyapatite + liquid (variable composition) and a three-phase region hydroxyapatite–liquid–vapor are significant features. The positions of the tie lines in two-phase areas were not determined exactly, but represent approximate compositions. The same system, showing polythermal equilibria with estimated liquidus and vaporus field boundaries at 1000 bar, is given

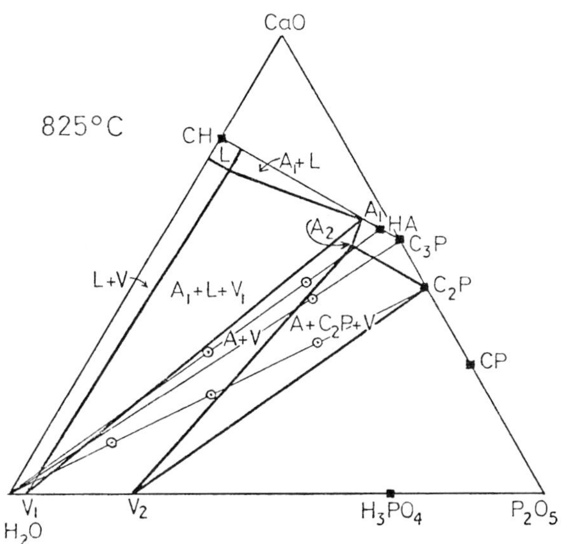

Fig. 8. Estimated isothermal section at 825°C for parts of the join $CaO-P_2O_5-H_2O$ at 1000 bars. Data for the region CH–HA–H_2O are taken from Figs. 7a and 7b and other experimental points are shown as circles. The compositions A_1 and A_2 are estimates but V_1 and V_2 are determined (subjected to the qualifying statements in the text) by the experimental points (from Biggar, 1966).

in Fig. 8; the important features are the positions of the binary eutectic E_2 in the system $Ca(OH)_2-H_2O$ at 785°C and $Ca(OH)_2$ 95%, H_2O 5%; and the ternary invariant point E_3 at 735°C, $Ca(OH)_2$ 92%, $Ca_3(PO_4)_2$ 4%, H_2O 4%. As described in Section IV, the phase equilibria as determined by Biggar were confirmed and utilized for growth of single crystals of hydroxyapatite from $Ca(OH)_2$-rich melts (Roy, 1971; Eysel and Roy, 1973).

A schematic representation of a proposed isothermal section of the ternary system $CaO-P_2O_5-H_2O$ at 825°C and 1000 bar (Biggar, 1966) is given in Fig. 8, based largely on data obtained from the join $Ca_3(PO_4)_2-H_2O$, $Ca_2P_2O_7-H_2O$, and $Ca(OH)_2$–OHAp (hydroxyapatite)–H_2O joins. The points A_1 and A_2 are schematic only, suggested by changes in XRD spacings of the apatite. For compositions higher in CaO than represented by the join A_1-V_1, a single vapor composition V_1 (approximately, since all vapor contains some CaO) exists in equilibrium with liquid and OHAp of composition A_1 (approximately); but to the phosphorus-rich side of this join more P_2O_5-rich vapors are in equilibrium.

(2) *Intermediate temperature equilibria.* Skinner (1973) reported equilibrium studies of the same system in the range 300–600°C. The equilibria, represented as isothermal sections from 600 to 300°C are shown in Fig. 9. Skinner

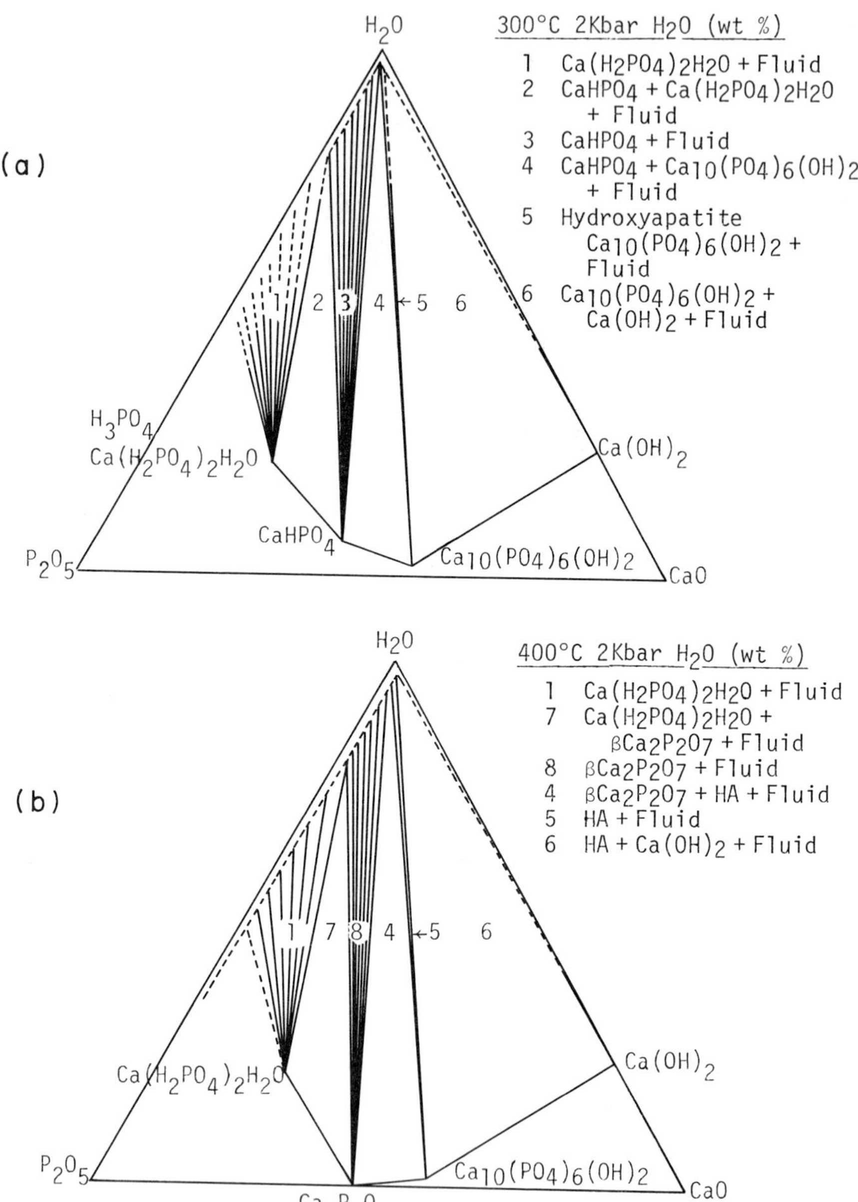

Fig. 9. Experimentally determined equilibrium phase diagrams CaO–P_2O_5–H_2O. (a) 300°C, (b) 400°C, (c) 500°C, (d) 600°C (from Skinner, 1973).

(c)
H2O
500°C 2Kbar H2O (wt %)
1 Ca(H2PO4)2H2O + Fluid
7 Ca(H2PO4)2H2O + βCa2P2O7 + Fluid
8 βCa2P2O7 + Fluid
4 βCa2P2O7 + HA + Fluid
5 HA + Fluid
6 HA + Ca(OH)2 + Fluid
1 7 8 4 ←5 6
H2PO4
Ca(H2PO4)2H2O
Ca(OH)2
P2O5
βCa2P2O7
Ca10(PO4)6(OH)2
HYDROXYAPATITE
CaO

(d)
H2O
600°C 2Kbar H2O + Fluid
1 Ca(H2PO4)2H2O + Fluid
7 Ca(H2PO4)2H2O + βCa2P2O7 + Fluid
8 βCa2P2O7 + Fluid
4 βCa2P2O7 + HA + Fluid
5 HA + Fluid
6 HA + Ca(OH)2 + Fluid
1 7 8 4 ←5 6
Ca(H2PO4)2H2O
Ca(OH)2
Ca10(PO4)6(OH)2
P2O5
Ca2P2O7
CaO

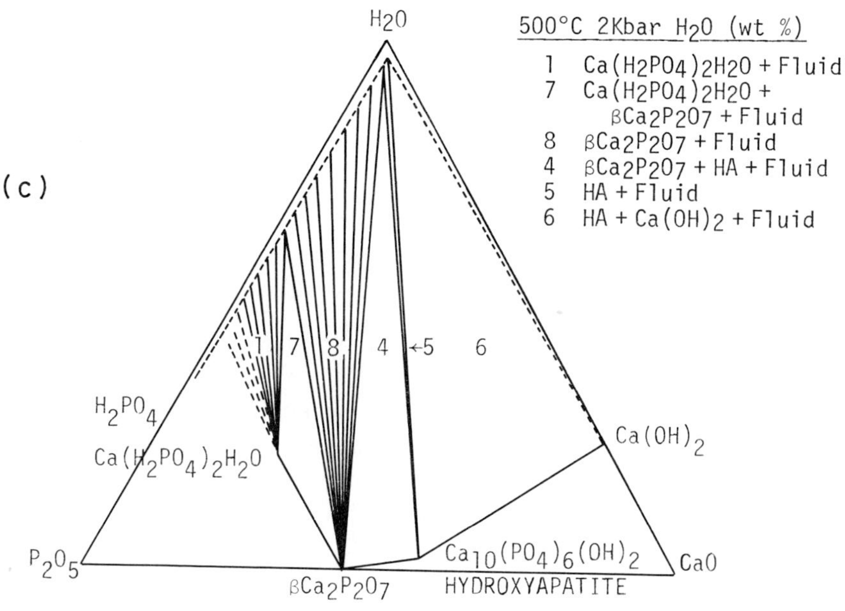

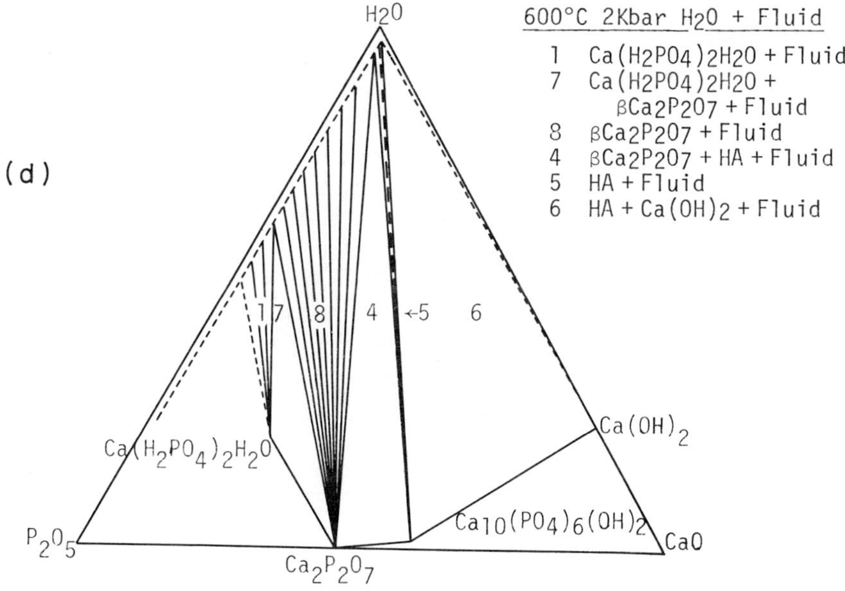

Fig. 9 (*cont.*)

TABLE IX

UNIT CELL PARAMETERS OF APATITES FROM THREE PHASE FIELDS

Apatite + monetite + fluid					Apatite + portlandite + fluid				
	a	Å	*c*	Å		*a*	Å	*c*	Å
300-121	9.417	(1)	6.879	(2)	CAP 70	9.421	(1)	6.882	(1)
300-133	9.419	(1)	6.881	(1)					
Apatite + $Ca_2P_2O_7$ + fluid									
400-CAP 12	9.4193	(1)	6.881	(1)	400-101	9.420	(1)	6.880	(1)
500-CAP 11	9.424	(1)	6.887	(1)	500-131	9.423	(1)	6.882	(1)
600-122	9.423	(1)	6.882	(1)	600-118	9.4129	(8)	6.879	(1)
600-CAP 45	9.4225	(5)	6.8825	(7)	600-CAP 44	9.4151	(7)	6.8819	(9)

[a] Skinner (1973).

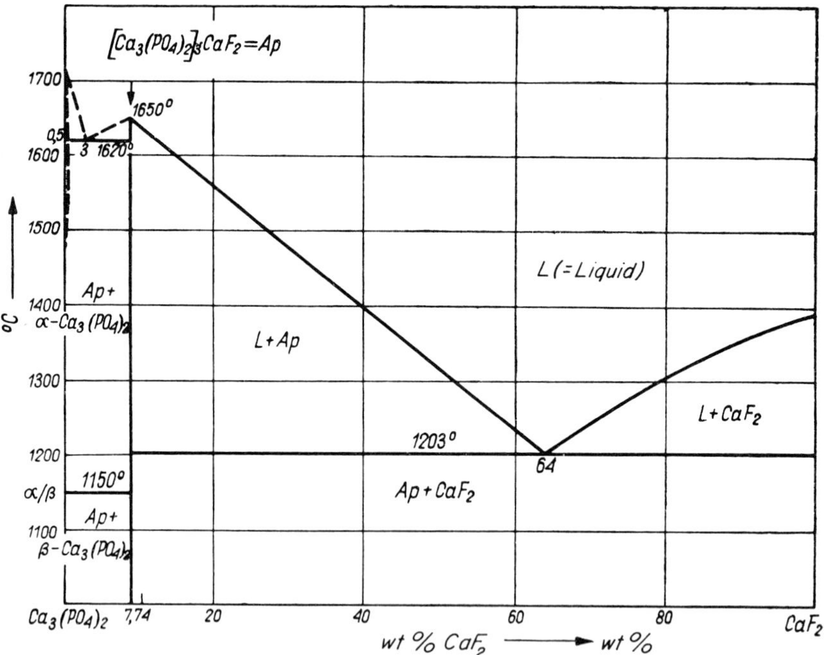

Fig. 10. Phase diagram of the system $Ca_3(PO_4)_2$–CaF_2 (from Berak, 1961a).

used measurement of pH of quenched fluid to estimate compositions of vapor near the H_2O–P_2O_5 edge. Hydroxyapatite crystallized from different phase fields had slightly different cell parameters. Table IX shows the range of variation found. Skinner represented the HA composition as a single point, in contrast to the deliberately exaggerated representation of Biggar. However, it is possible that the compositional variation is too small for representation in a diagram this size.

The equilibria at 600° differ from the 825° equilibria in that no melt phase is present at the lower temperature; and HA is also shown to exist in equilibrium with a much narrower compositional range fluid. Little change is observed in the lower temperature isothermal sections 500 and 400°C, but at 300°C monetite ($CaHPO_4$) is stable and coexists with assemblages of $Ca(H_2PO_4)_2H_2O$ + solution, and OHAp + solution; $Ca_2P_2O_7$ is no longer stable in the presence of high water vapor pressures.

2. CaO–P_2O_5–CaF_2

a. $Ca_3(PO_4)_2$–CaF_2 *Section*

The original phase diagram for the section $Ca_3(PO_4)_2$–CaF_2 of the ternary system CaO–P_2O_5–CaF_2 was published by Nacken (1912) and has been changed relatively little to date. Berak (1961a) extended the range of compositions studied and represented the complete section $Ca_3(PO_4)_2$–CaF_2, reproduced here as Fig. 10. The important features are congruent melting of $Ca_{10}(PO_4)_6F_2$ at ~1650°C, a eutectic with $Ca_3(PO_4)_2$ at 1620°C, and a second with CaF_2 at 1203°C. Although according to Berak (1976) further refinements in the equilibria are being investigated, the determination was sufficiently precise to provide the necessary information for flux growth of fluorapatite (Prener, 1967), crystals of which were only slightly deficient in fluorine from theoretical. The problem concerning the possible stable existence of spodiosite, $Ca_2(PO_4)F$, analogous to the naturally occurring mineral, remains unsettled, but it appears unlikely that it is stable at liquidus temperatures.

b. CaO–$Ca_3(PO_4)_2$–CaF_2 *System*

This ternary system was investigated by Berak (1961b) and is reproduced here as the phase diagrams Figs. 11 and 12. The section $Ca_4P_2O_9$–Apatite (Ap) is quasi-binary, with a eutectic at 58% $Ca_4P_2O_9$, 42% Ap at 1580°C. The section CaO–Ap is also quasi-binary with a eutectic at 13% CaO, 87% Ap at 1585°C. The temperatures of the three ternary eutectics are shown in Fig. 11. No crystalline solubility was reported between apatite and the other phases.

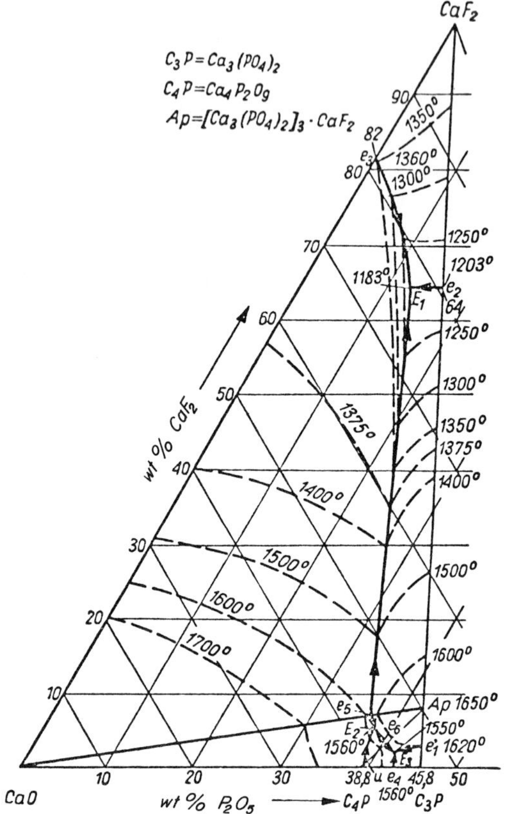

Fig. 11. Phase diagram of system CaO–$Ca_3(PO_4)_2$–CaF_2 (from Berak, 1961b).

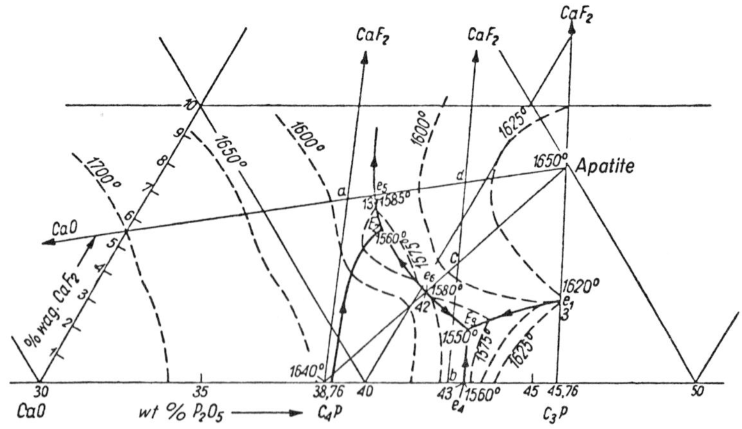

Fig. 12. Phase diagram of system CaO–$Ca_3(PO_4)_2$–Apatite (from Berak, 1961b).

3. $CaO–P_2O_5–CaCl_2$

Nacken (1912) determined the phase diagram for the section

$$Ca_{10}(PO_4)_6Cl_2–CaCl_2$$

of the system $CaO–P_2O_5–CaCl_2$, and his diagram for the system is reproduced as Fig. 13. According to Prener (1967) chlorapatite crystallized from melts of its own composition is highly deficient in Cl, while at lower temperatures near 1040°C stoichiometric chlorapatite can be crystallized. Nacken's original study shows the uncertainty in melting behavior of the apatite compound, by use of dashed lines at the liquidus. An additional feature is the presence of an intermediate compound that melts incongruently, chlorspodiosite $Ca_2(PO_4)Cl$ (Kingsley *et al.*, 1965).

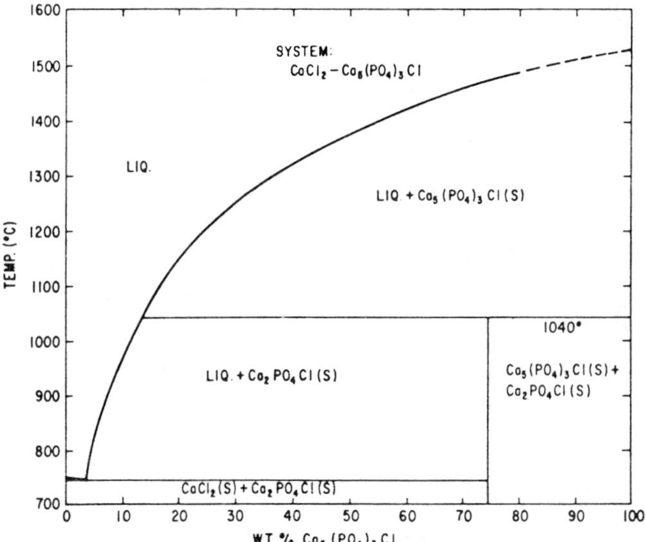

Fig. 13. Crystallization of chlorapatite and $CaCl_2 \cdot Ca_3P_2O_8$ from calcium chloride melts (from Nacken, 1912).

4. $CaO–P_2O_5–H_2O–CO_2–CaF_2$

a. $CaO–P_2O_5–CO_2$

No systematic research has been reported, to the authors' knowledge on detailed phase equilibria in this system in the absence of H_2O. The authors synthesized carbonate apatites almost free from H_2O by use of high CO_2

gas pressures at elevated temperatures, but reaction was sluggish and the synthesis was very difficult in the absence of a preexisting OH-apatite structure or of H_2O otherwise present in the system to exert a partial pressure of water vapor.

b. CaO–P_2O_5–H_2O–CO_2

Extensive literature exists on the synthesis of mixed carbonate-hydroxy-apatites from solutions at room temperature and slightly above. The high temperature data are mainly those of Biggar (1969), who presented a number of isobaric (1000 bar pressure) T–X sections in the range 600–900°C for the join $Ca(OH)_2$–$CaCO_3$–$Ca_3(PO_4)_2$, and the system with H_2O added. The position of the ternary eutectic $Ca(OH)_2 + CaCO_3 +$ apatite $=$ liquid was about 654°, 53% $Ca(OH)_2$, 1% $Ca_3(PO_4)_2$, and 46% $CaCO_3$. Crystalline solubility of CO_2 in the apatite was believed to be present to a small extent, but was difficult to measure. The join $Ca(OH)_2$–$CaCO_3$–$Ca_3(PO_4)_2$ in projection is given in Fig. 14, and isothermal sections in the system $Ca(OH)_2$–$Ca_3(PO_4)_2$–$CaCO_3$–H_2O–CO_2 are shown in Fig. 15.

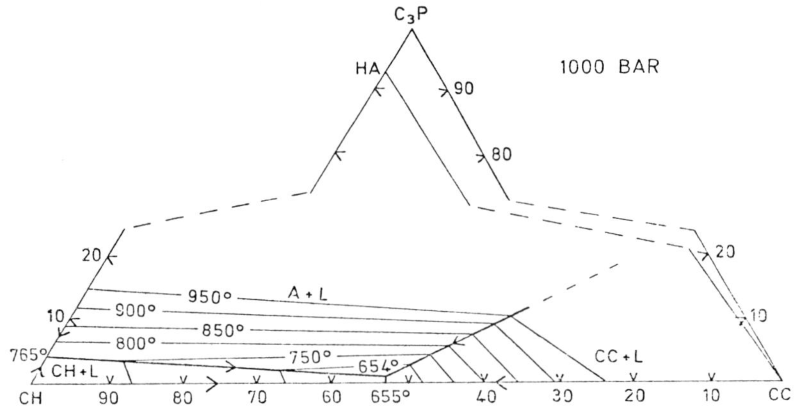

Fig. 14. The join CH–CC–C_3P in projection at 1000 bar showing liquidus field boundaries and thermal contours for liquidus surfaces (from Biggar, 1969).

c. CaO–P_2O_5–CaF_2–H_2O

Phase equilibria involving solids, liquids, and vapors on the join $Ca(OH)_2$–CaF_2–$Ca_3(PO_4)_2$–H_2O were investigated at pressures up to 1000 bar in the range 600–950°C (Biggar, 1967). One of the most useful representations is the series of isothermal sections for the join $Ca(OH)_2$–CaF_2–FAp–OHAp

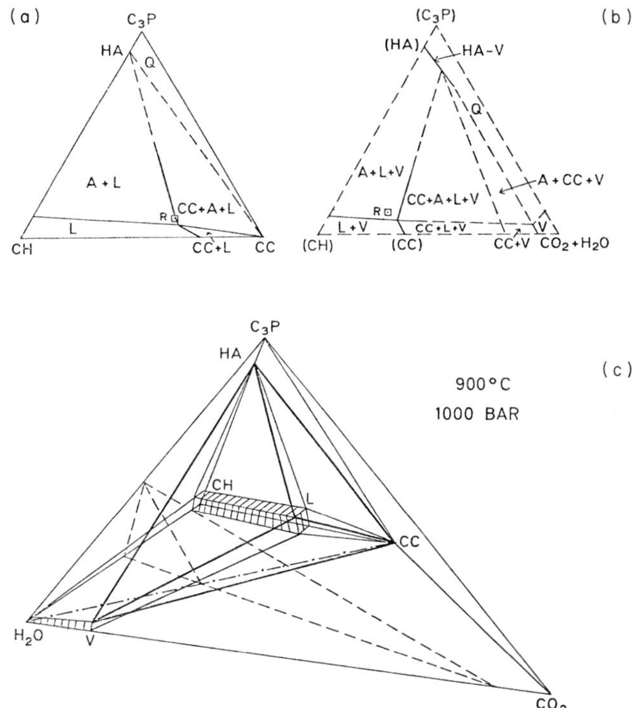

Fig. 15. Isothermal sections at 900°C. (a) The join CH–CC–C_3P. (b) The section at 50% water of the join CH–CC–C_3P–H_2O, but distorted and plotted in terms of the system CH–C_3P–CC–H_2O–CO_2, the plane of the plot being shown as dashed lines (c) and the parentheses (b) indicate points that contain 50% H_2O. *Q* indicates phase fields that were not studied and were presumed to contain defect apatites. (c) Partial isothermal section for the join CH–CC–C_3P–H_2O–CO_2 showing clearly the departure of the vapor from pure water. The liquidus surfaces L(A), L(CC), L(V) are shown shaded and the vaporous surface which coexists with liquid V(L) is shown; other vaporous surfaces are omitted (from Biggar, 1969).

at 1000 bar, shown in Fig. 16 at temperatures from 675 to 950°C. The lowest melting liquid was at 675°; and the changing composition of FAp–OHAp solid solutions could be measured from the changing spacings of chosen *x*-ray diffraction peaks, to locate the positions of phase boundaries.

Experiments were also repeated along the join $Ca(OH)_2$–CaF_2–FAp–OHAp with water added to make total compositions containing 50% H_2O. No significant changes in apatite composition were noted in the presence of excess water, although the isobaric invariant point in the presence of vapor was found to be 10° lower than the corresponding one in equilibrium with liquid only.

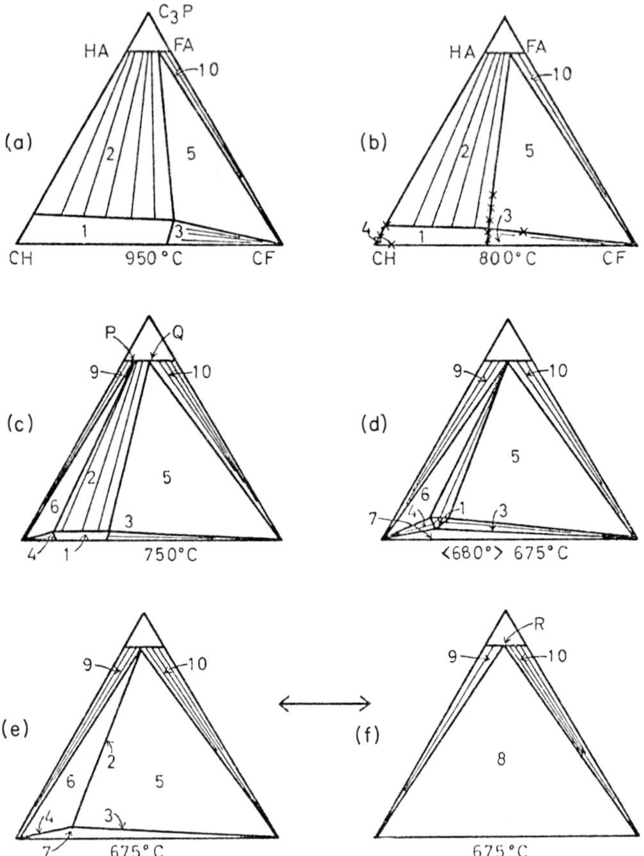

Fig. 16. Isothermal sections for the join CH–CF–FA–HA at 1000 bar. The experimentally determined points from which the sections are determined are illustrated as crosses in (b) only. The phase fields are numbered as follows: 1, L; 2, A + L; 3, CF + L; 4, CH + L; 5, CF + A + L; 6, CH + A + L; 7, CH + CF + L; 8, CH + CF + A; 9, CH + A; 10, CF + A. The reaction CH + CF + A → L, E_6, is illustrated by (e) and (f) and the liquid composition involved is $CH_{73}CF_{24}C_3P$ 3% (from Biggar, 1967).

d. $CaO–P_2O_5–CaF_2–H_2O–CO_2$

The study of the quaternary system $Ca(OH)_2–CaF_2–CaCO_3–Ca_3(PO_4)_2$, which is a section of the quinary system, was done by inference only (Biggar, 1969) and is given here in Fig. 17 as his interpretation of the probable equilibria through combining the bounding ternary systems. Calcite, fluorite, and apatite coexist with liquid in this system at temperatures below 600°C, and a low melting quaternary liquid, about 575°C at 1000 bars is postulated as existing at a composition near the ternary $Ca(OH)_2–CaCO_3–CaF_2$ eu-

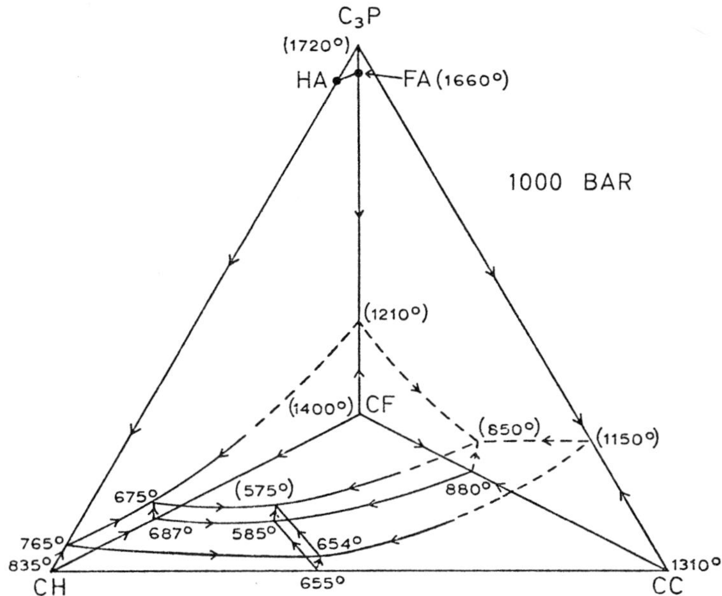

Fig. 17. Schematic diagram of the liquidus phase relationships in the system CH–CC–CF–C_3P at 1000 bar; estimated temperatures are shown in parentheses (from Biggar, 1969).

tectic. Coexisting water vapor would lower the temperatures another 10–40°C.

5. Miscellaneous Systems

a. CaO–SrO–P_2O_5–H_2O

Collin (1959, 1960) synthesized a complete series of crystalline solutions of stoichiometric hydroxyapatites between Ca and Sr from hot basic solutions, and hydrothermal investigations of Schoenberg (1963) showed that these well crystallized phases were stable in the range 300–800°C at elevated pressures.

b. CaO–MgO–P_2O_5–H_2O

Dinger (1974) and Dinger and Roy (1977) investigated equilibria involving the apatite and whitlockite phases in the above system, in the temperature range 260–400°C at 1 kbar (98 MPa) pressure. They found less than ~1 at. % equilibrium substitution of Mg for Ca in apatite in this p–T range, while on the other hand, Mg was necessary for stable whitlockite formation, which correlated with structural observations made by Gopal and Calvo (1972) on the latter.

c. $CaO–PbO–P_2O_5–H_2O$

Engel *et al.* (1975) established the existence of an equilibrium series of crystalline solutions between the end member Ca–Pb hydroxyapatite phases at 350°C under elevated pressures (and apparently also at lower temperatures), although work remains to determine the degree of cation structural ordering (Merker and Wondratschek, 1959) in the individual members. A break in the lattice constants at the Pb_6Ca_4 composition is correlated with x-ray intensity evidence for ordering of Pb at the M_{II} and Ca at the M_I site. The high-Pb members have much lower thermal stabilities than Ca-rich members.

d. $3CaO \cdot P_2O_5–3CaO \cdot B_2O_3$

A defect apatite structure was found to be formed stably over a temperature range in the above system from a range of compositions up to 80 mole % $3CaO \cdot P_2O_5$, as shown in the phase diagram of Fig. 18 (Bauer, 1968). The lattice constants were $a_0 = 9.343$ Å and $c_0 = 6.949$ Å, for a composition $12CaO \cdot 3P_2O_5 \cdot B_2O_3$.

e. Miscellaneous Oxyapatites

Although generally the phase equilibrium data are sparse, some may be inferred from crystal growth studies for certain systems which are important for rare-earth doping of apatites (e.g., Brixner and Bierstedt, 1975).

Shirvinskaya and Bondar (1975a) determined phase equilibria in the systems $Sr_2GeO_4–Sr_3(PO_4)_2$ (Bondar *et al.*, 1975), $Ba_2GeO_4–Ba_3(PO_4)_2$

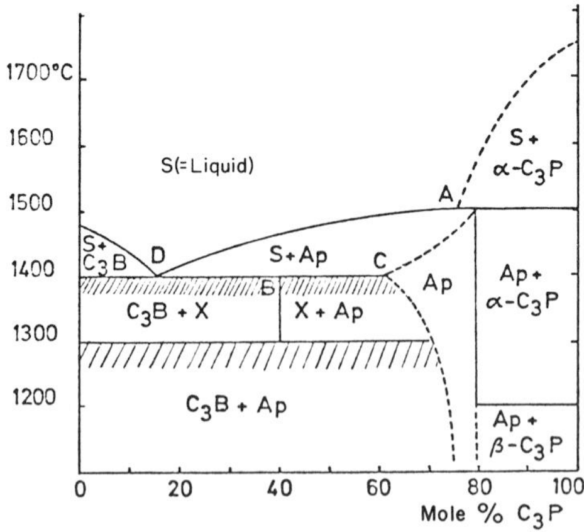

Fig. 18. The section $3CaO \cdot P_2O_5–3CaO \cdot B_2O_3$ in the ternary system $CaO–B_2O_3–P_2O_5$. C_3B, $3CaO \cdot B_2O_3$; C_3P, $3CaO \cdot P_2O_5$; Ap, apatite structure phase; X, second ternary phase.

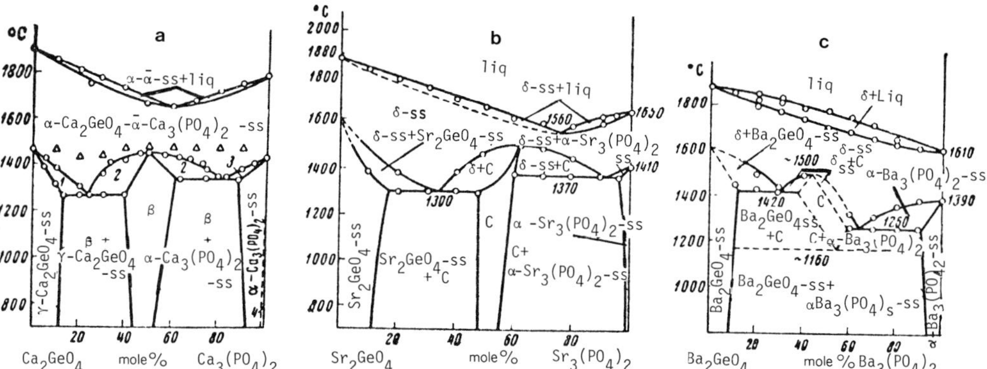

Fig. 19. The systems (a) Ca_2GeO_4–$Ca_3(PO_4)_2$; (b) Sr_2GeO_4–$Sr_3(PO_4)_2$, and (c) Ba_2GeO_4–$Ba_3(PO_4)_2$ [from Shirvinskaya and Bondar (1975a, p. 96)].

(Shirvinskaya and Bondar, 1975a), and Ca_2GeO_4–$Ca_3(PO_4)_2$ (Shirvinskaya and Bondar, 1975b), the latter analogous to the system Ca_2SiO_4–$Ca_3(PO_4)_2$. The subsolidus stability of an apatite structure compound having the formula $M_5(GeO_4)(PO_4)_2$ (M = Ca, Sr, or Ba), was established, possessing some extent of crystalline solution and stable up to 1400–1500°C. At higher temperatures they decompose to a continuous series of crystalline solutions (see Fig. 19).

C. Low Temperature Solution Equilibra

Considerable controversy still persists concerning phase equilibrium relations involving apatites from low temperature solutions, for which reactions are generally extremely sluggish, and conditions frequently involve supersaturation and metastability. Much of the earlier work from the standpoint of applications to fertilizers and biological systems has been summarized by Van Wazer (1958). The further difficulty concerning experimental studies has been that the apatites precipitated from low temperature solutions are poorly crystalline and usually nonstoichiometric with respect to the idealized formula $Ca_{10}(PO_4)_6(OH)_2$, indeed calcium deficient (Bjerrum, 1958; Berry, 1967).

1. CaO–P_2O_5–H_2O

The solubilities of apatite in solution in this system at temperatures up to 37°C have been studied very extensively, particularly for agricultural and biological applications. One means of expressing the equilibria is in the following solubility diagram (Fig. 20). It is seen that hydroxyapatite is the

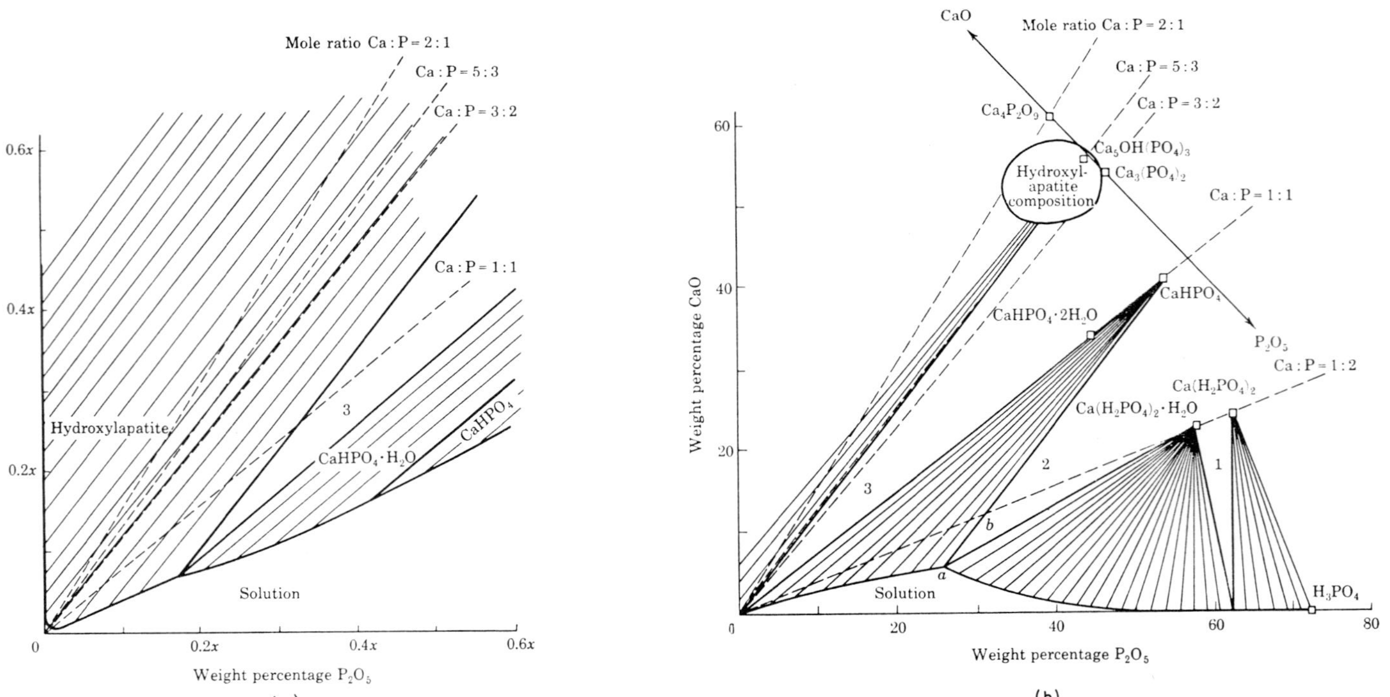

Fig. 20. The system $CaO–H_2O–P_2O_5$ for the calcium orthophosphates at 25°C. (a) Represents a one hundred- –four hundredfold magnification of the origin of (b) of this figure. (b) Is based on precise data, but (a) is only an approximation. In (a), $\chi = \sim\frac{1}{3}–1$ (from Van Wazer, 1958).

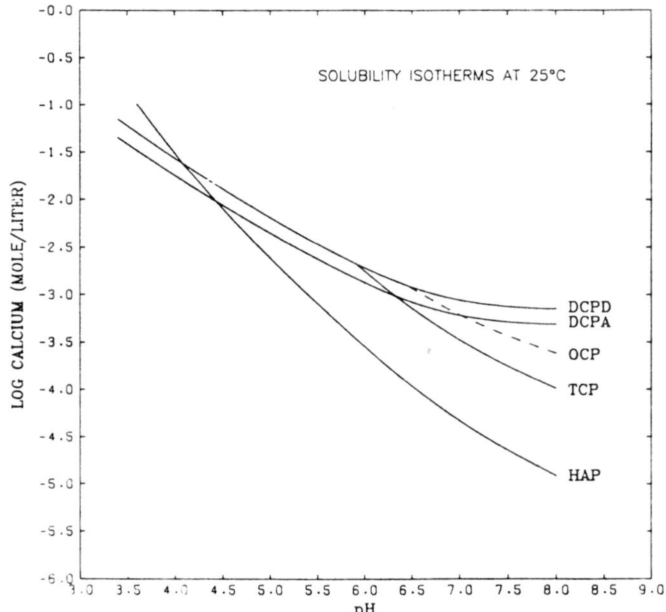

Fig. 21. Calculated isotherms at 25°C illustrating the relative stability of several calcium phosphates. DCPD, $CaHPO_4 \cdot 2H_2O$; DCPA, $CaHPO_4$; OCP, $Ca_4(PO_4)_3H \cdot 2.5H_2O$; TCP, β-$Ca_3(PO_4)_2$; HAP, $Ca_5(PO_4)_3OH$ (from Gregory *et al.*, 1974).

phase having lowest solubility. In addition to apatites, solubilities of $CaHPO_4$ (or $CaHPO_4 \cdot 2H_2O$) and OCP $[Ca_8H_2(PO_4)_6 \cdot 5H_2O]$ are of most concern, for they are in most closely adjacent positions compositionally. Other versions of solubility diagrams incorporate more recent and extensive solubility studies, such as in Fig. 21.

Other system bases may be selected for representation of the data, as in Fig. 22 where the solubilities are plotted as a function of $Ca(OH)_2$ and H_3PO_4 concentration.

Numerous investigators have found that it is useful to calculate solubilities (Madsen, 1975) and express them in terms of chemical potential diagrams. This approach is most recently summarized by Brown and Chow (1976). Mikhaylov (1968) also calculated solubilities from solubility products, and expressed them in terms of $-\log [PO_4^{3-}]$ and pH (Fig. 23). The thickened lines in the figure show regions of unsaturation for the other phases.

Certain problems are inherent in such phase studies, largely deriving from sluggishness of reactions, metastability, and crystal size. Sawatani and Roy (1977) investigated the transformation

$$\text{brushite } [CaHPO_4 \cdot 2H_2O] \rightleftharpoons \text{monetite } [CaHPO_4] + 2H_2O$$

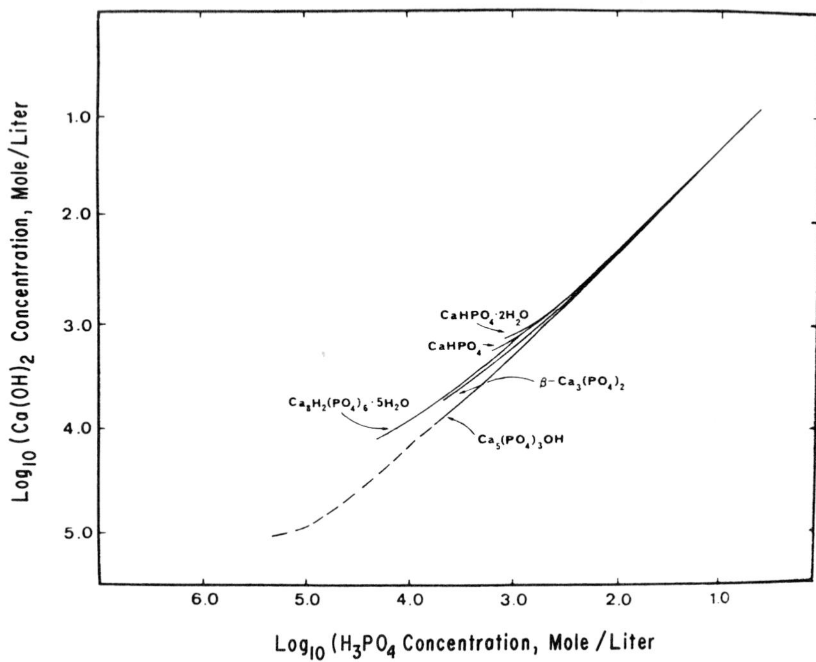

Fig. 22. The solubility phase diagram for crystalline calcium phosphates in the ternary system, $Ca(OH)_3$–H_3PO_4–H_2O at 25°C. Broken lines represent calculated values (from Brown, 1973).

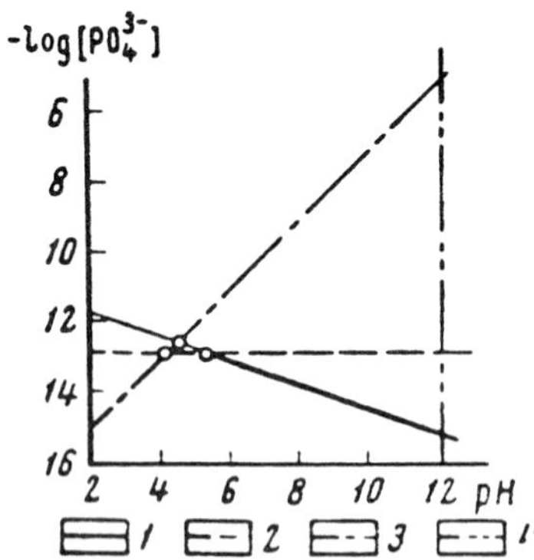

Fig. 23. Regions of existence (lines) of solid phases in the system CaO–P_2O_5–H_2O at $[Ca^{2+}] = 10^{-2}$ M as $[PO_4^{3-}]$ against pH: (1) hydroxylapatite, (2) tricalcium phosphate, (3) brushite, and (4) calcium hydroxide (from Mikhaylov, 1968).

as a function of pressure, which varied from 43°C at 15,000 psi to (probably) 25°C at 1 atm (Fig. 24). Thus, e.g., it is never quite clear which phase is stable when solubilities are measured at 25°C, 1 atm. A second example concerns $3CaO \cdot P_2O_5$, which precipitates from room temperature solution as a semi-amorphous phase (Boskey and Posner, 1973). High surface area materials

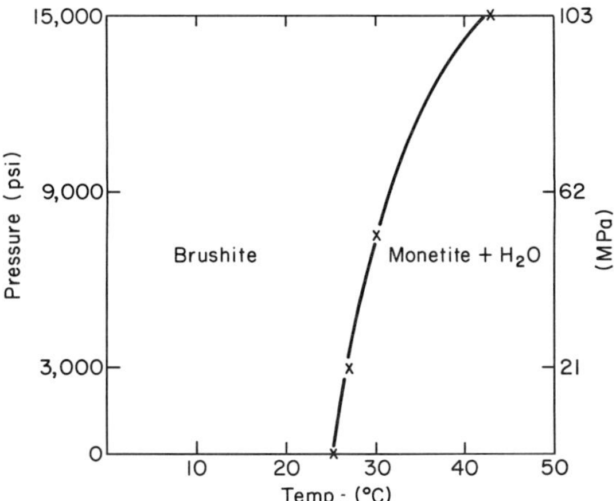

Fig. 24. The brushite–monetite equilibrium.

generally have greater apparent solubilities. In addition, the formation of defect hydroxyapatite from interaction of $3CaO \cdot P_2O_5$ with water (Berry, 1967) is a further complexity. The likelihood that anhydrous crystalline $3CaO \cdot P_2O_5$ is not stable in equilibrium with H_2O has already been discussed (Dinger, 1974; Dinger and Roy, 1977).

2. $CaO–P_2O_5$ + Aqueous Carbonate, $CaCl_2$ or CaF_2, and Other Solutions

The determination of low-temperature phase equilibria in the system $CaO–P_2O_5–H_2O$ with additions of fluoride, chloride, carbonate, or other salts introduces additional complexity to a system which is experimentally extremely difficult for equilibrium determinations. Primarily, the goals of understanding biological processes have spurred the various attempts. A recent review including some of this work has been completed (Brown and Young, 1977).

a. Carbonate

The substitution of carbonate for hydroxyl or phosphate groups in hydroxyapatite, reported by Wallaeys (1951), Le Geros *et al.* (1968), Montel (1971), and others, has already been discussed. The crystallinity of carbonate containing apatites prepared from room temperature solutions is generally even poorer than hydroxyapatite, and has made the determination of equilibria next to impossible. Somewhat better results have been obtained under low-temperature hydrothermal conditions (Eysel and Roy, 1973, 1975; Roy *et al.*, 1974; Dinger, 1974; Dinger and Roy, 1977) under which crystal growth of the apatite phase under certain conditions has been accelerated by the presence of carbonate. Systematic studies of the phase equilibria have not yet been performed, but indications have been found (Roy *et al.*, 1974; Dinger and Roy, 1977) that an equilibrium concentration of carbonate may be present in the OH-apatite phase for a given set of conditions.

b. Chloride and Fluoride

Precise solubility data for chlorapatite and fluorapatite in low temperature solutions are sparse. Valyashko *et al.* (1968) calculated heats of formation for hydroxyl- and fluorapatites, to be −3206.3 and −3262.5 kcal/mole, respectively. He also used the data of Gottschal (1958) for chlorapatite (−3172 kcal/mole) to calculate the relative activity products in water shown as a function of temperature in Fig. 25. All solubility products decrease with increasing temperature, and the order of solubility products is F < OH < Cl at all temperatures up to 350°C.

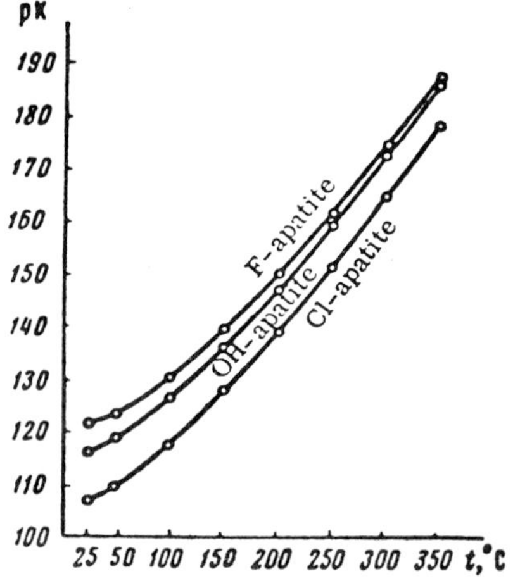

Fig. 25. Dependence of activity products of apatites on temperatures (from Valyashko *et al.*, 1968).

Duff (1972) calculated free energies of formation of chlorapatite, and of solid solutions of Cl^- and OH^- apatite as well as Cl^- and F^- apatites. Also, stability relations of solid solutions as a function of pH were calculated, and the various phase fields were defined as a function of activities of Cl^- and of Ca^{2+}. Notably, the F^- end member was found to extend considerably the apatite phase field stability under more acidic conditions. While theoretical, this corresponds to experimental observations in dentistry (Brudevold *et al.*, 1965). Dinger and Roy (1977) did not investigate equilibria involving F^- addition singly, but in preliminary studies at higher temperatures and pressures (260°C, 100 MPa) found that $Mg^{2+} + F^{-}$ extended the stability field of apatite with respect to whitlockite. This is the opposite from the effect observed with room temperature equilibria (Duff, 1971).

IV. CRYSTAL GROWTH

Much of the difficulty experienced in investigating apatites is attributed to the inability to grow pure stoichiometric single crystals which are needed for physical properties measurements. This section presents single-crystal growth studies by various techniques, most of which utilize data obtained from the phase equilibrium diagrams discussed previously. Examples are treated from each of the following growth methods: solution, gel, melt, hydrothermal, flux, and anion exchange.

A. Solution Growth

Precipitation of apatitic phases from solutions at low temperatures is a very common method of synthesis of powdered samples. The temperature range 25–37°C also is the most directly relevant for understanding biological processes. The apatites produced in this manner are usually nonstoichiometric where the Ca:P ratio departs from the theoretical value of 1.67, contain excess water and commonly carbonate, and have a low mean index of refraction. Apatites formed by direct precipitation characteristically have very fine crystallite size which yield poor x-ray diffraction patterns (Brown and Chow, 1976; Posner, 1969). Very limited success has been reported for growing crystals of apatite from aqueous solutions. Nancollas and Mohan (1970) studied the kinetics of growth of hydroxyapatite at 25°C by seeding supersaturated calcium phosphate solutions with crystals. The initial growth proceeds through the formation of a precursor phase with a Ca:P ratio of 1.45 which slowly reacts with the solution to attain hydroxyapatite composition, but only in crystals of small size.

B. Gel Growth

Scientific activity in the use of the gel method for research purposes was revived in these laboratories some years ago (Henisch *et al.*, 1965; McCauley, 1965; McCauley and Roy, 1974); in the latter two studies various phosphates were grown including calcium phosphates, probably not including true OHAp, but what was actually OCP several millimeters in length.

TABLE X

Gel CRYSTAL GROWTH RUNS

One-interface runs:	Bottom:	Gel + PO_4^{3-}
	Top:	Ca^{2+} in solution
Two-interface runs:	Bottom:	Gel + PO_4^{3-}
	Middle:	Gel + HAc
	Top:	Ca^{2+} in solution

Ca^{2+} source	PO_4^{3-} source	Acid
Cl^-, I^-, NO_3^-	H_3PO_4, $(NH_4)_2HPO_4$	H_3PO_4 or +HAc
	pH range 7–10	

Temp.	pH	Products
Room temp.	9.5–10	OHAp (very small)
Room temp.	8	Unknown
45°, 60°C	9.75	OHAp (very small)
45°C	8.20	$CaHPO_4 \cdot 2H_2O$, $CaHPO_4$ (1 cm)
60°C	8.20	$CaHPO_4$ (1 cm)
60°C	range (~6–7)	OCP (1–2 mm)
Room temp.	8.20	$CaHPO_4 \cdot 2H_2O$ (1 cm)

A more extensive study was therefore undertaken by the present authors in which pH, temperature, and reactants were varied. The typical experimental conditions are described in the papers referred to above. The results of our runs are summarized in Table X.

It will be noted that all crystalline phases in the ternary system $CaO–P_2O_5–H_2O$ have been grown, some into crystals larger than 1 cm in one or two directions. However, although the conditions for apatite formation have been found, the crystals have not grown beyond a few microns in size. Thus while the first stage has been reached, it is not self-evident that it will be possible to actually grow larger crystals of OHAp in gels because of its extremely low solubility product, and the general instability of gels under high pH conditions. However, the resulting apatites may be useful because they are well-crystallized fine apatites.

C. Melt Growth

Phase equilibria studies have shown that some apatites melt congruently and single crystals can be pulled from the stoichiometric melt. Mazelsky *et al.* (1968) used the Czochralski method for pulling pure and neodymium-doped fluorapatite. Vaporization of CaF_2 from the melt is reflected in the chemical analysis of the crystals which appear to be about 5% deficient in CaF_2. The phase equilibrium diagram for the system $Ca_3(PO_4)_2$–CaF_2 (Fig. 10) indicates that apatite with excess CaF_2 acting as a flux can be pulled from the melt at much lower temperatures where the vaporization of CaF_2 probably would not be as severe. A deficiency of CaF_2 in the crystals should also be eliminated with a CaF_2 rich melt. Fluorapatite crystals 5 mm long and 1 cm maximum diameter were pulled from the melt using the Kyropoulos method by Johnson (1961) who also used this technique to grow chlorapatite, but vaporization of $CaCl_2$ at the melting point severely limited the size of the crystals. By a cyclical process of heating and cooling of a chlorapatite melt which caused the larger crystals to grow at the expense of the smaller crystals, Johnson (1962) obtained crystals $3 \times 5 \times 10$ mm^3 but chemical analysis detected chloride ion deficiencies. Other melt grown crystals showed 5.73% chlorine content compared to the theoretical values of 6.81% (Prener, 1968). Segall *et al.* (1962) and Piper *et al.* (1965) found evidence that in melt grown fluorapatite, fluorine vacancies were present as well as various combinations of fluorine vacancies with oxygen.

Single crystals of the anion-deficient apatite of the formula $Pb_{8+x}Na_{2-x}\cdot(PO_4)_6O_{x/2}[\ ]_{2-x/2}$ ($x \approx 1$) as well as rare earth-doped compositions have been grown by the Czochralski technique by Brixner and Bierstedt (1975). The formula was based on extrapolation of lattice parameters given by Engel (1973) for both the pure oxyapatite $Pb_{10}(PO_4)_6O[\ ]$ and the pure sodium compound $Na_2Pb_8(PO_4)_6[\ ]_2$. Semiquantitative spectroscopic examination of the crystals indicated approximately 1% Na which is close to the 0.93% Na required for the composition $Pb_9Na(PO_4)_6O_{0.5}[\ ]_{1.5}$. They also reported that wet chemical analysis gave 75.1% Pb and 7.5% P compared to the calculated amounts of 75.63% Pb and 7.54% P. Attempts to grow pure $Pb_8Na(PO_4)_6[\ ]_2$ by using stoichiometrically prepared feed material produced crystals with a composition $Pb_{8.5}Na_{1.5}(PO_4)_6O_{0.25}[\ ]_{1.75}$. Brixner and Bierstedt (1975) also Czochralski pulled rare earth-doped crystals where the general composition of the starting material was $Pb_{7.75}Ln_{0.25}Na_2(PO_4)_6O_{0.125}[\ ]_{1.875}$, where Ln = La, Pr, Nd, Sm, Eu, Gd, Tb, Dy, Ho, and Er. They stated that due to segregation the true sodium concentration in the crystal was less than two moles per molecule and the above formulas were of only nominal compositions. It was also reported (Brixner and Bierstedt, 1975) that the lead orthophosphate of apatite structure $Pb_3(PO_4)_2$ is actually a sodium stabilized apatite of the formula $Pb_{8+x}Na_{2-x}(PO_4)_6[\ ]_{2-x/2}O_{x/2}$.

Federov *et al.* (1975) grew alkaline earth-rare earth oxysilicophosphate crystals by crystallization from stoichiometric melts. The general formula was $Ca_4Ln_6(SiO_4)_4(PO_4)_2O_2$ where Ln = La, Nd, Sm, Gd, Dy, Ho, Er,

Yb, and Y. The x-ray powder patterns denoted differences in the unit cell lattice parameters depending on the lanthanide atom.

D. Hydrothermal Preparation and Crystal Growth

This includes reports of growth in both relatively low temperature or elevated temperature solutions, as well as in high temperature $Ca(OH)_2$ flux solutions. An extensive literature exists on the preparation of apatite at temperatures up to 100°C. The hydrothermal work is somewhat more limited but includes that of Morey and Ingerson (1937) and Klement (1938) who prepared OHAp crystals by the hydrothermal method. Perloff and Posner (1956) prepared well-crystallized stoichiometric OHAp free from many impurities by hydrolysis of $CaHPO_4$ in a platinum-lined hydrothermal bomb. Hayek *et al.* (1958) grew crystals about 2×0.3 mm^2 using a small thermal gradient in a hydrothermal system. Collin (1959, 1960) reproducibly obtained stoichiometric OHAp from hot basic solutions, and increased the degree of crystallinity by heating such materials dry to 950°C. The latter work included extensive preparations of (Ca, Sr)–OHAp crystalline solutions. Schoenberg (1963) extended this work to the hydrothermal region at 300–800°C, carefully controlling the composition by using sealed tubes, and obtained well-crystallized hydroxyapatites, with much better crystallinity in the calcium-rich (relative to strontium) range of compositions. The phase equilibrium studies of Biggar were referred to in Section III. From compositions having high $Ca(OH)_2$ content, Biggar (1966, 1967) was able to crystallize OHAp crystals up to 0.25 mm in length. Kirn and Leidheiser (1968) reported hydrothermal synthesis of 2 mm length OHAp crystals. Roy (1971) grew crystals up to 3–4 mm long by slow cooling from the liquidus temperature under hydrothermal conditions at pressures up to 30,000 psi (Roy and Tuttle, 1956), using mixtures on the binary join $Ca_3(PO_4)_2$–$Ca(OH)_2$ and in the ternary system $Ca_3(PO_4)_2$–$Ca(OH)_2$–H_2O. Eysel and Roy (1973) extended this work to grow rod-shaped crystals up to 8 mm in length. Mixtures in the pure ternary system were studied, and also compositions with CO_2 present. OHAp as well as CO_2 saturated crystals containing 0.7% CO_2 were grown using a $Ca(OH)_2$ flux with oscillatory temperature from 750 to 880°C and pressures from 14,500 to 17,500 psi. In the case with CO_2 present the relevant equilibria are those of the $Ca_3(PO_4)_2$–$Ca(OH)_2$–$CaCO_3$–H_2O system (Biggar, 1966, Wyllie, 1967). Hydroxyapatite forms a binary eutectic at 765°C with $Ca(OH)_2$. The eutectic is lowered by the addition of water to 735°C which is the lowest temperature for flux growth of OHAp in the system $Ca_{10}(PO_4)_6(OH)_2$–$Ca(OH)_2$–H_2O (Fig. 6). With 50 wt % water the liquidus is very steep and growth from 900°C (upper temperature limit of cold seal vessels) to 735°C crystallizes only small amounts of hydroxyapatite. The oscillating temperature method, described by Scholz and Kluchkow (1967) and applied hydrothermally by Kolb and Laudise (1971), was used (Eysel and Roy, 1973) to reduce the number of nuclei due to the oscillation and enhance the growth of preferred nuclei depending on the temperature distribution in the tube. Relatively large crystals could therefore be grown from compositions containing only a small amount of phosphate. More recently Mengeot *et al.* (1973) grew single crystals of calcium hydroxyapatite $7 \times 3 \times 3\,mm^3$ in size in hydrothermal bombs. At pressures of 45,000–60,000 psi hydroxyapatite was found to exhibit retrograde solubility in water between 300 and 670°C. Hydroxyapatite powder was dissolved at the colder region of a sealed gold tube and recrystalized at the hotter region on seed crystals.

E. Flux Growth

Crystals grown from the melt at high temperatures usually are severely strained due to the large temperature gradients present during growth. Growth from flux solutions is possible at much lower temperatures under almost isothermal conditions. Fluor- and chlorapatite crystals have been grown by Prener (1967) from melts using CaF_2 and $CaCl_2$ fluxes, respectively. Chemical analysis of fluorapatite crystals about 5–6 mm in length and chlorapatite crystals 3–4 mm in length agreed to within 0.1% of the theoretical values. For fluorapatite growth, Prener (1967) used a mixture of CaF_2 flux (45 wt %) and $Ca_5(PO_4)_3F$ (55 wt %) which lowered the liquidus temperature to 1325°C (see Fig. 10). This is far below the 1650°C needed for pulling a crystal from the stoichiometric melt. In many cases inclusions of the flux were seen by microscope examination but Prener (1967) always found a large number of flux free crystals. The formation of apatites that are halide deficient was eliminated by using CaF_2 and $CaCl_2$ solvents which provide a halide-rich solution for the growing crystals.

Ito (1968) grew single crystals of oxyapatite by flux fusion. The sample and sodium tungstate flux were cooled from 1150 to 750°C at 1°C/hr and quenched from 750°C to room temperature. The transparent crystals were very small, 0.05 mm. Large euhedral transparent crystals of $Na_2La_8Si_6O_{24}F_2$ up to 8 mm were also grown using a sodium fluoride flux. The crystals were grown by a combination of slow cooling (2°C/hr) and flux evaporation from 1350 to 950°C. The crystals of Nd, Pr, Sm, and Gd analogs were also successfully grown.

Single crystals of bromapatite [$Ca_{10}(PO_4)_6Br_2$] 0.8 mm long were grown by Dykes (1974), using a calcium bromide flux. Equal weights of bromapatite powder and $CaBr_2$ flux were melted and slowly cooled in an hydrogen bromide atmosphere from 800 to 740°C at a rate of 0.68°C/hr. Based on chemical analyses, Dykes concluded that the crystals are stoichiometric with the space group $P6_3/m$.

F. Anion Exchange into Large Natural or Synthetic Crystals

In certain mineral species with suitable structures it has been shown that cations or anions may be exchanged with solutions or melts with high activities of other ions. Zeolites are, of course, the best known such cation exchangers. The feldspars have been shown to readily exchange alkali ions (e.g., Na^+ from NaCl melts exchanging directly with K^+ in a feldspar) even in fairly large crystals. By 1939, Schleede *et al.* (1939) had shown that finely divided natural FAp could be converted to OHAp by heating at 1360°C in steam. Wallaeys (1952) repeated the work with ClAp by heating at 800°C in steam. Elliott and Young (1967) likewise were successful with a somewhat larger crystal (0.4 mm) by heating to 1300°C for two weeks in steam. It was therefore argued that with much higher activities of OH^-, H_2O, or other ions it should be possible to effect the exchange in larger (centimeter size) crystals and at lower temperatures.

Experiments were conducted by Roy and Roy (1977) on samples $5 \times 3 \times 1$ mm^3 of natural FAp from Durango, Mexico, of clear gem quality. Their halogen content, F = 4.75% and Cl = 0.73%, were anomalously high. Spectrochemically they were found to contain 0.05% Mn, 0.08% Fe, 0.4% Si, 0.05% Mg, 1.2% Ce, 0.01% Yb, 0.8% La, 0.1% Y, 0.01% Ti, 0.05% Sr, and 0.2% Na, plus traces of Al, Ba, V, Ni, Cr, Mo, and Sn. The unit- cell dimensions were $a = 9.39$ and $c = 6.88$ Å and IR showed a small OH peak at 3530 cm^{-1}.

In order to make the Durango specimens into pure end-member FAp, molten-salt exchange was attempted with KF, LiF, MgF_2–CaF_2 eutectic, MnF_2, AgF, PbF_2, and CaF_2 (solid state) at the lowest temperatures feasible (450–900°C). Concentrated fluoride solutions were also tried at temperatures from 25 to 200°C. In most cases it was difficult to find conditions where the crystals did not dissolve; in only a few were there any substantial changes in lattice parameters, F composition as determined by electron-probe analysis, or IR absorption spectra. The most successful exchange was at 900°C with the single crystal packed with powdered CaF_2 which gave the pure F end member. Melts with various chlorides produced color and x-ray evidence for some changes in composition but this could not be determined quantitatively with the electron probe.

In the attempt to exchange OH for F, it was decided to avoid alkali hydroxides since Na^+ might also be exchanged. Several runs were made at

830°C at 15,000 psi to react FAp with pure water for 2, 4, 6, and 8 weeks. The OH content was followed by IR absorption; the peak at 3530 cm^{-1} increased by a factor of 5 to 6 in 8 weeks; this however was only a partial replacement of the F. The refractive index also increased regularly from an N_w of 1.635 to 1.644 in 8 weeks. Thus the exchange of OH for F is extremely slow under these exchange conditions.

Further interesting details in the OH exchange are revealed in other hydrothermal exchange experiments. Thus if the exchange is carried out at 915°C and 1000 psi for 3 weeks, an OH band much broader than normal appears centered at 3420 instead of 3530 cm^{-1} indicating that the OH groups are in abnormal sites in this material. Likewise, when a ClAp is used as starting material, the peak appears at 3580 cm^{-1}. As Young *et al.* (1968) and Kay *et al.* (1964) have shown, the F, Cl, and OH do not occupy identical sites. The above data would suggest two possible alternatives when OH ions are substituted in small amounts for halogen ions: (1) that they may (at first) accept the sites of the previous anion occupant rather than their own equilibrium sites; or (2) that interaction with the major anion causes a shift in the frequency.

A tentative explanation for our apparent failure to achieve as rapid exchange as have other workers is the following. In previous work where steam was reacted with a FAp at 1300–1350°C, the actual reaction was, possibly as follows:

$$\text{FAp} \xrightarrow{1350^\circ\text{C}} \begin{matrix}\text{F-deficient apatite}\\ \text{with open channels}\end{matrix} + \text{HF} \xrightarrow{+H_2O} \text{OHAp}.$$

The actual expulsion of F ions makes possible a much faster reaction. Under our current conditions true thermally activated diffusion is the only force driving the reaction and this at a very low temperature. It is interesting to note that Riboud (1968) found a decomposition temperature for OHAp of 1325°C at low H_2O pressures (in his case 4–6 Torr).

V. SUMMARY AND CONCLUSIONS

The apatite structure (either ideal or slightly distorted) defines a large family of crystalline phases, made additionally complex through the multiplicity of sites available for ionic substitution, either of individual ions or through coupled substitutions. Although knowledge of the crystal chemistry of apatite is rather extensive, more detailed information regarding syntheses and equilibria throughout a broad range of temperatures (and pressures) is required to delineate the potential of this family of materials. The pressure variable, in particular, has been utilized only to a minor extent, and this was principally for hydroxyl containing phases. The vast majority of reported experimental work on apatites is limited to synthesis, much of this at a single temperature.

The existing phase equilibrium data for some systems have been applied to the solution of crystal growth problems, achieving excellent results, and in others, materials syntheses have been performed most reliably for systems having such data available. Quite clearly the available coverage of phase equilibrium studies is not sufficient to enable a full exploration of the potential of this complex mineral family.

References

Ahrens, L. H. (1952). *Geochim. Cosmochim. Acta* **2**, 155–169.
Akhavan-Niaki, A. N. (1961). *Ann. Chim.* (*Paris*) **6**, 51–79.
Ames, L. L., Jr. (1959). *Econ. Geol.* **54**, 829–841.
Baker, W. E. (1966). *Am. Min.* **51**, 1712–1721.
Balz, W. (1961). Dissertation (Tech) Hoch. Karlsruhe.
Bauer, H. (1959). *Angew. Chem.* **71**, 374.
Bauer, H. (1968). *Bull. Soc. Chim. France*, 1703–1705.
Berak, J. (1961a). *Rocz. Chem.* **35**, 23–29.
Berak, J. (1961b). *Rocz. Chem.* **35**, 69–77.
Berak, J. (1976). Personal communication.
Berry, E. E. (1967). *J. Inorg. Nucl. Chem.* **29**, 317–327.
Biggar, G. M. (1966). *Min. Mag.* **35**, 1110–1122.
Biggar, G. M. (1967). *Min. Mag.* **36**, 539–564.
Biggar, G. M. (1969). *Min. Mag.* **37**, 75–82.
Bjerrum, Niels (1958). *K. Dan. Vidensk. Selsk. Mat. Fys. Medd.* **31**, 1–79.
Blasse, G. (1975). *J. Solid State Chem.* **14**, 181–184.
Bondar, I. A., Shirvinskaya, A. K., Kanclir, E., Kapralik, I., and Panek, Z. (1975). *Chem. Zvesti* **29** (3), 366–372.
Boskey, A. L., and Posner, A. S. (1973). *J. Phys. Chem.* **77**, 2313.
Bredig, M. A., Franck, H. H., and Fuldner, H. (1932). *Z. Elektrochem.* **38**, 158.
Brenner, P., Engel, G., and Wondratschek, H. (1970). *Z. Kristallogr.* **131**, 206–212.
Brixner, L. H., and Bierstedt, P. E. (1975). *J. Solid State Chem.* **13**, 24–31.
Brown, W. E. (1973). *In* "Environmental Phosphorus Handbook" (E. J. Griffith, A. Belton, J. M. Spencer, and D. T. Mitchell, eds.), pp. 203–239. Wiley, New York.
Brown, W. E., and Chow, L. C. (1976). *Annu. Rev. Mater. Sci.* **6**, 213–236.
Brown, W. E., and Young, R. A. (Eds.) (1977). "Structural Properties of Hydroxyapatite and Related Compounds" (in preparation).
Brudevold, V., McCann, H. G., and Gron, P. (1965). *In* "Caries Resistant Teeth," CIBA Symp. Churchill, London.
Cockbain, A. G. (1968). *Min. Mag.* **36**, 654–660.
Cockbain, A. G., and Smith, G. V. (1967). *Min. Mag.* **36**, 411–421.
Collin, R. L. (1959). *J. Am. Chem. Soc.* **81**, 5275–5278.
Collin, R. L. (1960). *J. Am. Chem. Soc.* **82**, 5067–5069.
Cooray, D. G. (1970). *Am. Mineral.* **55**, 2038–2041.

Dinger, D. R. (1974). M.S. Thesis, The Pennsylvania State Univ., University Park, Pennsylvania.
Dinger, D. R., and Roy, D. M. (1977). Manuscript in preparation.
Drafall, L., and Roy, D. M. Unpublished.
Duff, E. J. (1971). *Chem. Ind.* (*London*), 16 Oct. 1971, 1191–1193.
Duff, E. J. (1972). *J. Inorg. Nucl. Chem.* **34**, 95–100.
Dykes, E. (1974). *Mater. Res. Bull.* **9**, 1227–1236.
Elliott, J. C. (1969). *Calcif. Tissue Res.* **3**, 293–307.
Elliott, J. C. (1971). *Nature* (*London*) **230**, 72.
Elliott, J. C., Mackie, P. E., and Young, R. A. (1973). *Science* **180**, 1055–1057.
Elliott, J. C., and Young, R. A. (1967). *Nature* **214**, 904–907.
Engel, G. (1970). *Z. Anorg. Allg. Chem.* **378**, 49–61.
Engel, G. (1973). *J. Solid State Chem.* **6**, 286–292.
Engel, G., Krieg, F., and Reif, G. (1975). *J. Solid State Chem.* **15**, 117–126.
Eysel, W., and Roy, D. M. (1973). *J. Crystal Growth* **20**, 245–250.
Eysel, W., and Roy, D. M. (1975). *Z. Kristallogr.* **141**, 11–24.
Fedorov, N., Andreev, I. F., and Maliksetyan, N. S. (1975). *Kristallografiya* **20**, 455–457.
Felsche, J. (1972). *J. Solid State Chem.* **5**, 266–275.
Fisher, D. J., and McConnell, D. (1969). *Science* **164**, 551–553.
Fowler, B. O. (1974). *Chemistry* **13**, 207.
Gay, (1957). *Min. Mag.* **31**, 455.
Gopal, R., and Calvo, C. (1972). *Nature* (*London*) Phys. Sci. **237**, 30–32.
Gottschal, A. I. (1958). *J. Chem. Inst.* **11**, 45.
Gregory, T. M., Moreno, E. C., Patel, J. M., and Brown, W. E. (1974). *J. Res. Nat. Bur. Stand. Sect. A* **78A**, 667–674.
Grisafe, D. A., and Hummel, F. A. (1970a). *J. Solid State Chem.* **2**, 160–166.
Grisafe, D. A., and Hummel, F. A. (1970b). *Am. Mineral.* **55**, 1131–1145.
Harada, K., Nagashima, K., Nakao, K., and Kato, A. (1971). *Jpn. Am. Min.* **56**, 1507–1518.
Hayek, W., Bohler, W., Lechlutner, J., and Petter, H. (1958). *Z. Anorg. Chem.* **295**, 241.
Hendricks, S. B., Jefferson, M. E., and Mosley, V. M. (1932). *Z. Kristallogr.* **81**, 352.
Henisch, H. K., Dennis, J. C., and Hanoka, J. I. (1965). *J. Phys. Chem. Solids* **26**, 493.
Hill, W. L., Faust, G. T., and Reynolds, D. S. (1944). *Am. J. Sci.* **242**, 457–477.
Hounslow, A. W., and Chao, G. Y. (1970). *Can. Mineral.* **10**, 252.
Ito, J. (1968). *Am. Mineral.* **53**, 890–907.
Jensen, A. T., and Rowles, S. L. (1957a). *Nature* **179**, 912–913.
Jensen, A. T., and Rowles, S. L. (1957b). *Acta Odontol. Scand.* **15**, 121–139.
Johnson, P. D. (1961). *J. Electrochem. Soc.* **108**, 159–162.
Johnson, P. D. (1962). *In* "Luminescence of Organic and Inorganic Materials," pp. 563–575. Wiley, New York.
Kay, M. I., Young, R. A., and Posner, A. S. (1964). *Nature* (*London*) **204**, 1050–1052.
Kingsley, J. D., Prener, J. S., and Segall, B. (1965). *Phys. Rev.* **137**, A189.
Kirn, J. F., and Leidheiser, H. (1968). *J. Crystal Growth* **2**, 111–112.
Klement, R. (1938). *Z. Anorg. Allg. Chem.* **237**, 161.
Klement, R. (1939). *Naturwissenschaften* **27**, 57–58.
Klement, R., and Dihn, P. (1941). *Naturwissenschaften* **26**, 20, 301.
Klement, R., and Harth R. (1961). *Chem. Ber.* **94**, 1452–1456.
Klement, R., and Haselbeck, H. (1965). *Z. Anorg. Allg. Chem.* **336**, 113–128.
Kolb, E. D., and Laudise, R. A. (1971). *J. Crystal Growth* **8**, 191.
Kreidler, E. R., and Hummel, F. A. (1967). *Inorg. Chem.* **6** (5), 891.
Kreidler, E. R., and Hummel, F. A. (1970). *Am. Mineral.* **55**, 170–184.

LeGeros, R. Z., Trautz, O. R., LeGeros, J. P., and Klein, E. (1968). *Bull. Soc. Chim. Fr.*, 1712–1718.
Lux, H. (1946). *Z. Naturforsch.* **1**, 281–283.
Machatschki, F. (1939). *Centralbl. Mineral.* 161–164.
Mackie, P. E., Elliott, J. C., and Young, R. A. (1972). *Acta Crystallogr. Sect. B* **28**, 1840–1848.
Mackie, P. E., and Young, R. A. (1973). *J. Appl. Crystallogr.* 6, 26–31.
Madsen, H. E. L. (1975). *Acta Chem. Scand. Ser. A* **29**, 745–748.
Mayer, I., and Makogon-Loewry, V. (1969). *Isr. J. Chem.* **7**, 717–719.
Mayer, I., Roth, R. S., and Brown, W. E. (1974). *J. Solid State Chem.* **11**, 33–37.
Mazelsky, R., Ohlmann, R. C., and Steinbrugge, K. (1968). *J. Electrochem. Soc.* **115**, 68–70.
McCarthy, G. J., and Davidson, M. T. (1975). *Am. Ceram. Soc. Bull.* **54**, 783–786.
McCauley, J. W. (1965). M.S. Thesis, The Pennsylvania State Univ., University Park, Pennsylvania.
McCauley, J. W., and Roy, R. (1974). *Am. Mineral.* **59**, 947–963.
McClellan, G. H., and Lehr, J. (1969). *Am. Mineral.* **54**, 1374–1391.
McConnell, D. (1937). *Am. Mineral.* **22**, 977–986.
McConnell, D. (1938). *Am. Mineral.* **23**, 1–19.
McConnell, D. (1952). *Bull. Soc. Fr. Mineral. Cristallogr.* **75**, 428–445.
McConnell, D. (1973). "Apatite: Its Crystal Chemistry, Mineralogy, Utilization, and Geologic and Biological Occurrences." Springer-Verlag, New York and Wien.
Mehmel, M. (1930). *Z. Kristallogr.* **75**, 323–331.
Mehta, S., and Simpson, D. R. (1975). *Am. Mineral.* **60**, 134–138.
Mengeot, M., Harvill, M. L., and Gilliam, O. R. (1973). *J. Crystal Growth* **19**, 199–203.
Merker, L., and Wondratschek, H. (1957). *Z. Kristallogr.* **109**, 110–114.
Merker, L., and Wondratschek, H. (1959). *Z. Anorg. Allg. Chem.* **300**, 41–50.
Mikhaylov, A. S. (1968). *Geokhimiya* **7**, 851–859.
Montel, G. (1971). *Bull. Soc. Fr. Mineral. Cristallogr.* **94**, 300–313.
Mooney, R. W., and Aia, M. A. (1960). *J. Am. Chem. Soc.* **82**, 433–462.
Morey, G. W., and Ingerson, E. (1937). *Econ. Geol.* **32**, 607–761.
Nacken, R. (1912). *Centralbl. Mineral.* 545–559.
Nadal, M., Trombe, J. C., Bonel, G., and Montel, G. (1970). *J. Chim. Phys.* (Fr.) **67**, 1161.
Nancollas, G. H., and Mohan, M. S. (1970). *Arch. Oral Biol.* **15**, 731–745.
Náray-Szabó, St. (1930). *Z. Kristallogr.* **75**, 387–398.
Nurse, R. W., Welch, J. H., and Gutt, W. (1958). *Nature (London)* **182**, 1230.
Perloff, A., and Posner, A. S. (1956). *Science* **124**, 583–584.
Piper, W. W., Kravitz, L. C., and Swank, R. K. (1965). *Phys. Rev.* **138**, A1802–A1814.
Posner, A. S. (1969). *Physiol. Rev.* **49**, 760–792.
Posner, A. S., Perloff, A., and Diorio, A. F. (1958). *Acta Crystallogr.* **11**, 308–309.
Prener, J. S. (1967). *J. Electrochem. Soc.* **114**, 77–83.
Prener, J. S. (1968). Communicated to Princeton Univ. Conf. Phys. Biol. Properties of Apatite.
Prener, J. S. (1971). *J. Solid State Chem.* **3**, 49–55.
Riboud, P. W. (1968). *Bull. Soc. Chim. Fr.*, 1701–1703.
Roy, D. M. (1971). *Mater. Res. Bull.* **6**, 1337–1340.
Roy, D. M., Eysel, W., and Dinger, D. (1974). *Mater. Res. Bull.* **9**, 35–40.
Roy, D. M., and Roy, R. (1977). *In* "Structural Properties of Hydroxyapatite and Related Compounds" (W. E. Brown and R. A. Young, eds.), in preparation.
Roy, R. (1975). *Am. Ceram. Soc. Bull.* **54**, 459 (1975).
Roy, R., and Tuttle, O. F. (1956). "Physics and Chemistry of the Earth," Vol. I, pp. 138–180. Pergamon, Oxford.
Sawatani, S., and Roy, D. M. (1977). Unpublished.
Schleede, A., Meppen, B., and Zorgensen, O. (1939). *Angew. Chem.* **52**, 316.

Schoenberg, H. P. (1963). *Biochim. Biophys. Acta* **75**, 96–103.

Scholz, H., and Kluckow, R. (1967). *In* "Crystal Growth" (H. S. Preiser, ed.), p. 475. Pergamon, Oxford.

Schwarz, H. (1967a). *Z. Anorg. Allg. Chem.* **356**, 29–35.

Schwarz, H. (1967b). *Z. Anorg. Allg. Chem.* **356**, 36–45.

Schwarz, H. (1967c). *Inorg. Nucl. Chem. Lett.* **3**, 231–236.

Schwarz, H. (1968). *Z. Anorg. Allg. Chem.* **357**, 43–53.

Segall, B., Ludwig, G. W., Woodbury, H. H., and Johnson, P. D. (1962). *Phys. Rev.* **128**, 16–19.

Shannon, R. D., and Prewitt, C. T. (1969). *Acta. Crystallogr. Sect. B* **25**, 925–946.

Shirvinskaya, A. K., and Bondar, I. A. (1975a). *Dokl. Akad. Nauk. SSSR* **225** (1), 95–98.

Shirvinskaya, A. K., and Bondar, I. A. (1975b). *Inorg. Mater.* (*SSSR*) **11** (11), 2026–2032 (in Russian).

Silverman, S. R., Fugat, R. K., and Weiser, J. D. (1952). *Am. Mineral.* **37**, 211–222.

Simpson, D. R. (1964). *Am. Mineral.* **49**, 363–376.

Simpson, D. R. (1965). *Science* **147**, 501–502.

Simpson, D. R. (1968). *Am. Mineral.* **53**, 432–444.

Skinner, H. C. W. (1973). *Am. J. Sci.* **273**, 545–560.

Strunz, H., Freigang, H., and Contag, B. (1960). *Neues Jahrb. Mineral. Monatsh.*, 47.

Sudarsanan, K., and Young, R. A. (1968). *Am. Cryst. Assoc. Meet. Buffalo, August 11–16*, paper LL2.

Sudarsanan, K., Young, R. A., and Donnay, J. O. H. (1973). *Acta Crystallogr. Sect. B* **29**, 808–814.

Trautz, O. R., LeGeros, J. P., and LeGeros, R. Z. (1964). *J. Dent. Res.* **43**, 751.

Trombe, J. C., and Montel, G. (1975). *C. R. Acad. Sci.* **280**, 567–570.

Trömel, G. (1932). *Z. Phys. Chem.* (*ABA*) **158**, 422–432.

Trömel, G. (1943). *Stahl Eisen* **63**, 21.

Trömel, G., and Eitel, W. (1957). *Z. Kristallogr.* **109**, 231–239.

Trömel, G., and Fix, W. (1961). *Archiv. Eisenhüttenwes.* **32**, 209–213.

Trömel, G., Harkort, H.-J., and Hotop, W. (1948). *Z. Anorg. Chem.* **256** (5–6), 257.

Urusov, V. S., and Khudolozhkin, V. O. (1974). *Geokhimiya* **10**, 1509–1515.

Valyashko, V. M., Kogarko, L. N., and Khodakovskiy, I. L. (1968) *Geokhimiya* No. 1, 26–36.

Van Rees, H. B., Mengeot, M., and Kostiner, E. (1973). *Mater. Res. Bull.* **8**, 1307–1310.

Van Wazer, J. R. (1958). "Phosphorus and its Compounds," Vol. I. Wiley (Interscience), New York.

Wallaeys, R. (1952). *Ann. Chim.* (*Paris*) **7**, 808–846.

Wallaeys, R. (1954). *In* "Silicon, Sulphur, Phosphates," pp. 183–190. Int. Union Pure Appl. Chem., Münster, Verlag Chemie, Weinheim.

Welch, J. H., and Gutt, W. (1961). *J. Chem. Soc.* **874**, 4442–4444.

Wondratschek, H. (1963). *Neues Jahrb. Mineral. Abh.* **99**, 113–160.

Wyllie, P. J. (1967). *J. Am. Ceram. Soc.* **50**, 43.

Young, R. A. (1974). *J. Dent. Res.*, Supple. to No. 2, **53**, 193–203.

Young, R. A., and Elliott, J. C. (1966). *Arch. Oral. Biol.* **11**, 699–707.

Young, R. A., Sudarsanan, K., and Mackie, P. E. (1968). *Bull. Soc. Chim. Fr.* 1760–1763.

V

The Relationship of Phase Diagrams to Research and Development of Sialons*

K.H. JACK

WOLFSON RESEARCH GROUP FOR HIGH-STRENGTH MATERIALS
CRYSTALLOGRAPHY LABORATORY
THE UNIVERSITY, NEWCASTLE UPON TYNE
ENGLAND

* The work at Newcastle has been supported by Joseph Lucas Limited, by the Wolfson Foundation and, more recently, by the European Research Office of the U.S. Army.

ISBN 0-12-053205-0

I. INTRODUCTION

A. Silicon Nitride

Silicon nitride is one of the special ceramics at present being developed for gas turbines and other high-temperature engineering applications. Its unique combination of properties—high strength, wear resistance, high decomposition temperature, oxidation resistance, excellent thermal shock resistance, low coefficient of friction, and resistance to corrosive environments—should make it ideal. Indeed, it has been called the ceramic of the decade since its introduction as an engineering material by Parr *et al.* (1959) two decades ago! A major difficulty has been the fabrication of precisely shaped components with these desirable properties. Because it is a covalently bonded solid, the self-diffusivity is very small and so pure silicon nitride cannot be sintered to maximum density merely by firing. The temperature required to allow sufficient material transport is high enough to give an appreciable nitrogen decomposition pressure. Three methods of shaping are used giving "reaction-bonded," "hot-pressed," and "pressureless-sintered" products.

In reaction bonding, the required shape is first made from compacted silicon powder which is then nitrided in molecular nitrogen or nitrogen–hydrogen mixtures at about 1400°C to give a product of mixed alpha and beta silicon nitrides with about 25% porosity. The original dimensions of the silicon compact remain virtually unchanged during nitriding and so complex shapes can be obtained but their high porosity and the consequent high internal surface area impair strength and oxidation resistance.

The second route is to nitride powdered silicon to give α-silicon nitride powder. With suitable oxide additives, for example magnesia, yttria, ceria or zirconia, this can be hot pressed at 1–2 ton in.$^{-2}$ in a graphite die at 1700–1800°C to give a fully dense β-silicon nitride with a high strength at temperatures up to about 1000°C.

Pressureless-sintered silicon nitride with less than 5 vol % porosity is obtained by using larger amounts of oxide additives (e.g., 5 wt % MgO) and heating for long periods at lower temperatures (1500°C) or short periods at higher ones (1750°C).

Without additive, the densification of silicon nitride by either hot pressing or pressureless sintering is negligible below 1700°C. The function of the added oxide is to provide conditions for liquid-phase sintering (Rae *et al.*, 1978) by solution of α-Si_3N_4 and reprecipitation of β. Magnesium oxide, the first widely used additive, reacts with the silica that is always present as a surface layer on the silicon nitride powder to give what was thought (Wild *et al.*, 1972) to be a silicate liquid near the $MgSiO_3$–SiO_2 eutectic composition and which, on cooling, gives a low softening-temperature glass. Impurities, particularly alkali and alkali-earth oxides, lower the glass viscosity still further and so the strength and the creep resistance of the high-density product decrease rapidly above 1000°C. It is now well established that the liquid phase and the glass that forms from it are oxynitrides (Terwilliger and Lange, 1974; Drew and Lewis, 1974; Jack, 1974). With yttria as a hot-pressing additive, an Y–Si–O–N liquid also allows liquid-phase densification with α–β transformation, but the improved properties of the product (Gazza, 1975) are due to the further reaction of the liquid with silicon nitride to give one or more highly refractory grain-boundary bonding phases. These accommodate in solid solution the impurities which would otherwise form a low softening-point glass (Rae *et al.*, 1975). Again however, as with magnesia, one advantage is at the expense of another, and some compositions of the high-strength, creep-resistant silicon nitride hot pressed with yttria fail catastrophically at about 1000°C in an oxidizing environment. The extensive cracking that occurs is due to the oxidation of the grain-boundary yttrium–silicon oxynitrides, giving oxides with a markedly different specific volume from that of the starting material.

These limitations and problems of silicon nitride technology have stimulated the current search for improved materials. It has been argued (Jack, 1974) that densification of ostensibly "pure" silicon nitride either with or without an additive can never give a homogeneous, single-phase product because of the surface silica on each powder particle. Where the second phase is vitreous it will soften at high temperatures and second-phase inclusions, whether they are crystalline or vitreous, will act as stress raisers or will initiate cracks because of their differences in thermal expansion relative to the matrix.

An alternative approach is to use the principles of "ceramic alloying" inherent in the production of "sialons," the acronym given (Jack, 1973) to phases in the Si–Al–O–N and related systems that were discovered concurrently in Japan (Oyama and Kamigaito, 1971; Oyama, 1972a) and in England (Jack and Wilson, 1972). Silicon nitride is merely the first of what is now known to be a very wide field of nitrogen-containing ceramics that include vitreous as well as crystalline materials.

B. Analogies between Nitrogen Ceramics and the Silicates

In each of its structural modifications α and β, Si_3N_4 is built up of SiN_4 tetrahedra joined in a three-dimensional network by sharing corners; each nitrogen corner is common to three tetrahedra. In a similar way, the three-dimensional framework silicates are built up of SiO_4 tetrahedra. Indeed, the atomic arrangement of β-Si_3N_4 (see Fig. 1) is the same as that of beryllium

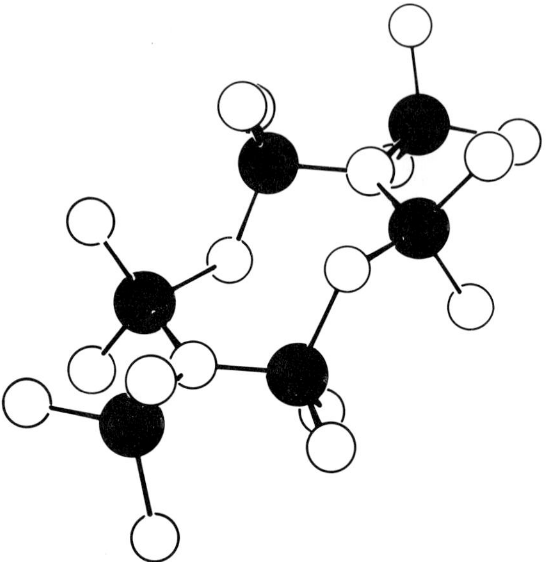

Fig. 1. The crystal structure of β-silicon nitride (Si_3N_4) (from Jack, 1976).

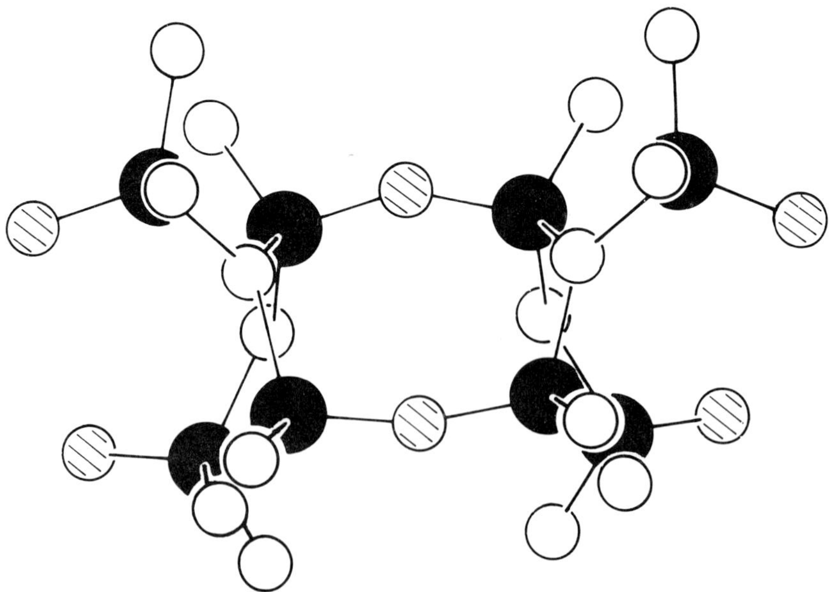

Fig. 2. The crystal structure of silicon oxynitride (Si_2N_2O) (from Jack, 1976).

silicate (Be_2SiO_4) and of zinc silicate (Zn_2SiO_4); note that the metal:non-metal atom ratio in the three solids is 3M:4X.

Silicon oxynitride (Si_2N_2O) in the Si–O–N system is potentially as good a ceramic as silicon nitride (Washburn, 1967). It is built up of SiN_3O tetrahedra [Idrestedt and Brosset (1964); see Fig. 2] and consists of parallel sheets of silicon–nitrogen atoms joined by Si–O–Si bonds. Alternatively, it can be regarded as a distorted wurtzite-type structure with one-third of the metal-atom sites vacant.

In all mineral silicates the building unit is the $[SiO_4]^{4-}$ tetrahedron carrying four negative charges; see Fig. 3. These units may occur separately or in pairs, or they may be joined together in chains or rings, in two-dimensional sheets or in three-dimensional networks. Aluminum plays a very special role in the silicates because the $[AlO_4]^{5-}$ tetrahedron—this time with five negative charges—is about the same size as $[SiO_4]^{4-}$ and can replace it in the chains, rings, sheets, or networks, provided that the necessary charge or valency compensation is made elsewhere in the structure. For example, in the amphibole silicate tremolite, the tetrahedra are joined in double chains of composition $[Si_4O_{11}]^{6-}$ that are held together by Ca^{2+} and Mg^{2+} cations. Additional anions, hydroxyl or fluorine, are present outside the double chains and the simplest tremolite composition is represented by

$$(Ca_2Mg_5)^{14}[Si_8O_{22}]^{\overline{12}}(OH)_2^{\overline{2}}$$

However, up to one-fourth of the SiO_4 tetrahedra may be replaced by AlO_4 tetrahedra without any change in structure and electrical neutrality can then be maintained by replacing $2Mg^{2+}$ by $2Al^{3+}$ outside the chains or by introducing additional alkali or alkali-earth cations. At the same time, Mg^{2+} can be replaced by Fe^{2+}, Ca^{2+} by $2Na^+$, and $2OH^-$ by $2F^-$ or O^{2-}. A typical tremolite has a simple, invariant structure but a wide and complex

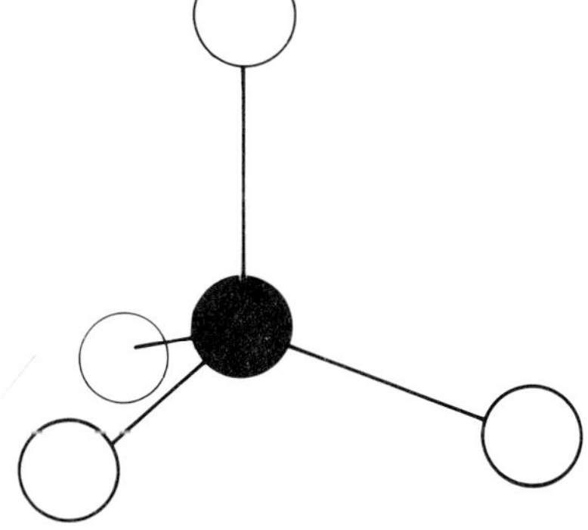

Fig. 3. The tetrahedral unit in the silicates and in nitrogen ceramics (from Jack, 1976).

range of chemical composition that must be represented as

$$(\mathrm{Ca, Na})_2^{2-4}(\mathrm{Na, K})_{0-2}^{0-2}(\mathrm{Mg, Fe})_1^2(\mathrm{Mg, Fe; Al, Fe})_4^{8-12}$$
$$[(\mathrm{Al, Si})_2\mathrm{Si}_6\mathrm{O}_{22}]^{\overline{12}-\overline{14}}(\mathrm{O, OH, F})_2^{\overline{2}-\overline{4}}$$

By analogy with the silicates it was thought that it might be possible to replace N^{3-} in Si_3N_4 by O^{2-} if at the same time Si^{4+} were replaced by Al^{3+}. Charge compensation might also be feasible by introducing other atoms and it was suggested (Wild *et al.*, 1968) that a variety of new materials, vitreous as well as crystalline, could be built up with the (Si, Al)(O, $N)_4$ tetrahedron as a structural unit in the same way that the almost infinite range of silicates is built up of (Si, $Al)O_4$ tetrahedra.

Since then, the exploration of the reversible replacement

$$Si^{4+}N^{3-} \rightleftharpoons Al^{3+}O^{2-} \tag{1}$$

has been widened and it is now clear that the sialons are essentially silicates or aluminosilicates in which oxygen is partly replaced by nitrogen. So far, structure types have been characterized in which the (Si, Al)(O, $N)_4$ tetrahedra occur as isolated units, as rings, as sheets, and as three-dimensional networks. The mutual replacement of oxygen and nitrogen in vitreous and crystalline silicates gives an additional degree of freedom that is well worth scientific and technological exploration. It also adds to the complexity of phase relationships in silicate-type systems and so it is not surprising that mistakes were made in the early interpretation of experimental observations. Until recently the exploration of the sialons has been superficial, but reliable preparative methods have now been established and products are now being characterized more precisely.

The biggest single factor in this progress has been the establishment of phase limits and phase relationships in the appropriate quaternary and quinary systems and by their representation in so-called "behavior diagrams."

II. THE SIALONS

A. Early Preparative Work

1. The Reaction of Silicon Nitride with Alumina

In spite of the predicted replacement of Si–N by Al–O according to Eq. (1), the first results of reacting silicon nitride with alumina by hot pressing at 1700–2000°C were surprising. As shown by Fig. 4, the x-ray diffraction patterns of products containing up to 70 wt % of Al_2O_3 were identical with β-Si_3N_4 except that the reflections moved to slightly lower Bragg angles

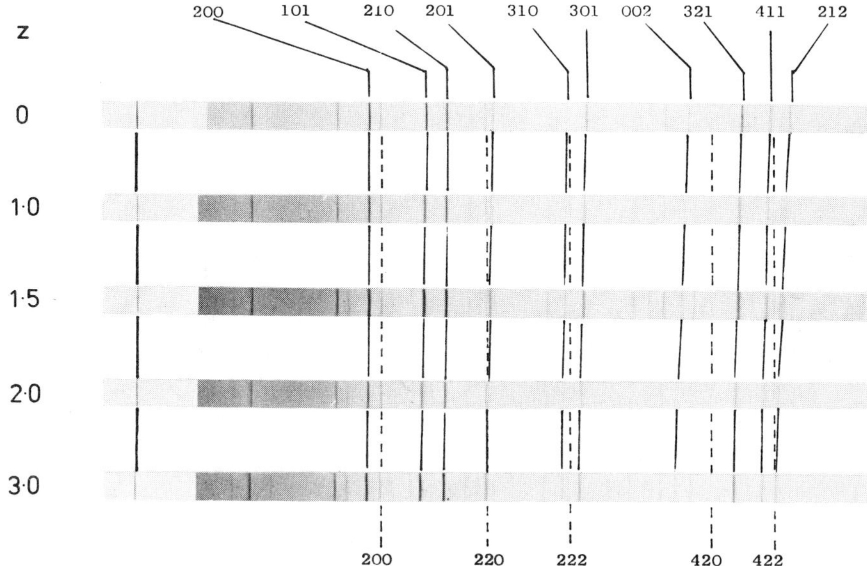

Fig. 4. X-ray diffraction patterns of β'-sialons prepared by reaction of Si_3N_4 and Al_2O_3 at 1700°C.

as the alumina content increased. The homogeneous sialon solid solution was therefore termed β' and the following alternative compositions, both of which satisfy valency requirements, were considered:

$$\begin{array}{ll} \beta & Si_6^{24}N_8^{\overline{24}} \\ \downarrow & \\ \beta' & Si^{24-3x}_{6-0.75x}Al^{2x}_{0.67x}O^{\overline{2x}}_{x}N^{\overline{24-3x}}_{8-x} \\ \searrow & \\ & Al^{16}_{5.33}O^{\overline{16}}_{8} \equiv Al_2O_3 \end{array} \tag{2}$$

$$\begin{array}{ll} \beta & Si_6^{24}N_8^{\overline{24}} \\ \downarrow & \\ \beta' & Si^{24-4z}_{6-z}Al^{3z}_{z}O^{\overline{2z}}_{z}N^{\overline{24-3z}}_{8-z} \\ \searrow & \\ & Al^{18}_{6}O^{\overline{12}}_{6}N^{\overline{6}}_{2} \equiv Al_2O_3\cdot AlN \end{array} \tag{3}$$

On the basis of limited chemical analyses and the observation of only one single-phase crystalline product it was concluded that the sequence shown by Eq. (2) represented the reaction and that the β' product was $Si_{6-0.75x}Al_{0.67x}O_xN_{8-x}$ with x, the number of nitrogens in the unit cell replaced by oxygen, reaching a maximum of about six. It was not appreciated that products represented by Eq. (3) might be obtained with (i) volatilization of silicon monoxide and nitrogen or, in the reducing environment of the graphite hot-pressing die, (ii) volatilization of carbon and silicon monoxides, or (iii) with simultaneous formation of a silica-rich glass. For example, possible reactions to produce a β' composition $Si_{6-z}Al_zO_zN_{8-z}$ with $z = 4$ are given by Eqs. (4)–(6):

$$4Si_3N_4 + 6Al_2O_3 = 3Si_2Al_4O_4N_4 + 6SiO + 2N_2 \tag{4}$$

$$Si_3N_4 + 2Al_2O_3 + C = Si_2Al_4O_4N_4 + SiO + CO \tag{5}$$

$$Si_3N_4 + 2Al_2O_3 = Si_2Al_4O_4N_4 + SiO_2 \tag{6}$$

X-ray photographs of some of the β' products showed traces of an additional phase ascribed to reaction with the silica always present on the silicon nitride powder. Deliberate reaction of increasing amounts of silica with β' or with silicon nitride:alumina mixtures gave increasing quantities of the new phase which, because its structure could not be determined, was called "X." Just as silica reacts with silicon nitride to give silicon oxynitride

$$Si_3N_4 + SiO_2 = 2Si_2N_2O \tag{7}$$

it was also found to react with β' in an appropriate ratio to give an oxynitride-type sialon designated as O′. The reactions between silicon nitride, alumina, and silica were represented by the diagram shown in Fig. 5.

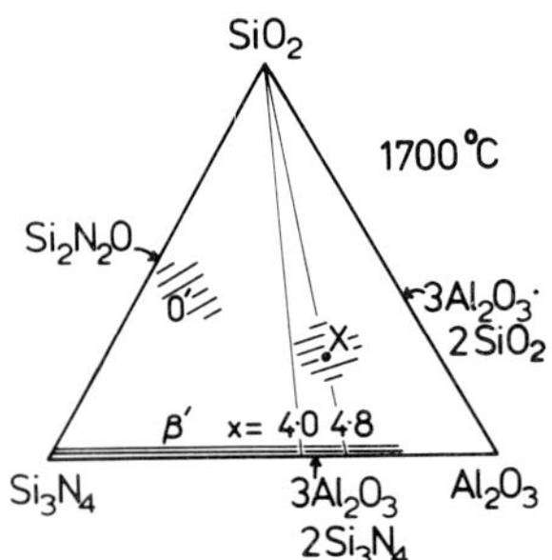

Fig. 5. Early representation of phase relationships in the SiO_2–Si_3N_4–Al_2O_3 system at 1700°C (now known to be incorrect) (from Jack, 1973).

2. Reactions of Mixed Oxides with Silicon Nitride

Magnesium–silicon nitride ($MgSiN_2$), manganese–silicon nitride ($MnSiN_2$), lithium–silicon nitride ($LiSi_2N_3$), and other metal–silicon nitrides have structures based on the wurtzite-type aluminum nitride, AlN, in which the structural units are AlN_4 tetrahedra. In $MgSiN_2$ there are equal numbers of MgN_4 and SiN_4 tetrahedra and in $LiSi_2N_3$ there are twice as many SiN_4 tetrahedra as there are LiN_4 units. It seemed probable that Mg, Mn, Li, and perhaps other metals might be incorporated into sialon structures.

By reaction of magnesia and alumina or of magnesium spinel ($MgAl_2O_4$) with silicon nitride a range of Mg–Si–Al oxynitrides with the β' structure was prepared and with higher MgO concentrations other oxynitride phases

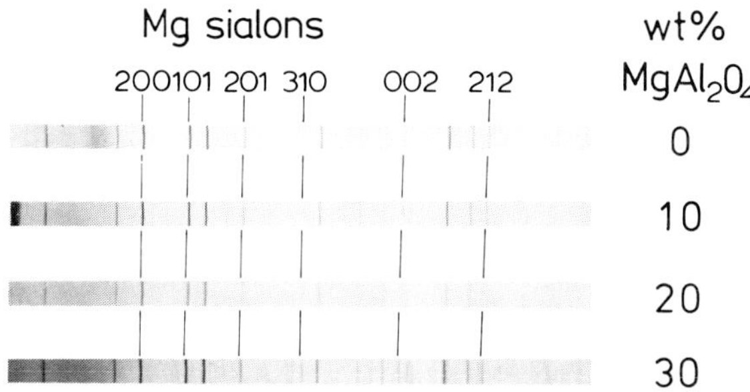

Fig. 6. X-ray photographs of β'-magnesium sialons prepared by reaction of $MgAl_2O_4$ and Si_3N_4 at 1700°C (from Jack, 1976).

"Y" and "W" with distorted AlN-type structures were obtained. X-ray photographs of β'-magnesium sialons prepared by reaction of $MgAl_2O_4$ and Si_3N_4 at 1700°C are shown in Fig. 6 and the phase relationships first proposed in the MgO–Si_3N_4–Al_2O_3 system are given in Fig. 7.

By reaction of $LiSi_2N_3$ with alumina or of lithium aluminum spinel ($LiAl_5O_8$) with silicon nitride β'-lithium sialons were found as well as lithium- and nitrogen-containing analogs of α-silicon nitride, silicon oxynitride, and cristobalite. Phase relationships were thought to be shown by Fig. 8.

Preliminary work (Jack, 1973) suggested a large region of β'-homogeneity in the Be_2SiO_4–Si_3N_4–Al_2O_3 system (see Fig. 9) and at about the same time Oyama (1972b) observed an extended β'-phase area in the Ga_2O_3–Si_3N_4–Al_2O_3 system (Fig. 10).

3. The Si_3N_4–Al_2O_3–AlN System

The $Si_{6-0.75x}Al_{0.67x}O_xN_{8-x}$ formulation for β'-sialon requires $\frac{1}{12}x$ vacant metal-atom sites in the hexagonal unit cell whereas there are no vacancies in the range $Si_{6-z}Al_zO_zN_{8-z}$. The terminal composition at $z = 6$ is $Al_2O_3 \cdot AlN$ with a metal: nonmetal atom ratio 3M:4X as in silicon nitride. Both in Japan (Oyama, 1974) and Newcastle Si_3N_4 was reacted with equimolecular mixtures of Al_2O_3 and AlN (i.e., the equivalent of the spinel $Al_3O_3N \equiv Al_2O_3 \cdot AlN$) and then with varying ratios of the aluminum oxide and nitride. It was concluded that the β'-phase field extended not only along the joins Si_3N_4–Al_2O_3 and Si_3N_4–Al_3O_3N corresponding to both the "x" and "z" compositions but also covered the region between these limits. The phase diagrams deduced for the Si_3N_4–Al_2O_3–AlN system in Japan and Newcastle were almost identical (see Fig. 11) but both were incorrect.

Doubts about the wide range of homogeneity for β' were raised by Lumby *et al.* (1975) who suggested (from the results of creep measurements) that

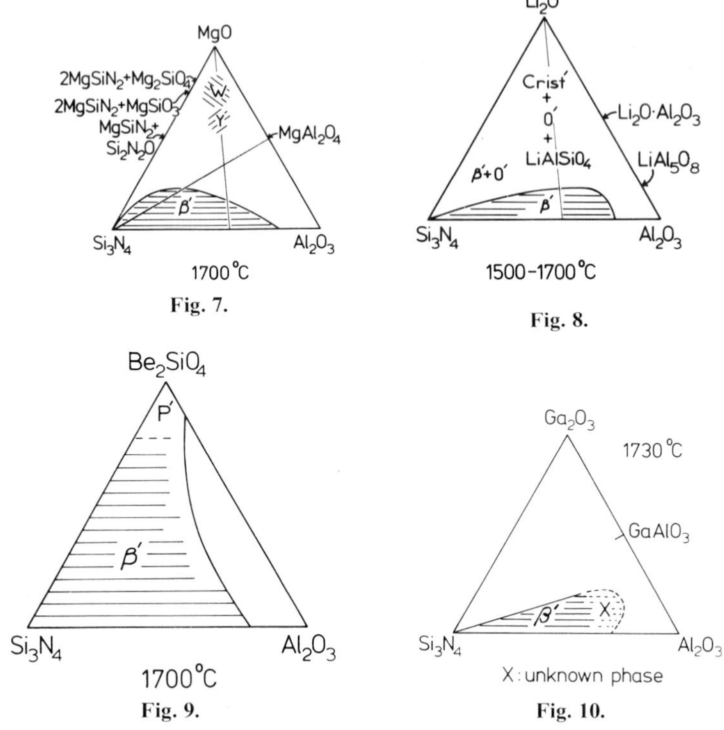

Fig. 7.

Fig. 8.

Fig. 9.

Fig. 10.

Fig. 7. Phase relationships in the $MgO–Si_3N_4–Al_2O_3$ system at 1700°C (now known to be incorrect) (from Jack, 1973).

Fig. 8. Phase relationships in the $Li_2O–Si_3N_4–Al_2O_3$ system at 1500–1700°C (now known to be incorrect) from Jack, 1973).

Fig. 9. Phase relationships in the $Be_2SiO_4–Si_3N_4–Al_2O_3$ system at 1700°C (now known to be incorrect) from Jack, 1973).

Fig. 10. Phase relationships in the $Ga_2O_3–Si_3N_4–Al_2O_3$ system at 1730°C, X: unknown phase (proposed by Oyama, 1972; from Jack 1973).

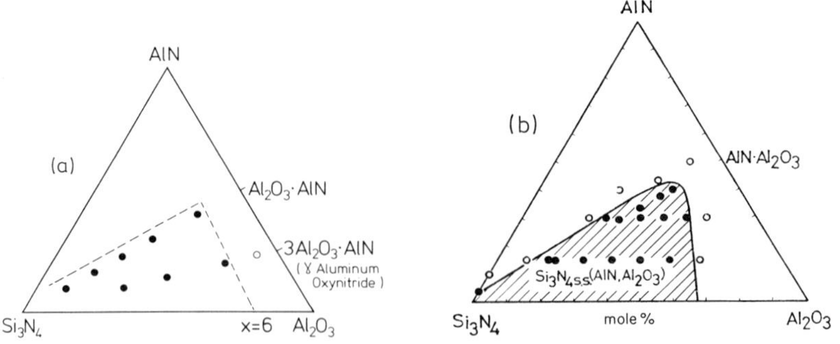

Fig. 11. Phase diagrams for the $Si_3N_4–Al_2O_3–AlN$ system above 1700°C (both are now known to be incorrect). (a) From results at Newcastle; (b) proposed by Oyama (1974). ●, solid solution; ○, solid solution and other phases (from Jack, 1976).

sialons with overall compositions along the Si_3N_4–Al_2O_3 join (the "x" formulation) contained glassy phases whereas the "z" compositions with 3M:4X showed minimum high-temperature creep. Since then, careful preparative work by Gauckler *et al.* (1975), by Lumby *et al.* (1977), and at Newcastle [see Jack (1976)] has shown that β' extends along the join Si_3N_4–Al_3O_3N from $z = 0$ to about $z = 4.2$ at 1750°C. The range of homogeneity with M:X unequal to 3:4 is quite limited.

B. The Representation of the Si–Al–O–N System

The Si–Al–O–N system has apparently four components and so should be represented by a regular tetrahedron (Fig. 12), each of the vertices of which represent one atom (or one gram atom) of the respective elements. A point within the tetrahedron represents one atom of composition

$$Si_aAl_bO_cN_{1-(a+b+c)}$$

However, in any condensed phase and by analogy with the silicates it is unlikely that the elements will have other than their accepted fixed valencies Si^{IV}, Al^{III}, O^{II}, and N^{III}. Thus, one degree of freedom is lost and because the sum of the positive valencies must equal that of the negative ones

$$4a + 3b = 2c + 3(1 - a - b - c) \tag{8}$$

From Eq. (8) it follows that the composition of a solid phase is given by

$$Si_aAl_bO_{3-7a-6b}N_{6a+5b-2}$$

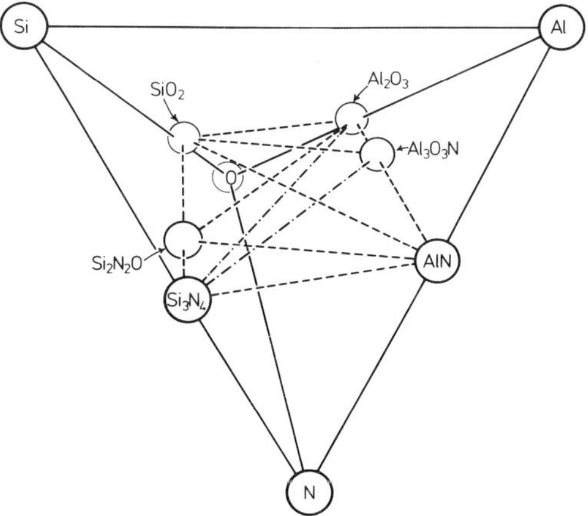

Fig. 12. The tetrahedral representation of the Si–Al–O–N system.

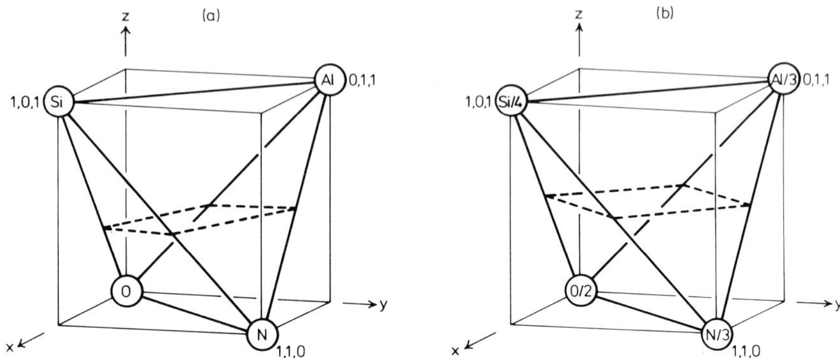

Fig. 13. Depiction of a tetrahedron in terms of three orthogonal axes. In (a) corners represent atomic units and in (b) equivalent units.

and the system becomes a pseudoternary one; any composition is defined by only two variables *a* and *b*.

The easiest depiction of a regular tetrahedron is the figure obtained by joining the four alternate corners of a cube (see Fig. 13) with orthogonal edges *x*, *y*, and *z*. In terms of these axes the four tetrahedron corners have coordinates:

	u	*v*	*w*
oxygen at	0	0	0
silicon at	1	0	1
aluminum at	0	1	1
and nitrogen at	1	1	0

The coordinates of a composition

$$Si_aAl_bO_{3-7a-6b}N_{6a+5b-2}$$

are

$$u = 7a + 5b - 2, \qquad v = 6a + 6b - 2, \qquad \text{and} \qquad w = a + b$$

which, because $hu + kv + lw + 2 = 0$, always lies on one plane with indices $(hkl) = (01\bar{6})$. This irregular quadrilateral plane which cuts the edges of the tetrahedron at the compositions $SiO_2/3$, $Al_2O_3/5$, $AlN/2$, and $Si_3N_4/7$ is shown in Fig. 13a and in a [001] projection by Fig. 14. It is a representation of the Si_3N_4–AlN–Al_2O_3–SiO_2 system and includes compositions of all possible condensed Si–Al–O–N phases. The only assumption is that the four combining elements retain their usual fixed valencies. The gaseous compound silicon monoxide in which silicon is divalent is therefore outside the plane. The irregular quadrilateral representation of the system has been

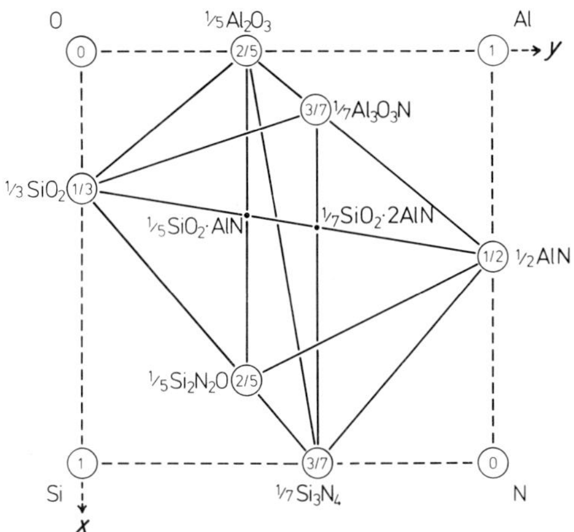

Fig. 14. Irregular quadrilateral plane representing the Si_3N_4–AlN–Al_2O_3–SiO_2 system. Figures within circles are heights above the plane of the paper along the cube axis *z*.

used recently (Wills *et al.*, 1977) to illustrate phase relationships but it is inconvenient for plotting compositions and a much simpler representation is obtained by expressing concentrations not in atoms or gram atoms but in equivalents.

Each corner of the Si–Al–O–N tetrahedron is then one equivalent of an element and, because one equivalent of any element or compound always reacts with one equivalent of any other species, the compositions of the compounds Si_3N_4, AlN, Al_2O_3, and SiO_2—expressed in equivalents–are located at the midpoints of the tetrahedron edges; see Fig. 13b. The square obtained by joining these midpoints is the simplest representation of the Si_3N_4–AlN–Al_2O_3–SiO_2 system.

The same conclusion can be reached in another way. Since the composition of any Si–Al–O–N compound is determined by only two variables, these can be plotted perpendicular to each other. Using equivalent concentrations, i.e., where the unit of concentration is the atom divided by valency, the ratios

$$\frac{[\mathrm{Al}/3]}{[\mathrm{Si}/4] + [\mathrm{Al}/3]} \quad \text{and} \quad \frac{[\mathrm{O}/2]}{[\mathrm{N}/3] + [\mathrm{O}/2]}$$

when plotted in this way give a square. This is exactly the same as for a reciprocal salt pair [see Zernicke (1955), and Findlay (1927)], e.g., Na_2SO_4–KCl–NaCl–K_2SO_4, where the equilibrium is written

$$\mathrm{Na} \cdot (\mathrm{SO_4})/2 + \mathrm{KCl} \rightleftharpoons \mathrm{NaCl} + \mathrm{K} \cdot (\mathrm{SO_4})/2$$

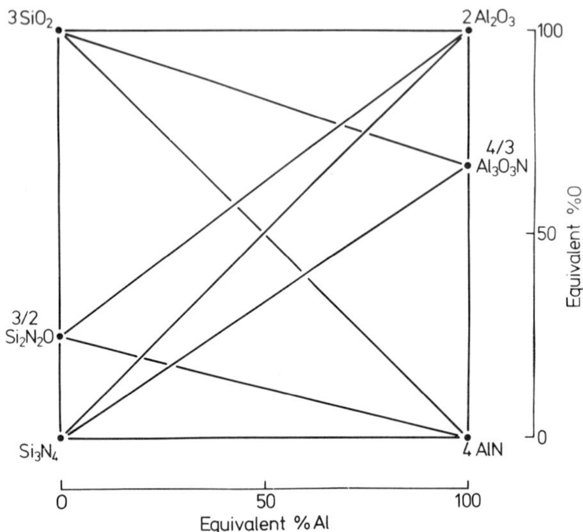

Fig. 15. The square representation of the Si–Al–O–N system, using equivalent concentrations.

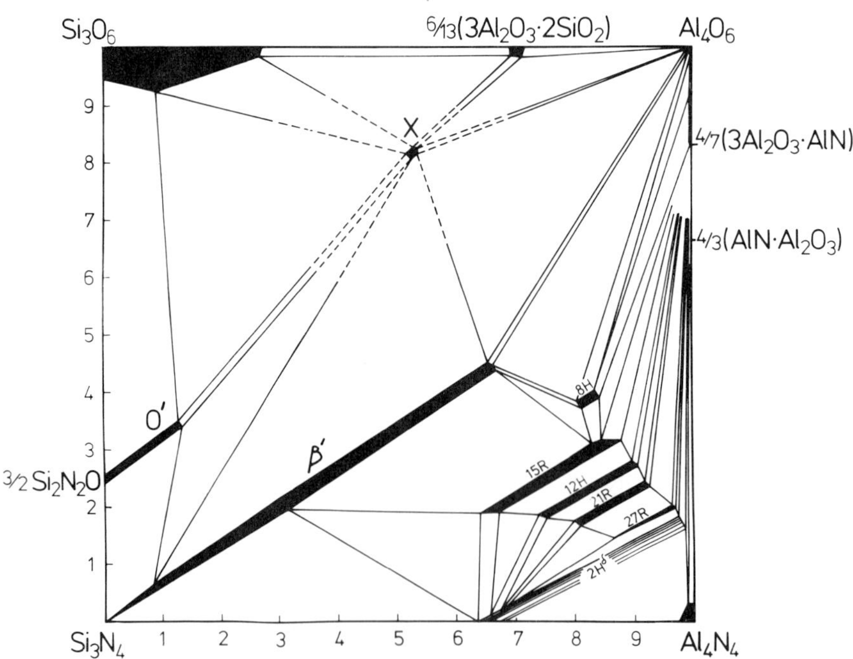

Fig. 16. The Si_3N_4–AlN–Al_2O_3–SiO_2 behavior diagram at 1750°C based on research at Newcastle.

and where, in the same square, the ratios

$$\frac{[K^+]}{[Na^+]+[K^+]} \quad \text{and} \quad \frac{[SO_4/2^-]}{[Cl^-]+[SO_4/2^-]}$$

are plotted against each other.

It is convenient, although not mandatory, to let the bottom left-hand corner of the square represent one mole of Si_3N_4; the other three corners then represent Al_4N_4, Al_4O_6, and Si_3O_6 as in Figs. 15 and 16.

It should be emphasized that all possible phases or mixtures of phases in which the combining elements Si, Al, O, and N have their normal valencies lie within this diagram. It is the same as the irregular quadrilateral plane of Fig. 14 except that concentrations are expressed in equivalents instead of atomic units. Any point on the square diagram is a combination of 12 +ve and 12 −ve valencies, i.e., it is convenient to regard compounds in ionic terms even though the interatomic bonding is predominantly covalent. In going from left to right, $3Si^{4+}$ is gradually replaced by $4Al^{3+}$; and from bottom to top, $4N^{3-}$ is replaced by $6O^{2-}$. The center of the square represents a composition $Si_{1.5}Al_2O_3N_2$. It should also be noted that the number of atoms changes with the change of position in the diagram but the number of equivalents remains constant. Because of the $12^+:12^-$ "composition" it is often convenient to scale the sides of the square 0–12; each unit is then one valency.

C. The Phases of the Si–Al–O–N System

1. Experimental Methods

The results of reacting together appropriate mixtures of Si_3N_4, AlN, Al_2O_3, SiO_2, and Si_2N_2O by pseudoisostatic hot pressing in a graphite die, using boron nitride as a powder vehicle (Colquhoun *et al.*, 1973), are shown by Fig. 16. Hot pressing was usually at 1750°C but higher and lower temperatures (1550–2000°C) were also used and, when possible, the same product was synthesized from different mixtures to check compositions and reproducibility. Thus, a mixture $4AlN:3SiO_2$ should give the same product as a mixture $Si_3N_4:2Al_2O_3$. There is often some weight loss due to volatilization at high temperature according to Eqs. (4) and (5) and in making up mixtures compensation must be made for the surface oxides present on the powder particles of the two nitrides. Recrushing, remilling, and hot pressing a second time after making minor composition adjustments are often necessary in order to prepare a pure phase. Moreover, to make these adjustments the limits of the required phase and its relationship with other phases in the system must be known and so the synthesis of sialons and the derivation of a phase diagram are related iterative processes.

Figure 16 is a behavior diagram for the particular experimental conditions used at Newcastle and it is not claimed to be an equilibrium phase diagram in a thermodynamic sense. However, recent calculations by Torre and Mocellin (1977) suggest that it is at least self-consistent and compatible with equilibrium at 1 atm pressure.

2. Phase Relationships

Each phase in the sialon system extends in a direction of constant metal: nonmetal atom ratio M:X along which Si–N is replaced increasingly by Al–O. In the direction in which the M:X ratio changes most rapidly, i.e., parallel to the $AlN–SiO_2$ join, the homogeneity range of each phase is small. These features are due, at least in part, to the similarity of the two bond lengths

$$\text{Si–N } (1.75\ \text{Å}) \approx \text{Al–O } (1.75\ \text{Å})$$

and the difference between the other two bonds

$$\text{Al–N } (1.87\ \text{Å}) \neq \text{Si–O } (1.62\ \text{Å})$$

3. The β'-Sialon Phase

The β'-phase has the same crystal structure as β-silicon nitride with a hexagonal unit cell containing M_6X_8. It extends along the 3M:4X line of the behavior diagram with a homogeneity range $Si_{6-z}Al_zO_zN_{8-z}$ where the maximum z value exceeds 4 at 1750°C but decreases with decreasing temperature to about 2 at 1450°C. The variation of unit-cell dimensions with composition is given by Fig. 17. Until recently, and prior to the establishment of a reliable behavior diagram, specimens usually contained small amounts of

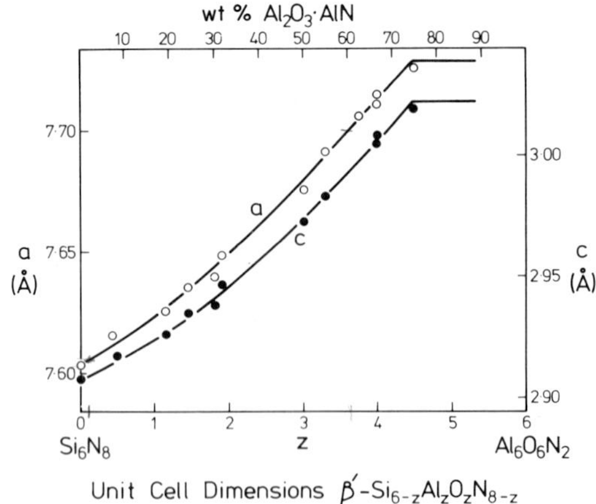

Fig. 17. Variation of unit-cell dimensions of β'-sialon with composition.

other vitreous or crystalline phases and so it is not certain whether the intrinsic properties of β' have yet been evaluated. Measurements on compositions containing about equal concentrations of silicon and aluminum, $Si_3Al_3O_3N_5$ with $z = 3$, show that because of its structure the physical and mechanical properties are similar to those of β-silicon nitride, but chemically it is closer to aluminum oxide. The thermal expansion coefficient (2.7×10^{-6}/°C) is less than that of β-Si_3N_4 (3.5×10^{-6}) and, although thermal conductivity is lower, it has good thermal shock properties [see Arrol (1974)]. Oxidation resistance is better than for silicon nitride and compatibility with molten metals is excellent. As described in Section V.C the highest cold strength so far found for any ceramic is observed with β'-sialon bonded with a yttrium aluminosilicate glass. The use of such materials for holding and conveying molten metals and also for cutting tools has recently been emphasized (Lumby *et al.*, 1978) and these applications are perhaps of more immediate importance—and are more easily realized—than the very exacting ones for ceramic turbine components.

A potential advantage of β'-sialon over silicon nitride is its easier densification by both pressureless sintering and hot pressing (Rae *et al.*, 1978). The usual ceramic techniques of extrusion, pressing, and slip casting can be employed to produce shapes from the mixed powder components together with a suitable additive, and then these can be fired to near theoretical density in an inert atmosphere at 1600–1700°C.

β' is much more stable than β-Si_3N_4 despite some reports to the contrary. Gauckler *et al.* (1977) have recently stated that a $z = 4$ sialon mix of $2SiO_2$:4AlN heated at 1950°C for 2 min gave β' corresponding only to $z \approx 2$ plus a glass, and the product on subsequent treatment at 1300°C for 8 hr consisted of mullite, alumina, two different AlN-polytypes, an unknown phase, and a greatly reduced amount of β'. The authors therefore suggested that $z = 4$ β' is unstable and that the isothermal sections of the sialon phase diagram must be markedly different at 1760 and 1300°C. Attempts to repeat these observations at Newcastle failed, and specimens of $z = 4$ β'-sialon remained unchanged after heating up to 2200°C in nitrogen. These discrepancies were resolved by an even more recent report (Gauckler, 1978) that the apparent thermal decomposition of β' was, in fact, a reaction of finely divided powder with major amounts of silica impurity; the formation of a silica-rich glass under these circumstances is exactly what might be expected from the behavior diagrams of Figs. 16 and 19.

4. The O′-Phase

The O′-phase with the structure of silicon oxynitride (Si_2N_2O) extends to a limited extent along the 2M:3X line toward Al_2O_3. No detailed examination of its properties has been made.

5. The X-Phase

The X-phase was originally thought to have a composition $SiAlO_2N$ and was described as a "nitrogen-mullite" (Jack, 1973). Recent research at Newcastle shows that it exists in "high" and "low" modifications depending on whether it is cooled rapidly or slowly from its melting temperature ($\approx 1720°C$). Both modifications have triclinic unit cells, simply related to one another and with dimensions:

	a (Å)	b (Å)	c (Å)	$\alpha(°\delta)$	$\beta(°\delta)$	$\gamma(°\delta)$
High-X	9.658	2.842	11.168	90.0	124.4	98.5
Low-X	9.684	8.560	11.211	90.0	124.5	98.5

Low-X is a superlattice of high-X, both having compositions near $Si_7Al_9O_{23}N_3$. Although the complete crystal structures are not yet resolved, they are built up of MX_4 tetrahedra and MX_6 octahedra as in mullite and so the original description does not seem unreasonable.

6. Tetrahedral AlN-Polytype Phases

Of the six uncharacterized phases reported by Gauckler *et al.* (1975) the one that occurs most frequently is observed as a minor phase in the hot pressing of AlN-rich β' compositions. It was designated in early work at Newcastle as "Y" and occurs also in the Mg–Si–Al–O–N system (Jack, 1973; Hendry *et al.*, 1975). The diffraction pattern of each of the phases near the AlN corner of the Si–Al–O–N diagram has been interpreted in terms of a new kind of polytype based upon the AlN structure but containing excess nonmetal atoms (Thompson, 1977a; Roebuck and Thompson, 1977; Jack, 1976). The structures are directly related to their compositions M_mX_{m+1} and are described by the Ramsdell symbols 8H, 15R, 12H, 21R, 27R, and $2H^{\delta}$. Thompson (1976) has described a similar series of polytypes with compositions $M_{m+1}X_m$ in the Be–Si–O–N system.

D. Other Si–Al–O–N Phase Diagrams

The isothermal section of the Si–Al–O–N system at 1760°C proposed by Gauckler *et al.* (1975) is reproduced as Fig. 18 and is superficially similar to Fig. 16. As shown by the comparison of phase compositions in Table I, the greatest discrepancies concern the Al–N polytypes. The Newcastle compositions are in excellent agreement with the measured densities and the proposed crystal structures (Thompson, 1977a; see Jack, 1976) the general features of which have been recently confirmed by direct lattice imageing

TABLE I

COMPOSITIONS OF Si–Al–O–N PHASES

Gauckler *et al.* (1975)		Published and unpublished research at Newcastle, 1976–1978	
Phase designation	M:X ratio	M:X ratio	Phase designation
β'	3:4	3:4	β'
—	—	2:3	O′
X_1	8:13	8:13	X
X_4	3:4	4:5	8H
X_2	4:5	5:6	15R
X_5	5:6	6:7	12H
X_6	6:7	7:8	21R
—		9:10	27R
X_7	8:9	>9:10	$2H^{\delta}$

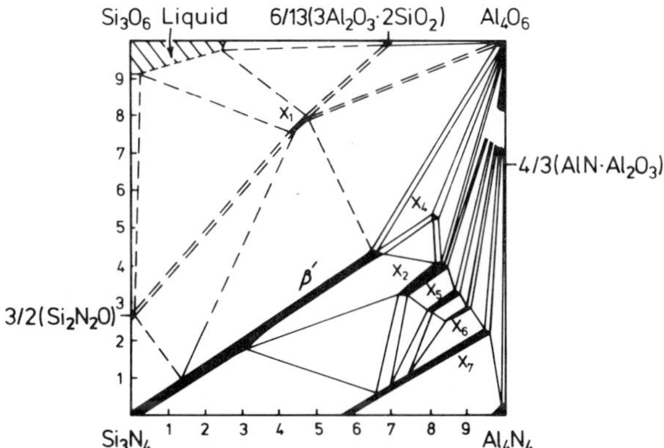

Fig. 18. Isothermal section of the Si–Al–O–N system at 1760°C (after Gauckler *et al.*, 1975). X_2 = 15R, X_4 = 8H, X_5 = 12H, X_6 = 21R, X_7 = 27R (Newcastle).

(Clarke *et al.*, 1978). It is beyond reasonable doubt that the compositions of the polytypes are close to those shown by Fig. 16 and not by Fig. 18.

The incomplete diagrams by Layden (1976) given as Fig. 19 are important in showing the extent of liquid phase and are in agreement with the Newcastle observation that X-phase melts at about 1720°C. The small liquid field at 1650°C near the silica-rich corner of the diagram extends considerably to include X-phase at 1750°C. At this temperature and above, there exists a large two-phase field of β' + liquid.

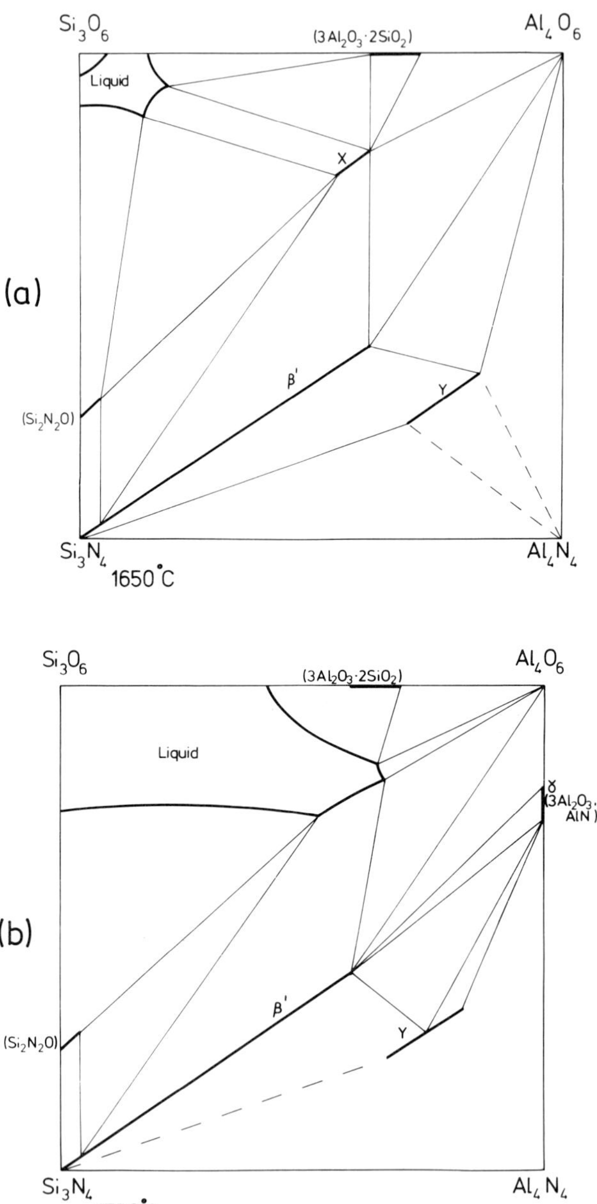

Fig. 19. Phase diagram for the Si–Al–O–N system (a) at 1650°C and (b) at 1750°C (after Layden, 1976).

E. Sialon Processing

It is clear that any process for the production of β' or indeed any other sialon material requires a precise knowledge of the phase limits and phase relationships in the Si–Al–O–N system over a wide range of temperature. The early "phase diagrams" reproduced as Figs. 5 and 7–11 were almost completely misleading, and rapid progress has been possible only since the more reliable but still incomplete behavior diagram of Fig. 16 was established. Mixtures of Si_3N_4, Al_2O_3, and AlN or other appropriate starting powders can be reacted to give a pure, theoretical density β'-sialon provided, as previously stated, that compensation is made for the surface oxide on the nitrides. Densification starts at about 1450°C due to particle rearrangement accompanying the chemical reaction and is complete above 1750°C after the formation of liquid. Even if the overall composition of the mix corresponds to β', the distribution of surface oxide ensures that liquid is produced initially. Complete reaction to β' requires about one hour at 1750–1800°C, and if the composition is perfectly balanced all liquid phase should disappear just when the reaction is finished. Compositions on the SiO_2-rich side of the 3M:4X line give β' + liquid and, although the latter should produce X-phase at equilibrium, crystallization is seldom complete and the product cools to give a glass that impairs creep and oxidation resistance. Small amounts of 15R polytype do not degrade properties and so a final composition on the AlN-rich side of β' is preferable.

Densification and reaction are facilitated by addition of metal oxide to the sialon mix because, in general, liquid metal–aluminosilicates containing nitrogen are formed and so allow the transport of material by a solution-reprecipitation process (Rae *et al.*, 1978). For example, with the addition of magnesia a liquid oxynitride that promotes densification is first formed at about 1500°C and then, as shown in Section IV.C, can subsequently react to be incorporated finally in a homogeneous, single-phase β'-magnesium sialon. To interpret the reactions that occur when a third metal M is introduced, where M is Mg, Li, Y, Ce, or Zr, it is necessary to consider the representation of M–Si–Al–O–N phase relationships.

III. THE REPRESENTATION OF METAL–SIALON SYSTEMS

If the third metal M in any M–Si–Al–O–N system has a fixed valency, the apparent five components are effectively reduced to four in the same way as described in Section IIB and any composition is then defined by only three variables. The system is pseudoquaternary or, in the classification

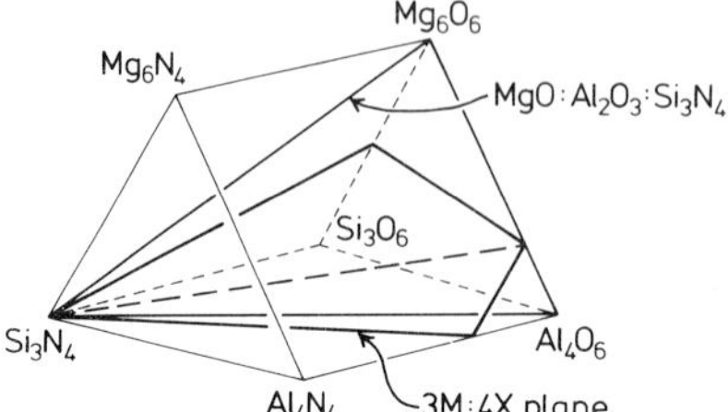

Fig. 20. The representation of the Mg–Si–Al–O–N system.

described by Zernike (1955), a "quaternary system of the third kind." This is represented by Jänecke's (1907) triangular prism in which all edges are equal. Figure 20 outlines this representation for the magnesium–sialon system; it is based on the standard Si_3N_4–Al_4N_4–Al_4O_6–Si_3O_6 square of Fig. 16 with Mg in equivalent units along the third dimension. The front triangular face of the prism represents nitrides and the rear face oxides. Thus, as shown by Fig. 21a the distance of any point *P* from the front face represents the concentration ratio in equivalents of O/(N + O). In the triangular section containing point *P* (see Fig. 21b) the metal-atom concentrations, again in equivalents, are represented in the usual way for any three-component system. Any point in the prism again represents a combination of 12 +ve and 12 −ve valencies and in Fig. 21 the edges of the

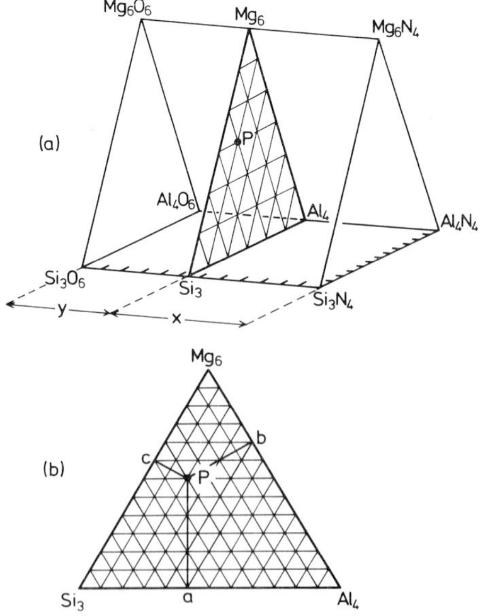

Fig. 21. Method of representing a composition by a point within the M–Si–Al–O–N triangular prism. (P represents $Mg_{3.00}Si_{1.00}Al_{0.67}O_{3.30}N_{1.80}$; see text.)

triangular prism are scaled in valency units; the point *P* thus has a composition in valency units

$$Mg^{6+}Si^{4+}Al^{2+}O^{6.60-}N^{5.40-}$$

and hence in atomic units

$$Mg_{3.00}Si_{1.00}Al_{0.67}O_{3.30}N_{1.80}$$

In the Mg–Si–Al–O–N system each of the phases of the basal square plane extend into the prism volume, often along planes of constant M:X value. Such planes are not parallel with any of the prism faces and they cut other triangular planes representing pseudoternary subsystems. Thus, in Fig. 20 the 3M:4X plane cuts the plane of the $Mg_6O_6:Si_3N_4:Al_4O_6$ subsystem along a single line joining Si_3N_4 with $\frac{3}{2}(MgAl_2O_4)$.

Other subsystems (i) $Si_3N_4:Y_2O_3:Al_2O_3$, (ii) $Si_3N_4:Y_2O_3$:"Al_3O_3N", (iii) $Si_3N_4:Y_2O_3$:AlN, and the 2M:3X and 3M:4X planes of the Y–Si–Al–O–N system are illustrated by Fig. 22. The complete investigation of

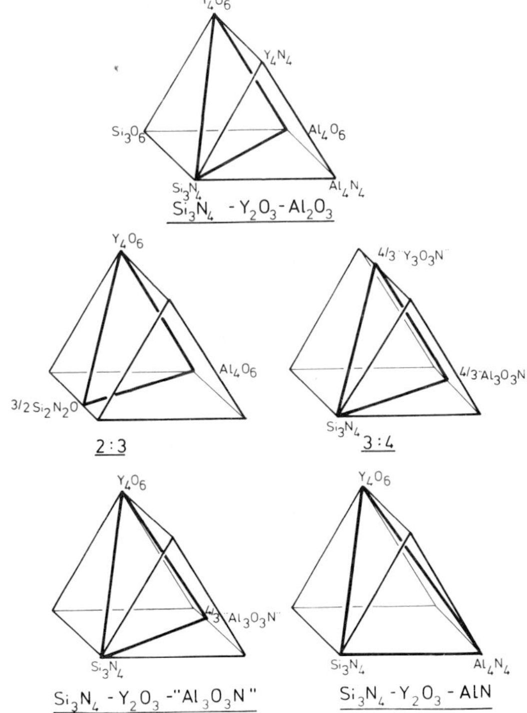

Fig. 22. Representative subsystems and planes of constant M:X value in the Y–Si–Al–O–N system.

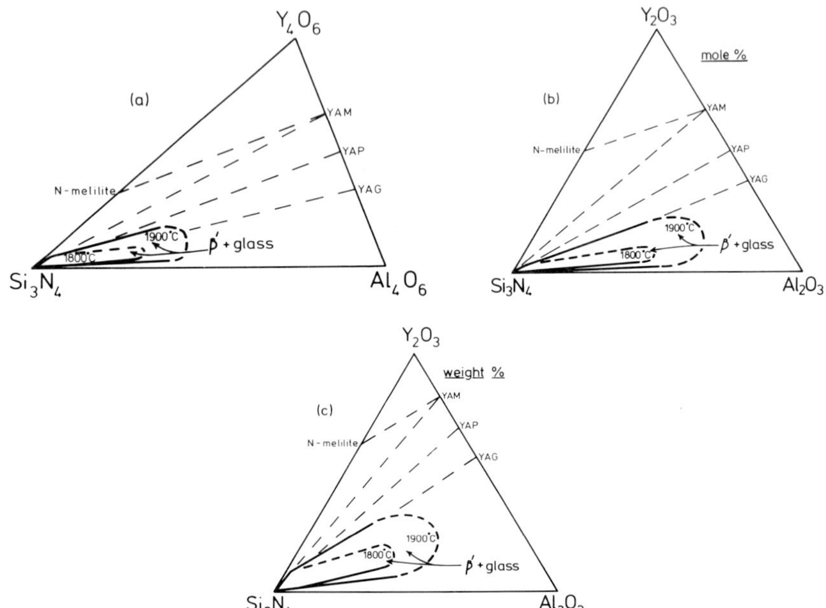

Fig. 23. The Y_2O_3–Si_3N_4–Al_2O_3 behavior diagram, showing glass formation after heating to 1900°C. (a) In equivalent concentration units, (b) in mole per cent, and (c) in weight per cent.

a whole system requires exploration of different subsystems the representations of which are subsequently fitted together in the Jänecke prism. The subsystems are therefore depicted by appropriately shaped triangles with compositions shown in equivalent units. Alternatively, any subsystem can be represented by an equilateral triangle with corners representing atomic or molecular compositions. A third method is, of course, to depict triangular subsystems with compositions in weight percent. Whatever method is used, it is recommended that the corners and the diagrams be marked appropriately. As an example, the Y_2O_3–Si_3N_4–Al_2O_3 behavior diagram, showing glass formation after heating to 1900°C, is represented in three different ways by Figs. 23a–c.

With the current increasing activity in sialon research and development it is worth urging the adoption of simple conventions in presenting phase diagrams. Accepting that Si_3N_4 is placed at the front, bottom, and left-hand corner of the prism and that the basal square of the Si–Al–O–N system is as shown by Figs. 16 and 18, the left- and right-hand square faces of the prism represent, respectively, the M–Si–O–N and M–Al–O–N systems while the front and back faces show the respective M–Si–Al–N and M–Si–Al–O systems. These five subsystems of the whole can be drawn simultaneously, as shown in Fig. 24, by opening the prism and allowing the

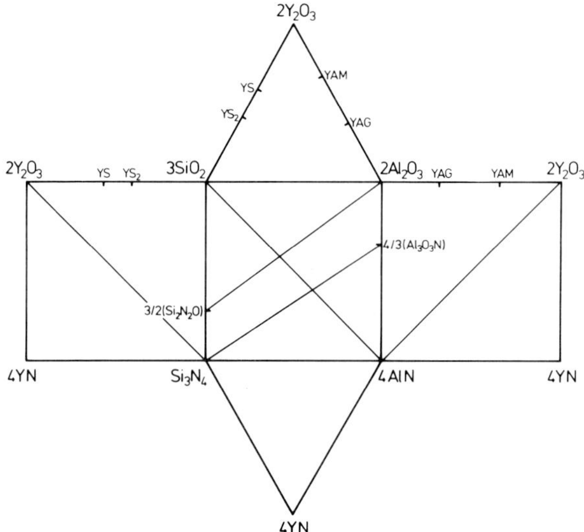

Fig. 24. Schematic and graphical representation of the Y–Si–Al–O–N system as an opened prism.

four nonbasal faces to fall parallel with the basal plane. Each subsystem is then viewed as from the *inside* of the prism.

It is less logical, but has been found more convenient in practice, to represent all the subsystems, except that of the basal plane, as if the prism faces are viewed from the *outside*. Examples of both methods of representing the Y–Si–O–N system, both in equivalent units, are given in Fig. 25. A convention by which the face is viewed as from the *outside* of the prism is recommended.

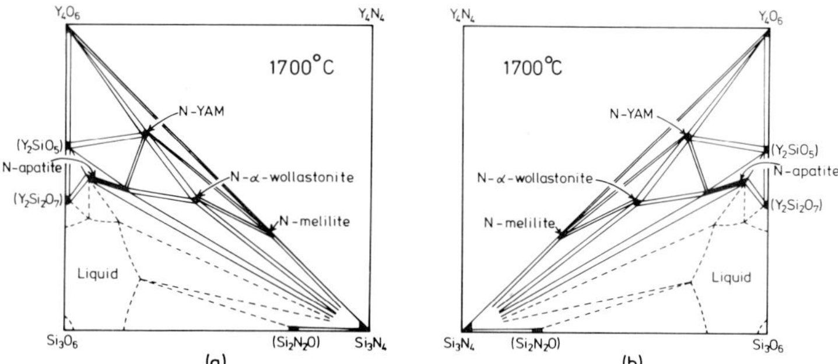

Fig. 25. The Y–Si–O–N system as the left-hand square face of the Y–Si–Al–O–N prism in equivalent concentration units. (a) As viewed from *outside* the prism and (b) as viewed from *inside* the prism.

IV. THE Mg–Si–Al–O–N AND RELATED SYSTEMS

A. Magnesium Sialons

Magnesium is tetrahedrally coordinated by nitrogen in magnesium nitride and magnesium silicon nitride, whereas it is most frequently coordinated octahedrally in oxides and silicates. Magnesium is therefore a "network former" in sialons and, as stated in Section III, extends the Si–Al–O–N phases from lines of constant M:X ratio into planes of the same ratio in the prism volume. The $MgO–Si_3N_4–Al_2O_3$ section shown by Fig. 26 cuts across planes of constant M:X and so shows very few extended single-phase regions except β'. In general, mixtures of magnesium-containing α', 12H, 15R, and nitrogen-spinel are observed.

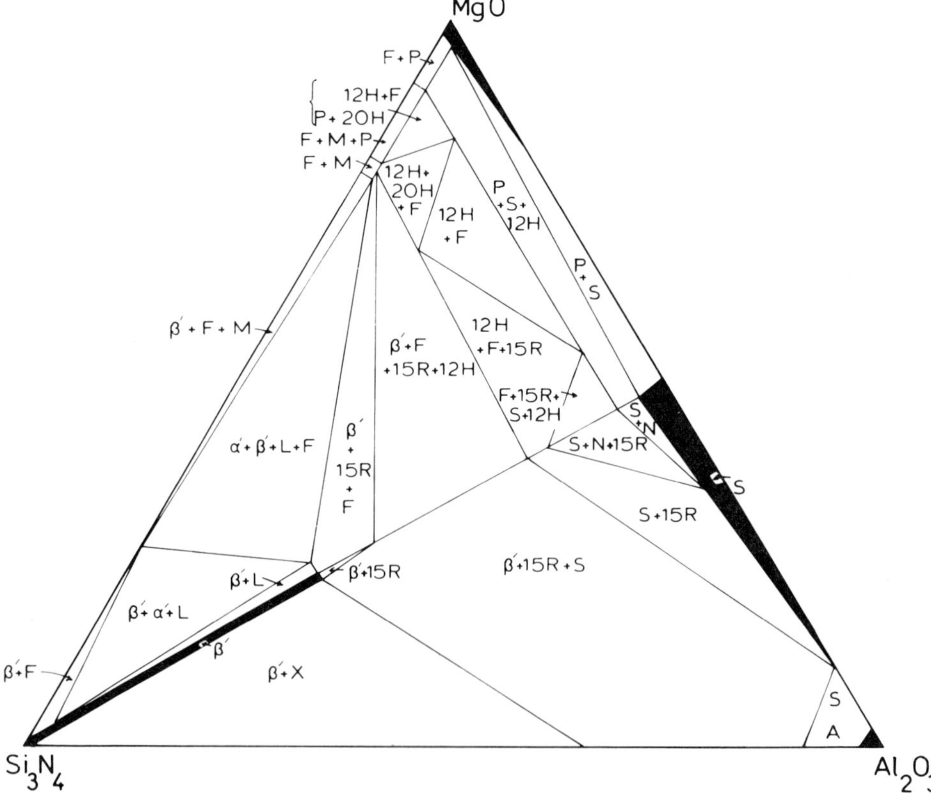

Fig. 26. The $MgO–Si_3N_4–Al_2O_3$ section of the Mg–Si–Al–O–N system at 1800°C. F, Forsterite; P, Periclase; S, Spinel; A, α-A_2O_3; L, Liquid; M, $MgSiN_2$. Reproduced from Jack (1974) by permission of The Chemical Society, London.

B. Magnesium Sialon Polytypes

All the polytypes observed in the basic Si–Al–O–N system are found in Mg–Si–Al–O–N at comparable compositions on the appropriate M:X planes but, in addition to the expected structures, new ones also appear. Figure 27 shows the region of homogeneity on the 6M:7X plane for the expected 12H magnesium sialon and for a modification 6H′ with the same 6M:7X ratio but for which structural details are not yet established. Similarly, on the 7M:8X plane (see Fig. 28) the expected 21R polytype and its 14H

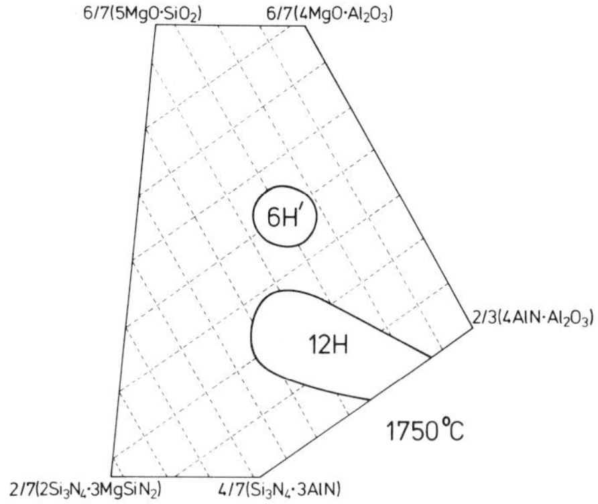

Fig. 27. The 6M:7X plane of the Mg–Si–Al–O–N system, showing two polytype modifications 12H and 6H′.

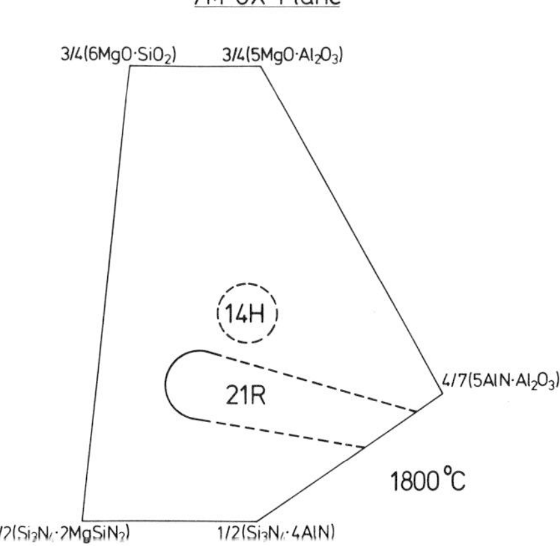

Fig. 28. The 7M:8X plane of the Mg–Si–Al–O–N system showing two polytype modifications 21R and 14H.

modification both have a fundamental periodicity of seven MX double layers one of which contains an extra nonmetal atom and becomes MX_2. It is assumed that the relatively minor structural differences between polytypes of the same M:X ratio are associated with differences in the atomic ordering of the atoms Mg, Si, and Al, or perhaps of O and N, as the relative numbers of these change within the overall constant M:X value. Such subtle structural changes can be correlated with changes in composition only if adequate representations of these latter changes can be shown on a phase diagram. Figures 27 and 28, and the many others like them that are used to represent different sections of the whole Mg–Si–Al–O–N system, are indispensable for the interpretation of the complex crystal chemistry of polytypes.

C. Hot-Pressed Silicon Nitride

In using magnesia as a hot-pressing additive to silicon nitride Lange (1976) states that the maximum strength and oxidation resistance at 1400°C occur with $MgO:SiO_2$ ratios that approach zero and infinity, that is, along the joins Si_3N_4–Si_2N_2O and Si_3N_4–MgO. This is explained by Fig. 29, representing the Mg–Si–O–N system on which the approximate region of liquid at 1700°C is given by a dashed outline. On cooling, all the liquid does not crystallize. In fact, glass formation in the system is remarkably easy as shown by the series of x-ray photographs of Fig. 30. When mixtures of 20 wt % MgO:80 wt % SiO_2 are pressed at 1700°C without Si_3N_4, the product, as expected, consists of crystalline protoenstatite and cristobalite. Addition of 10 wt % Si_3N_4 to this same mix gives, after the same treatment, a completely vitreous product which devitrifies at 1500°C to cristobalite, clinoenstatite, and silicon oxynitride. Corresponding optical micrographs are shown by Fig. 31. It is evident from the glass composition region shown in Fig. 29 that

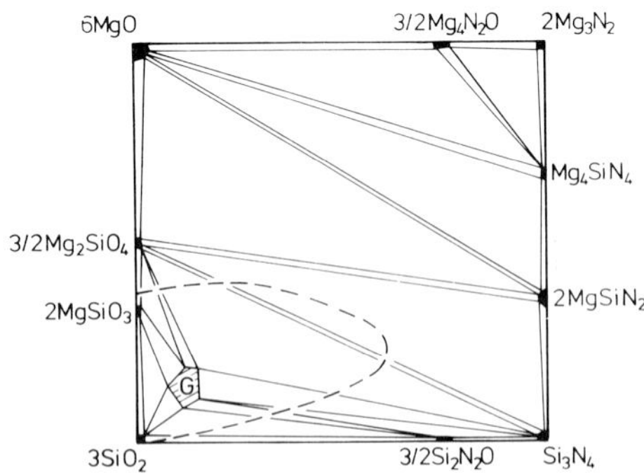

Fig. 29. The Mg–Si–O–N behavior diagram. ---, liquid at 1700°C.

p-e
c
c
wt % MgO SiO2 Si3N4
1
1700°C
20
80
0
2
19
77
4
3
18
72
10
1500°C
4
c
c-e
c
Si2N2O

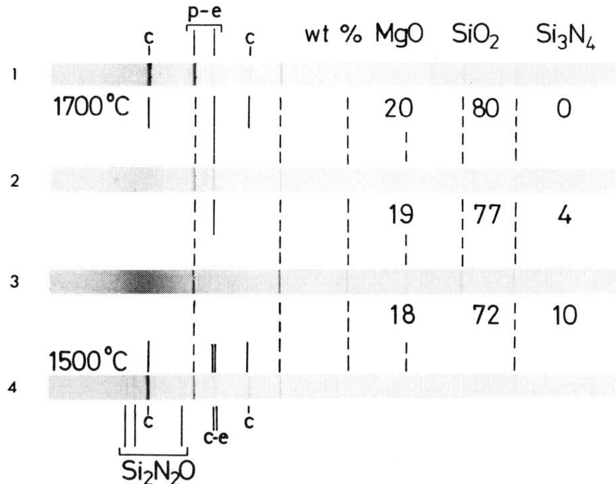

Fig. 30. X-ray photographs showing Mg–Si–O–N glass formation and devitrification. ---, KCl standard; c, cristobalite (SiO_2); p-e, protoenstatite; c-e, clinoenstatite.

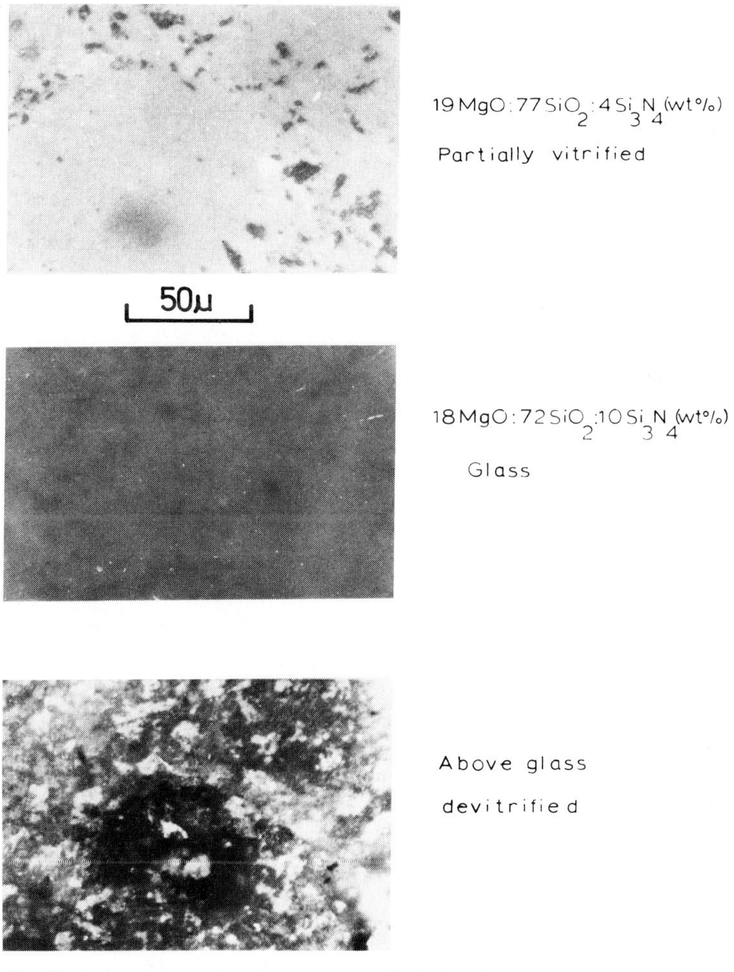

19 MgO : 77 SiO2 : 4 Si3N4 (wt%)
Partially vitrified
50μ
18 MgO : 72 SiO2 : 10 Si3N4 (wt%)
Glass
Above glass
devitrified

Fig. 31. Optical micrographs showing Mg–Si–O–N glass formation and devitrification ($\times 340$).

silicon nitride with its usual 4 wt % of surface silica will contain glass after hot pressing or sintering with added magnesia. Only when the MgO content is low enough to bring the overall composition within or near the Si_3N_4–Si_2N_2O two-phase region or is high enough to bring it into the Si_3N_4–$MgSiN_2$–Mg_2SiO_4 compatibility triangle, will glass be avoided.

D. Single-Phase Sialons

Liquid and glass-forming regions in Mg–Si–O–N extend into the Mg–Si–Al–O–N system. The 3M:4X plane (see Fig. 20) is important because, as shown by Fig. 32, the β'-sialon phase extends along this plane from the Si_3N_4–Al_3O_3N join toward forsterite [Mg_2SiO_4 ($\equiv$ $2MgO.SiO_2$)]. In densifying silicon nitride with magnesia the presence of a surface layer of silica prevents the production of a homogeneous single-phase material. Figures 32 and 33 show that by addition of the appropriate amounts of Al_2O_3 and AlN—equivalent to Al_3O_3N—with just sufficient MgO to react with the surface silica, a completely balanced composition can be obtained to give a homogeneous single-phase β'-magnesium sialon. Moreover, the magnesia and silica react initially with a small amount of silicon nitride to produce a Mg–Si–O–N liquid which allows liquid-phase sintering and then, by suitable heat-treatment after densification, the liquid is incorporated by further reaction into the β' solid solution.

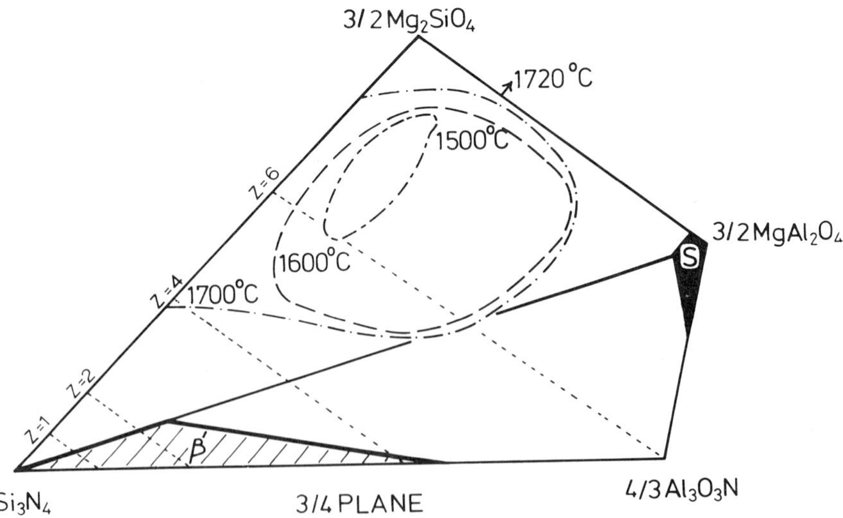

Fig. 32. The 3M:4X plane of the Mg–Si–Al–O–N system with liquid regions shown at 1500–1700°C. —·—, ——, —-—, liquid regions.

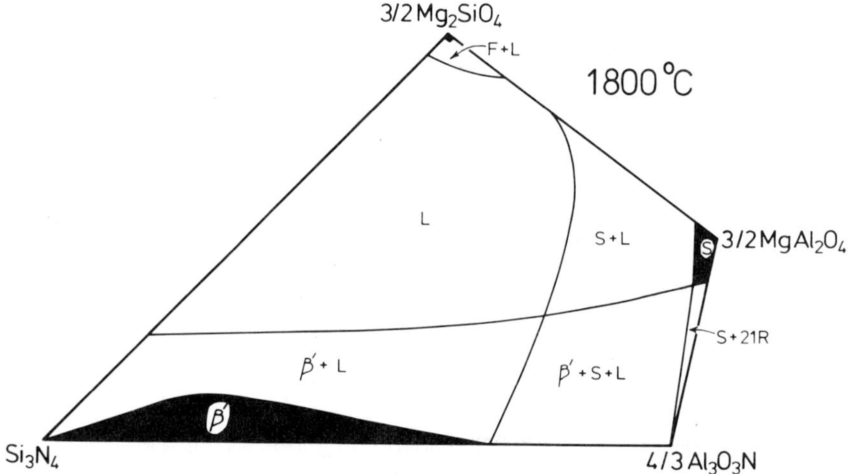

Fig. 33. The 3M:4X plane of the Mg–Si–Al–O–N system at 1800°C. F, forsterite; L, liquid; S, spinel.

E. MgO–BeO Additions

On chemical grounds, beryllia is expected to behave as an additive to silicon nitride and to sialons in a way similar to magnesia. However, because the silicate Be_2SiO_4 (phenacite) has the same atomic arrangement as β-Si_3N_4 it is much more soluble in β and β'-sialon than forsterite (Jack, 1973; Gauckler *et al.*, 1976). Thus, there is a very wide range of homogeneity of β' on the 3M:4X plane of the Be–Si–Al–O–N system; see Fig. 34. Oda *et al.* (1977) has reported near theoretical densities for silicon nitride pressureless sintered at 1760°C with mixed additions of (i) 3wt % BeO:3 wt % MgO, and (ii) 1.25 wt % BeO:3.75 wt % MgO:5 wt % CeO_2. The room-temperature rupture moduli

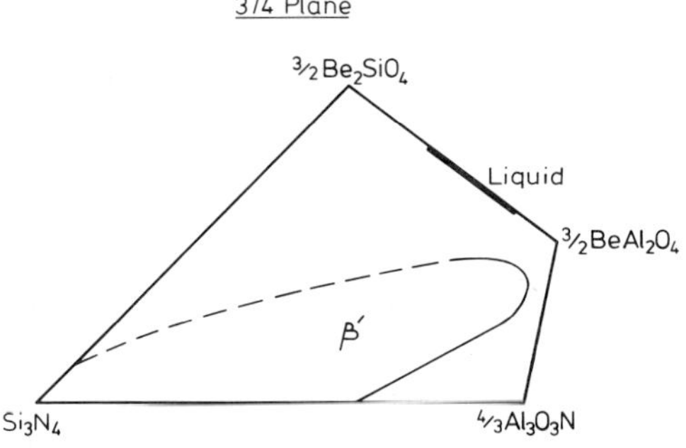

Fig. 34. The 3M:4X plane of the Be–Si–Al–O–N system (after Gauckler *et al.*, 1977).

are also high and, once again, the behavior is explicable in terms of the phase diagram. Mixtures of BeO and MgO added to silicon nitride and to β'-sialon compositions are expected to give lower eutectics than with MgO alone and so at a similar temperature they will extend the liquid regions. Moreover, the greater homogeneity range of β' should allow more compositional flexibility for the subsequent incorporation of all additives and impurities into a single-phase material.

The possible beneficial effect of using mixed additives in the densification of silicon nitride and sialons has not been explored systematically but to be effective this approach to ceramic alloying must be based upon a detailed knowledge of phase relationships.

F. Lithium Sialons

Lithium is another tetrahedrally coordinated "network former" in nitride and oxynitride systems. As with the magnesium sialons, completely new

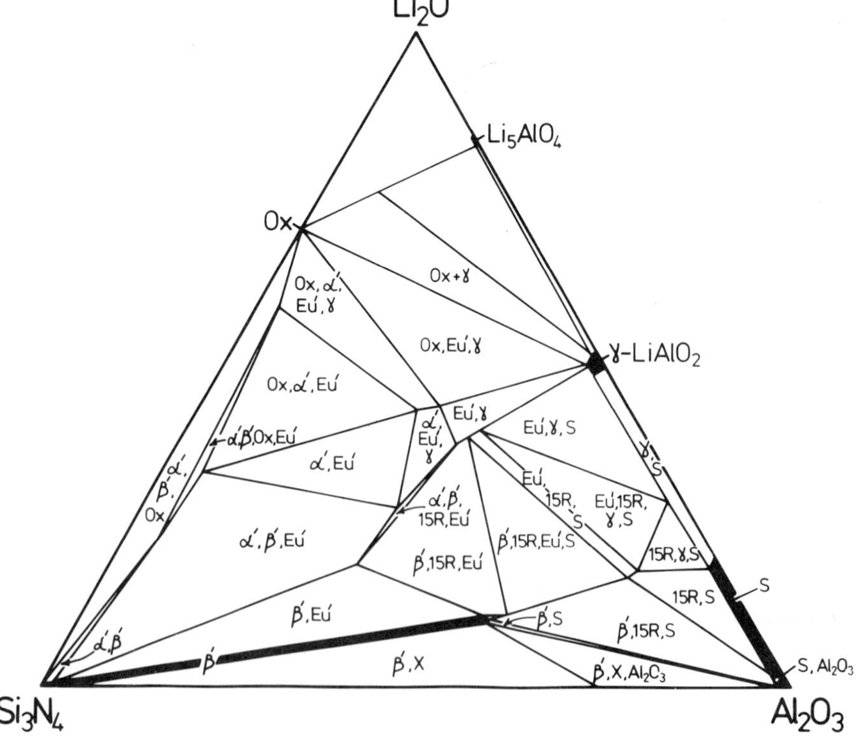

Fig. 35. The Li_2O–Si_3N_4–Al_2O_3 section of the Li–Si–Al–O–N system at 1550°C (from Jack, 1976).

phases occur in the Li–Si–Al–O–N system as well as those which extend from the basic Si–Al–O–N behavior diagram. Reaction of silicon nitride with lithium–aluminum spinel, $LiAl_5O_8$, gives β'-lithium sialon and with lithium aluminate, $LiAlO_2$, a sialon (α') with the structure of α-silicon nitride.

Other nitrogen-containing phases have structures based on β-eucryptite (Eu′), spinel (S), silicon oxynitride (Ox), and a 15R polytype. A behavior diagram of the Li_2O–Si_3N_4–Al_2O_3 section of the system is given by Fig. 35 but again it shows no extensive single-phase regions.

V. YTTRIUM SIALONS

A. The Y–Si–O–N System

In oxynitride systems yttrium is a "network modifier" presumably because its charge density is too small and its radius too large for tetrahedral coordination. It is important for two reasons. First, it was shown by Gazza (1973) that much improved high-temperature strengths can be obtained by hot pressing relatively impure silicon nitride with yttria instead of magnesia and second, the highest strength ceramic so far obtained is that with the addition of yttria and alumina to silicon nitride (Venables *et al.*, 1977).

The Y_2O_3–SiO_2–Si_3N_4 behavior diagram deduced from investigations at Newcastle (Rae *et al.*, 1975; Rae, 1976; Rae *et al.*, 1978) is given in equivalent units as one face of the Y–Si–Al–O–N system at 1700°C by Fig. 25. The following four quaternary phases have been fully characterized by analysis, synthesis, and crystal structure determination:

(i) N-melilite, $Y_2O_3 \cdot Si_3N_4 \equiv Y_2Si[Si_2O_3N_4]$,
(ii) N-apatite, with a range of homogeneity $(Y, Si, \square)_{10}[Si(O, N)_4]_6 (O, N, \square)_2$,
(iii) N-YAM, $2Y_2O_3 \cdot Si_2N_2O \equiv Y_4Si_2O_7N_2$ (J-phase), and
(iv) N-α-wollastonite, $Y_2O_3 \cdot Si_2N_2O \equiv 2YSiO_2N$ (H′-phase)

The existence of (i)–(iii) is confirmed by Lange *et al.* (1976) and Wills *et al.* (1976). More conventional phase diagrams are given in weight percent and mole percent by Fig. 36 and are similar, except in detail, to the less complete diagrams proposed by Lange and Wills.

N-melilite ($Y_2Si[Si_2O_3N_4]$) consists of parallel sheets of Si–O–N tetrahedra held together by yttrium cations sandwiched between them. It is isostructural with the melilite silicates akermanite ($Ca_2Mg[Si_2O_7]$), and gehlenite ($Ca_2Al[AlSiO_7]$) and with each of them it forms a complete series of high melting solid solutions. Thus, appreciable amounts of Ca, Mg, Al, and other

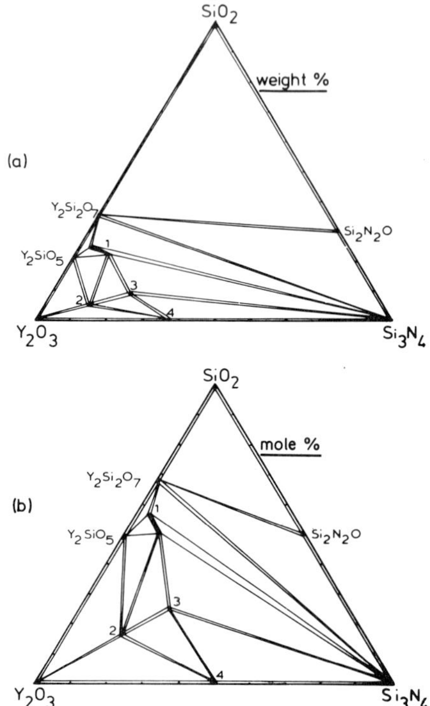

Fig. 36. The Y_2O_3–SiO_2–Si_3N_4 phase diagram at 1700°C from Rae (1976): (a) in weight percent; (b) in mole percent. (1) N-apatite; (2) N-YAM; (3) N-α-wollastonite; (4) N-melilite.

impurities in silicon nitride which would otherwise form a low softening-temperature glass are accommodated in the structure without impairing creep resistance and strength.

N-YAM, originally called "J-phase," is isostructural with the yttrium aluminate "YAM." N-α-wollastonite (originally "H′-phase") is isostructural with a yttrium aluminate "YAP" the low-temperature modification of which has a similar structure to α-wollastonite ($CaSiO_3$). These two yttrium–silicon oxynitrides lie on the Y_2O_3–Si_2N_2O join of the 2M:3X plane of the Y–Si–Al–O–N system (see Fig. 37) and they form solid solutions with the corresponding yttrium aluminates.

In the densification of silicon nitride with yttria, the latter reacts initially with surface silica and some of the nitride to give a liquid (see Fig. 25b) that provides conditions for liquid-phase sintering. As the reaction proceeds, the yttria- and silica-rich liquid combines with more Si_3N_4 to give one or more of the refractory bonding phases in which are dissolved the impurities that would otherwise degrade the creep properties.

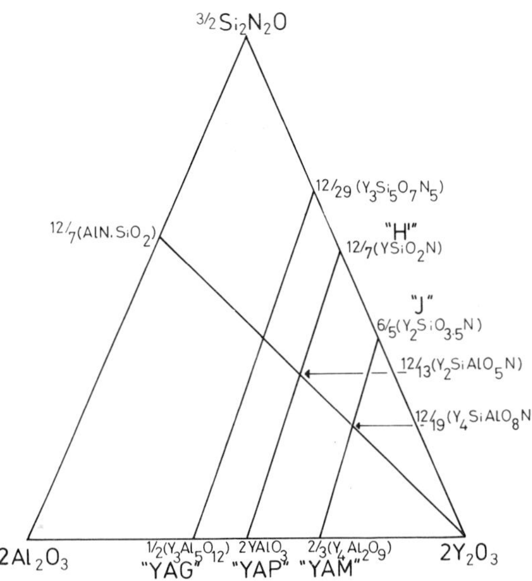

Fig. 37. The 2M:3X plane of the Y–Si–Al–O–N system. Reproduced from Jack (1974) by permission of The Chemical Society, London.

B. Oxidation Resistance of Y_2O_3–Si_3N_4 Ceramics

Yttria would seem to be the ideal hot-pressing additive. The product shows good oxidation resistance at 1400°C by forming a coherent, protective glaze. Unfortunately, oxidation at around 1000°C produces extensive cracking and this exposes fresh surfaces for further attack. Large quantities of y-$Y_2Si_2O_7$ and cristobalite are produced and the specific volume change initiates the cracks that make the process catastrophically progressive.

Lange *et al.* (1976) point out that materials within the compatibility triangle Si_3N_4–Si_2N_2O–$Y_2Si_2O_7$ show excellent oxidation resistance at both 1000 and 1400°C and this has been amply confirmed. Figure 38 shows two specimens after oxidation at 1000°C for 120 hr. The upper bar, showing no change from its initial appearance, had been hot pressed with additions of 5 wt % SiO_2 + 5 wt % Y_2O_3; the lower one, bent and extensively cracked, had been hot pressed with 15 wt % Y_2O_3.

Once again, the marked difference of behavior is due to the formation of different phases and of different proportions of phases under different processing conditions. Figure 39 is the Si_3N_4-rich corner of the Y_2O_3–SiO_2–Si_3N_4 system on which lines representing 2, 5, and 10 wt % surface silica are drawn. Clearly, the amount of added yttria necessary to bring the overall composition within the favorable Si_3N_4–Si_2N_2O–$Y_2Si_2O_7$ triangle,

Fig. 38. Si_3N_4 bars oxidized at 1000°C for 120h: (a) hot pressed with 5 wt % SiO_2 + 5 wt % Y_2O_3 (b) hot pressed with 15 wt % Y_2O_3.

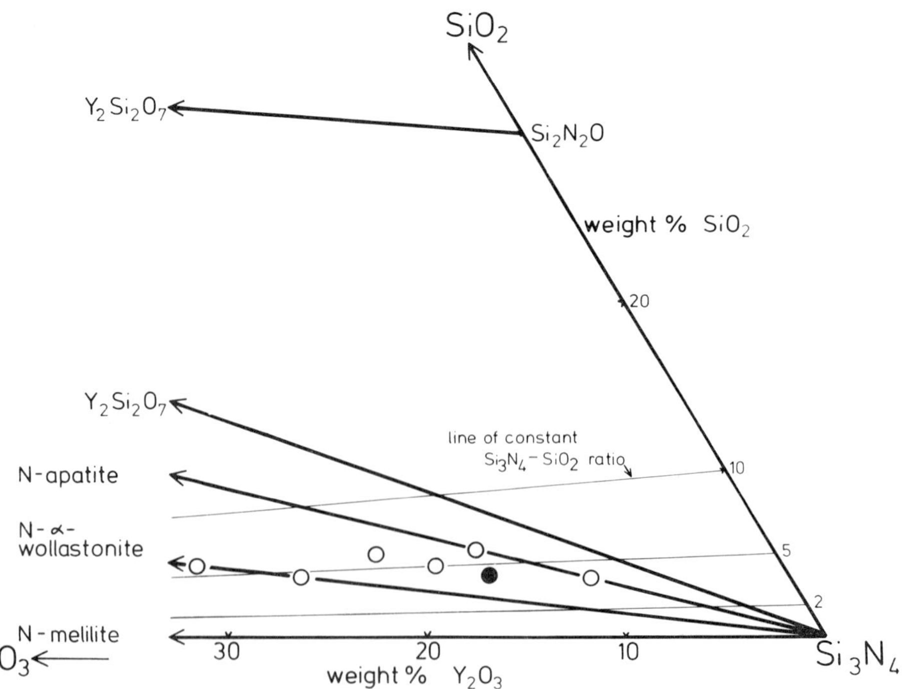

Fig. 39. The Si_3N_4 corner of the Y_2O_3–SiO_2–Si_3N_4 system showing 2, 5, and 10 wt % surface silica.

or to bring it into phase fields containing one or more of the easily oxidized quaternary phases, is very critically dependent on the amount of surface silica. The latter varies with particle size and with different commercial sources of silicon nitride.

By preventing the formation of quaternary oxynitrides, impurities such as calcium are no longer accommodated in a refractory grain-boundary phase and so, depending on the concentration of such impurities, the high-temperature creep resistance might be impaired by improving the oxidation resistance.

C. The Y–Si–Al–O–N System

Extensive glass formation is found in silicon nitride hot pressed and pressureless sintered with mixtures of yttria and alumina but no new nitrogen-containing crystalline phases occur other than those observed in the Si–Al–O–N and Y–Si–O–N systems; see Fig. 23. Venables *et al.* (1977) used 10 wt % of the eutectic Y_2O_3–Al_2O_3 composition as an additive and suggested that its excellence as a densification aid was due to the formation of a large volume of low-viscosity liquid that wetted the solid completely and so was able to migrate rapidly through the whole microstructure. The extremely high room-temperature flexual strength of the product (190,000 psi) has already been mentioned.

The diagrams of the subsystems AlN–Al_2O_3–Y_2O_3–YN and Si_3N_4–Al_2O_3–Y_2O_3 expressed in equivalent concentrations are shown, respectively, by Figs. 40 and 41. In large areas the tie lines run from the stable compounds

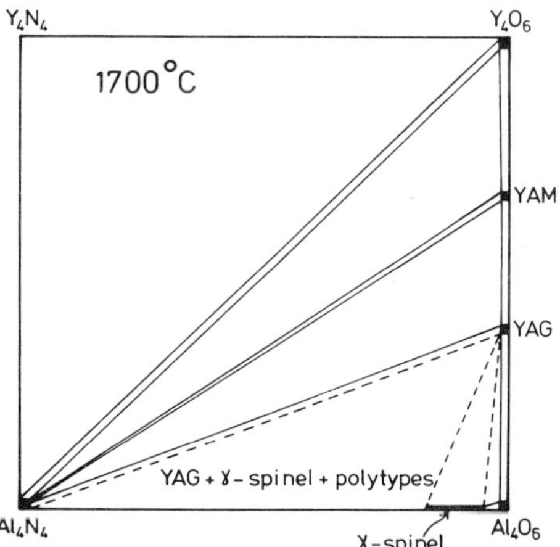

Fig. 40. The AlN–Al_2O_3–Y_2O_3–YN behavior diagram at 1700°C.

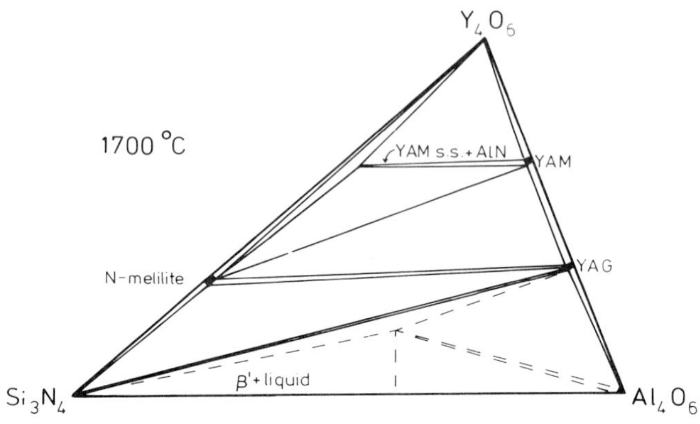

Fig. 41. The Si_3N_4–Al_2O_3–Y_2O_3 behavior diagram at 1700°C.

Si_3N_4, AlN, and β'-sialon to the oxide end of the system and hence nitride-rich compositions after densification can often contain oxide grain-boundary phases rather than the readily oxidizable yttrium–silicon or yttrium–silicon–aluminum oxynitrides.

As a consequence, both hot-pressed and pressureless sintered silicon nitride containing 5 wt % Y_2O_3:5 wt % Al_2O_3 as well as the usual surface silica show none of the catastrophic oxidation at 1000°C experienced by some of the Y_2O_3–Si_3N_4 compositions. Figure 42 shows specimens oxidized

Fig. 42. Si_3N_4 hot pressed with 5 wt % Y_2O_3:5 wt % Al_2O_3 and oxidized in air: (a) for 120 hr at 1000°C, (b) for 120 hr at 1400°C.

for 120 hr at 1000 and 1400°C. At 1000°C the appearance remains unchanged and the weight gain is negligible. However, at 1400°C the surface glaze is not completely protective and the parabolic oxidation rate constant ($\approx 10^{-1} mg^2 cm^{-4} hr^{-1}$) is nearly one thousand times greater than that of a hot-pressed Y_2O_3–Si_3N_4 composition within the favorable compatibility triangle. Yet again the explanation is obtained from a phase diagram. In the Y_2O_3–SiO_2–Al_2O_3 system the lowest liquidus is at approximately 1350°C and so oxidation of yttrium–silicon–aluminum phases at much above this temperature will eventually involve liquid formation. When this occurs, diffusivities increase, the transport of all species is enhanced, and the oxidation becomes more rapid.

VI. CERIUM AND ZIRCONIUM SIALONS

Ceria and zirconia are, like yttria, "network modifiers" in oxynitride systems. Very limited exploration has so far been made of sialon systems containing cerium and zirconium but the oxides are being used as densification additives for silicon nitride and their behavior in this role can be interpreted by the phase relationships of the respective metal–silicon–oxygen–nitrogen systems.

A. The Ce–Si–O–N System

1. Ceria as a Densifying Additive

Fully dense products are obtained by hot pressing silicon nitride with ceria, and with up to 10 wt % CeO_2 the only crystalline phase observed is β-Si_3N_4. As with yttria, the cerium oxide reacts with surface silica and some of the silicon nitride to give a liquid which, on cooling, forms a cerium–silicon oxynitride glass.

Indeed, similarities in the behavior of ceria and yttria might have been expected but in exploring the reaction of CeO_2 with Si_3N_4 Wills and Cunningham (1977) found only one quaternary phase which they claimed was $3Ce_2O_3 \cdot 2Si_3N_4$ having an orthorhombic unit cell and an unknown structure. However, a reinterpretation by Thompson (1977b) of Wills' and Cunningham's published diffraction data and his more detailed but still incomplete examination of the Ce–Si–O–N system at Newcastle show that the new cerium–silicon oxynitride is isostructural with the N-α-wollastonite yttrium–silicon oxynitride (see Section V.A). It has further been shown that cerium analogs of the other three quaternary Y–Si–O–N phases also occur by reaction of ceria with silicon nitride.

Ceric oxide decomposes to cerous oxide at about 1230°C but when silicon nitride is present the latter is oxidized with loss of nitrogen and with the formation of silica or silicon oxynitride according to Eqs. (9) and (10):

$$Si_3N_4 + 12CeO_2 \rightarrow 3SiO_2 + 6Ce_2O_3 + 2N_2 \tag{9}$$

$$2Si_3N_4 + 6CeO_2 \rightarrow 3Si_2N_2O + 3Ce_2O_3 + N_2 \tag{10}$$

The further reaction of silicon oxynitride with Ce_2O_3 or with Ce_2O_3 and SiO_2 allows the formation of the four quaternary Ce–Si–O–N phases that are isostructural with the corresponding yttrium phases (numbered as in Section V.A):

(i)	N-melilite,	$Ce_2O_3 \cdot Si_3N_4$
		$Y_2O_3 \cdot Si_3N_4$
(ii)	N-apatite,	$5Ce_2O_3 \cdot Si_2N_2O \cdot 4SiO_2$
		$5Y_2O_3 \cdot Si_2N_2O \cdot 4SiO_2$
(iii)	N-YAM,	$2Ce_2O_3 \cdot Si_2N_2O$
		$2Y_2O_3 \cdot Si_2N_2O$
(iv)	N-α-wollastonite,	$Ce_2O_3 \cdot Si_2N_2O$
		$Y_2O_3 \cdot Si_2N_2O$

A fifth quaternary phase, $Ce_2Si_6O_3N_8$, has no analog in the yttrium system and its structure has not yet been determined.

2. Properties of Cerium-Containing Nitrogen Ceramics

The almost exact correspondence between the Ce–Si–O–N and Y–Si–O–N systems suggests that the behavior and properties of silicon nitride and sialons densified by addition of ceria will be similar to those using yttria after allowance is made for the reduction of the ceric to cerous oxide. Thus, if the formation of quaternary cerium–silicon oxynitrides is avoided, the oxidation resistance of silicon nitride hot pressed with ceria should be good, and the advantage of ceria is that it has a built-in safeguard to assist in maintaining the overall composition within the oxidation-resistant Si_3N_4–Si_2N_2O–$Ce_2Si_2O_7$ compatibility triangle.

According to Eq. (9), 1 mole of SiO_2 is produced for every 2 moles of Ce_2O_3. By comparing Eqs. (11) and (12),

$$18SiO_2 + 9Y_2O_3 \rightarrow 9Y_2Si_2O_7 \tag{11}$$

$$18SiO_2 + 24CeO_2 + 2Si_3N_4 \rightarrow 12Ce_2Si_2O_7 + 4N_2 \tag{12}$$

it can be seen that for the same initial surface silica content nearly three times the molar concentration of CeO_2 (i.e., more than twice the weight percent) can be used compared with Y_2O_3 before any nitrogen-apatite phase appears in the final product (see Fig. 39). For the same concentrations by weight of additives, silicon nitride hot pressed with ceria should be much more oxidation resistant than when hot pressed with yttria.

B. Densification with Zirconia Additions

1. The Zr–Si–O–N System

From the Zr–Si–O–N behavior diagram at 1700°C (see Fig. 43) a small amount of liquid which dissolves silicon nitride will be formed locally at the silica-rich grain boundaries when silicon nitride is hot pressed with 5 wt % ZrO_2; this provides conditions for liquid-phase densification. The final product with a composition within the compatibility triangle Si_3N_4–Si_2N_2O–ZrO_2 will contain little glass and so should show good oxidation resistance. This agrees with the observation that after 120 hr at 1000 and 1400°C bar specimens show no detectable weight increase and are unchanged in appearance except for a very thin coherent surface glaze at the higher temperature.

It has been reported (Becher and Halen, 1978) that reaction of ZrO_2 with silicon nitride during brazing can lead to the formation of zirconium nitride (ZrN). Given an initial composition within the Si_3N_4–Si_2N_2O–ZrO_2 triangle of Fig. 43, loss of the volatiles silicon monoxide and nitrogen ($3SiO + N_2$ corresponds to the midpoint of the $3SiO_2$–Si_3N_4 join) will certainly move the composition of the remaining condensed material into the three-phase field Si_3N_4–ZrN–ZrO_2. In other words, the reaction

$$2Si_3N_4 + 2SiO_2 + 2ZrO_2 \rightarrow 2ZrN + 8SiO + 3N_2 \tag{13}$$

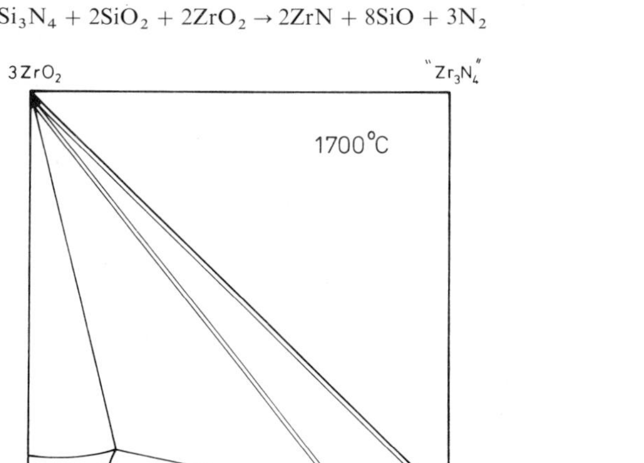

Fig. 43. The Zr–Si–O–N behavior diagram at 1700°C.

is thermodynamically possible if silicon monoxide and nitrogen are removed by volatilization.

2. Zirconium Sialons

Densification of silicon nitride with ZrO_2 and Al_2O_3 is more effective than with ZrO_2 alone and a recent Japanese investigation (Inomata *et al.*, 1976) suggests that the zirconia lowers the liquidus temperature in the Si–Al–O–N system as well as increases the liquid compositional range. Only one new phase, a zirconium–aluminum oxynitride, has been found in the system at Newcastle. Its structure is based on a superlattice of ZrN and so the interatomic bonding is probably nearer to that in an interstitial alloy than for a predominantly covalent ceramic; it oxidizes readily at 1000°C.

VII. Nitrogen Glasses

Figures 30 and 31 demonstrate the ease of glass formation in the Mg–Si–O–N system. Glasses occur in all the sialon systems so far investigated and it is perhaps not surprising that nitrogen can replace some of the oxygen in vitreous silicates, considering that it readily does so in the crystalline ones. The mechanical properties of nitrogen ceramics, particularly the high-temperature strength and creep resistance, depend markedly on the amount and characteristics of the grain-boundary glass but nitrogen glasses are also of interest as materials in their own right.

Additions of Al_2O_3 to Mg–Si–O–N compositions extend the vitreous region and mixtures of MgO, SiO_2, and AlN cold compacted, heated to 1650°C for 30 min and then cooled, give glasses containing up to 10 at. % N; a typical composition is $Mg_9Si_{21}Al_{10}O_{50}N_{10}$. In the yttrium sialons a transparent glass of composition $Y_9Si_{20}Al_9O_{53}N_9$ has a refractive index 1.76 and devitrifies by long heat treatment at 1200°C to give β-$Y_2Si_2O_7$, $Y_3Al_5O_{12}$ ("YAG") and Si_2N_2O. The maximum nitrogen content so far obtained in these glasses is about 12 at. %.

Even small concentrations of nitrogen not exceeding 1 at. % in oxide glasses are reported (Elmer and Nordberg, 1967) to increase the softening temperature, viscosity, and resistance to devitrification. Glasses with 10 at. % N or more might be expected to have even more unusual properties because if the nitrogen in the tetrahedral network is coordinated by three ligands, the structure should be more rigid and have a higher viscosity than the silicate glasses. The ease of shaping glasses, and the possibility of producing glass ceramics in which the crystalline phases are refractory nitrides and oxynitrides suggest that nitrogen-containing glasses are well-worth scientific and technological exploration. It widens still further the field of nitrogen materials and emphasizes the need for phase diagram exploration over a wide enough temperature range to include liquids as well as solids.

VIII. SUMMARY AND CONCLUSIONS

Sialons are phases in the Si–Al–O–N and related systems that are comparable in variety and diversity with the mineral silicates. They are built up of one-, two-, and three-dimensional arrangements of (Si, Al)$(O, N)_4$ and (Si, M)$(O, N)_4$ tetrahedra in the same way that the fundamental unit in the silicates is the (Si, Al)O_4 tetrahedron. They include vitreous as well as crystalline materials and are being explored for their thermal, mechanical, chemical, and electrical properties. The mutual replacement of oxygen and nitrogen gives an additional degree of freedom that offers good prospects for scientific investigation and technological development. Until recently the investigations have been superficial and frequent mistakes have been made in the interpretation of experimental observations. All processes for the production of sialons depend upon solid–state or solid–liquid reactions in multiphase systems and the properties of the products depend in many cases on the presence of secondary phases.

The conclusion is a simple one. In order to produce sialons and other nitrogen ceramics with desirable combinations of properties, the design of processing methods requires detailed information of phase limits and phase relationships over a wide temperature range. This information, presented in the form of phase diagrams, will continue to contribute to the solution of some of the immediate technological problems associated with applications of engineering ceramics. More important, the provision of reliable phase diagrams is probably the major factor in the scientific development of the new field of sialon materials.

References

Arrol, W. J. (1974). *Proc. Army Mater. Technol. Conf., Ceram. for High-Performance Appl., 2nd Hyannis, 1973*, p. 729.

Becher, P. F., and Halen, S. A. (1978). *Proc. Army Mater. Technol. Conf., Ceram. for High-Performance Appl. II, 5th, Newport, 1977*, p. 1077.

Clarke, D. R., Shaw, T. M., and Thompson, D. P. (1978). *J. Mater. Sci.* **13**, 217.

Colquhoun, I., Thompson, D. P., Wilson, W. I., Grieveson, P., and Jack, K. H. (1973). *Proc. Br. Ceram. Soc.* **22**, 181.

Drew, P., and Lewis, M. H. (1974). *J. Mater. Sci.* **9**, 261.

Elmer, T. H., and Nordberg, M. E. (1967). *J. Am. Ceram. Soc.* **50**, 275.

Findlay, A. (1927). "The Phase Rule and its Applications," 6th ed. Longmans Green, London.
Gauckler, L. J. (1978). *Proc. Army Mater. Technol. Conf., Ceram. for High-Performance Appl. II, 5th, Newport, 1977*, contribution to discussion.
Gauckler, L. J., Lukas, H. L., and Petzow, G. (1975). *J. Am. Ceram. Soc.* **58**, 346.
Gauckler, L. J., Lukas, H. L., and Tien, T. Y. (1976). *Mater. Res. Bull.* **11**, 503.
Gauckler, L. J., Boskovic, S., Petzow, G., and Tien, T. Y. (1977). *Proc. NATO Adv. Study Inst., Nitrogen Ceram., Canterbury, 1976*, p. 405.
Gazza, G. E. (1973). *J. Am. Ceram. Soc.* **56**, 662.
Gazza, G. E. (1975). *Bull. Am. Ceram. Soc.* **54**, 778.
Hendry, A., Perera, D. S., Thompson, D. P., and Jack, K. H. (1975). *In* "Special Ceramics" (P. Popper, ed.), Vol. 6, pp. 321–331. The Brit. Ceram. Res. Assoc., Stoke-on-Trent, England.
Idrestedt, I., and Brosset, C. (1964). *Acta Chem. Scand.* **18**, 1879.
Inomata, Y., Hasegawa, Y., Matsuyama, T., and Yajima, Y. (1976). *Yogyo Kyokai Shi* **84**, 600.
Jack, K. H. (1973). *Trans. Br. Ceram. Soc.* **72**, 376.
Jack, K. H. (1974). *Proc. Army Mater. Technol. Conf., Ceram. for High-Performance Appl., 2nd, Hyannis, 1973*, p. 265.
Jack, K. H. (1976). *J. Mater. Sci.* **11**, 1135.
Jack, K. H., and Wilson, W. I. (1972). *Nature (London) Phys. Sci.* **238**, 28.
Jänecke, E. (1907). *Z. Anorg. Chem.* **53**, 319.
Lange, F. F. (1976). *Basic Sci. Fall Meet. Am. Ceram. Soc., San Francisco* (presented paper).
Lange, F. F., Singhal, S. C., and Kuznicki, R. C. (1976). Westinghouse Electr. Corp. Res. Dev. Cent., Tech. Rep. No. 6.
Layden, G. K. (1976). *Basic Sci. Fall Meet. Am. Ceram. Soc., San Francisco.*
Lumby, R. J., North, B., and Taylor, A. J. (1975). *In* "Special Ceramics" (P. Popper, ed.), Vol. 6, pp. 283–298. The Br. Ceram. Res. Assoc., Stoke-on-Trent, England.
Lumby, R. J., North, B., Taylor, A. J., Lewis, M. H., Powell, B. D., and Drew, P. (1977). *J. Mater. Sci.* **12**, 61.
Lumby, R. J., North, B., and Taylor, A. J. (1978). *Proc. Army Mater. Technol. Conf., Ceram. for High-Performance Appl. II, 5th, Newport 1977*, p. 893.
Oda, I., Kaneno, M., and Yamamoto, N., (1977). *Proc. NATO adv. Study Inst., Nitrogen Ceram., Canterbury, 1976*, p. 359.
Oyama, Y. (1972a). *Jpn. J. Appl. Phys.* **11**, 760.
Oyama, Y. (1972b). *Jpn. J. Appl. Phys.* **11**, 1572.
Oyama, Y. (1974). *Yogyo Kyokai Shi* **82**, 351.
Oyama, Y., and Kamigaito, O. (1971). *Jpn. J. Appl. Phys.* **10**, 1637.
Parr, N. L., Martin, G. F., and May, E. R. W. (1959). Admiralty Mater. Lab. Rep. No. A/75(s).
Rae, A. W. J. M. (1976). Ph.D. Thesis, The University of Newcastle upon Tyne, England.
Rae, A. W. J. M., Thompson, D. P., Pipkin, N. J., and Jack, K. H. (1975). *In* "Special Ceramics" (P. Popper, ed.), Vol. 6, pp. 347–360. The Br. Ceram. Res. Assoc., Stoke-on-Trent, England.
Rae, A. W. J. M., Thompson, D. P., and Jack, K. H. (1978). *Proc. Army Mater. Technol. Conf., Ceram. for High-Performance Appl. II, 5th, Newport, 1977*, p. 1039.
Roebuck, P. H. A., and Thompson, D. P. (1977). *In* "High Temperature Chemistry of Inorganic and Ceramic Materials" (F. P. Glasser and P. E. Potter, eds.), pp. 222–228. The Chemical Society, London.
Terwilliger, G. R., and Lange, F. F. (1974). *J. Am. Ceram. Soc.* **57**, 25.
Thompson, D. P. (1976). *J. Mater. Sci.* **11**, 1377.
Thompson, D. P. (1977a). *Proc. NATO Adv. Study Inst., Nitrogen Ceram., Canterbury, 1976*, p. 139.

Thompson, D. P. (1977b). *J. Mater. Sci.* **12**, 2344.

Torre, J. P., and Mocellin, A. (1977). *Proc. NATO Adv. Study Inst., Nitrogen Ceram., Canterbury, 1976*, p. 63.

Venables, J. D., McNamara, D. K., and Lye, R. G. (1977). *Proc. NATO Adv. Study Inst., Nitrogen Ceram., Canterbury, 1976*, p. 391.

Washburn, M. E. (1967). *Am. Ceram. Soc. Bull.* **46**, 667.

Wild, S., Grieveson, P., and Jack, K. H. (1968). *Prog. Rep. No. 1*, Ministry of Defence Contract N/CP.61/9411/67/4B/MP387. See Wild *et al.* (1972, pp. 289–297).

Wild, S., Grieveson, P., Jack, K. H., and Latimer, M. J. (1972). *In* "Special Ceramics" (P. Popper, ed.), Vol. 5, pp. 377–384. The Br. Ceram. Res. Assoc., Stoke-on-Trent, England.

Wills, R. R., and Cunningham, J. A. (1977). *J. Mater. Sci.* **12**, 208.

Wills, R. R., Holmquist, S., Wimmer, J. M., and Cunningham, J. A. (1976). *J. Mater. Sci.* **11**, 1305.

Wills, R. R., Stewart, R. W., and Wimmer, J. M. (1977). *Am. Ceram. Soc. Bull.* **56**, 194.

Zernike, J. (1955). "Chemical Phase Theory." Kluwer, Deventer, Netherlands.

VI

The Use of Phase Diagrams in Development of Silicates for Thermal Shock Resistant Applications

*R. N. KLEINER** AND S. T. BULJAN*[†]

CERAMICS DEPARTMENT
GTE SYLVANIA INCORPORATED
CHEMICAL AND METALLURGICAL DIVISION
TOWANDA, PENNSYLVANIA

I. INTRODUCTION

Ceramic materials offer the potential for improved commercial products as a result of their high-temperature properties. In most of the applications where high temperatures are involved, the products are cycled between

* Present address: Coors Porcelain Company, 17750 32nd Avenue, Golden, Colorado 80401.
[†] Present address: GTE Laboratories, 400 Sylvan Road, Waltham, Massachusetts 02154.

ISBN-0-12-053205-0

room temperature and some elevated temperatures. Under these cyclic conditions, depending on the cycle rate, the thermal shock resistance of the ceramic becomes a major factor. One of the most severe and challenging applications for ceramics requiring high thermal shock resistance is in the heat exchanger for gas turbine engine applications.

Automotive companies have been engaged in the development of vehicular gas turbine engines for many years. The efficiency of the gas turbine engine and its ability to compete with piston engines are directly related to its temperature of operation and its ability to utilize heat effectively. The purpose of the heat exchanger is to recover waste heat losses and to preheat incoming air in order to decrease fuel consumption. The gas turbine engine environment is an extremely difficult one for materials to survive in because of mechanical loads, chemical attack, and thermal stresses.

The mechanical, chemical, and thermal shock-resistant properties required of materials for this stringent application will be described. Emphasis will be placed on maximizing the thermal shock resistance of silicate ceramics through lowering thermal expansion to meet the required properties of this application. The use of phase diagrams in the development of low thermal expansion materials for these applications will be discussed in detail.

II. HEAT EXCHANGER REQUIREMENTS

Regenerators are subjected to mechanical stresses, chemical corrosion, and rapid thermal cycles. The source of these stresses is described briefly in this section. Details of how the material properties can be altered to minimize these stresses are given in the following sections.

Mechanical stresses are transmitted to the regenerator through the regenerator drive system. A 28 in. diameter rotary regenerator showing the exhaust product and air inlet paths is given in Fig. 1. Drive systems which have been utilized are a hub drive and support, a rim drive and support, and a hub support and rim drive. Each of these systems creates different mechanical stresses on the regenerator. Mechanical designers have done a lot to minimize these stresses through elastomer stress absorbing layers and stress-relieving slots (Anderson *et al.*, 1975a). Torque from the rubbing seal creates additional stresses. Distortion of metal components such as the housing and cover also adds stress to the core. An enlarged view of a honeycomb matrix is shown in Fig. 2. The wall thickness is approximately 0.005 in. Matrices with walls as thin as 0.0035 in. have proven to have ade-

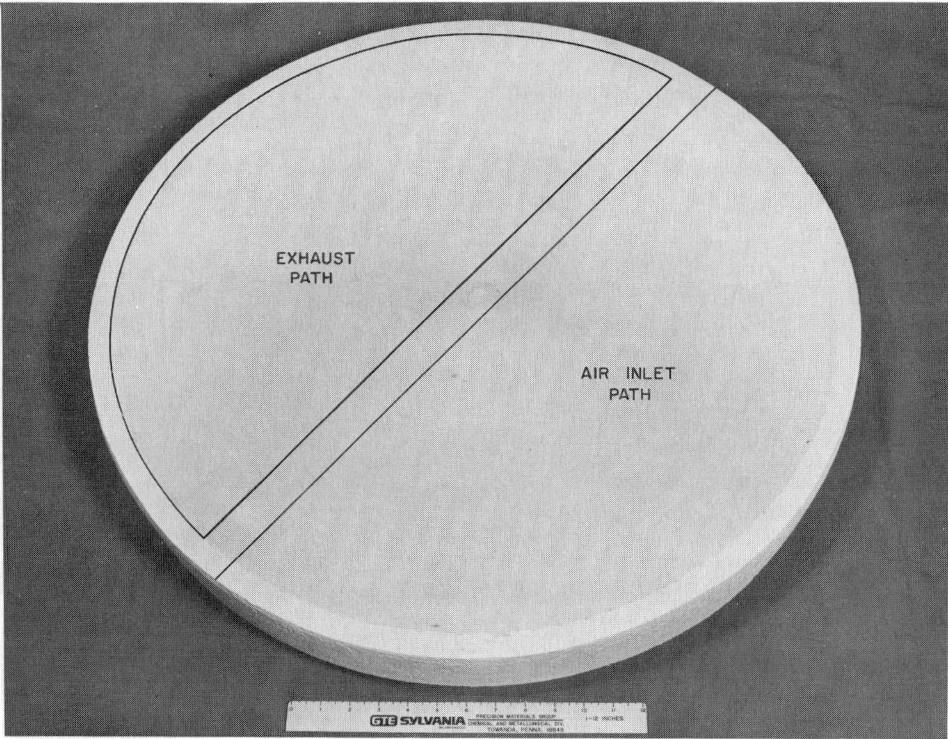

Fig. 1. Ceramic rotary regenerator showing exhaust product and air inlet paths.

quate strength to withstand the mechanical stresses of the gas turbine engine.

Sulfuric acid is formed during the combustion process from the sulfur present in the fuel. When the sulfuric acid condenses on the cold face of the regenerator, it attacks the regenerator material. The most promising regenerator material for several years has been a lithium–aluminum–silicate (LAS) glass-ceramic honeycomb material principally because of its low thermal expansion. The sulfuric acid was found to leach the lithium from the LAS matrix. This changed the thermal expansion of the matrix, creating large tensile stresses that resulted in cracks. Sodium, which is present in some fuels, salt water, or road salt attacks the hot face of the core. Sodium reacts with the LAS and the resulting material has a high thermal expansion. The difference in thermal expansion between the hot and cold sides creates internal radial cracks (Anderson *et al.*, 1975b).

Thermal shock resistance is an important parameter in turbine heat exchanger materials because of rapid engine starts and shut downs. The ambient temperature can be below 0°F and the turbine inlet temperature

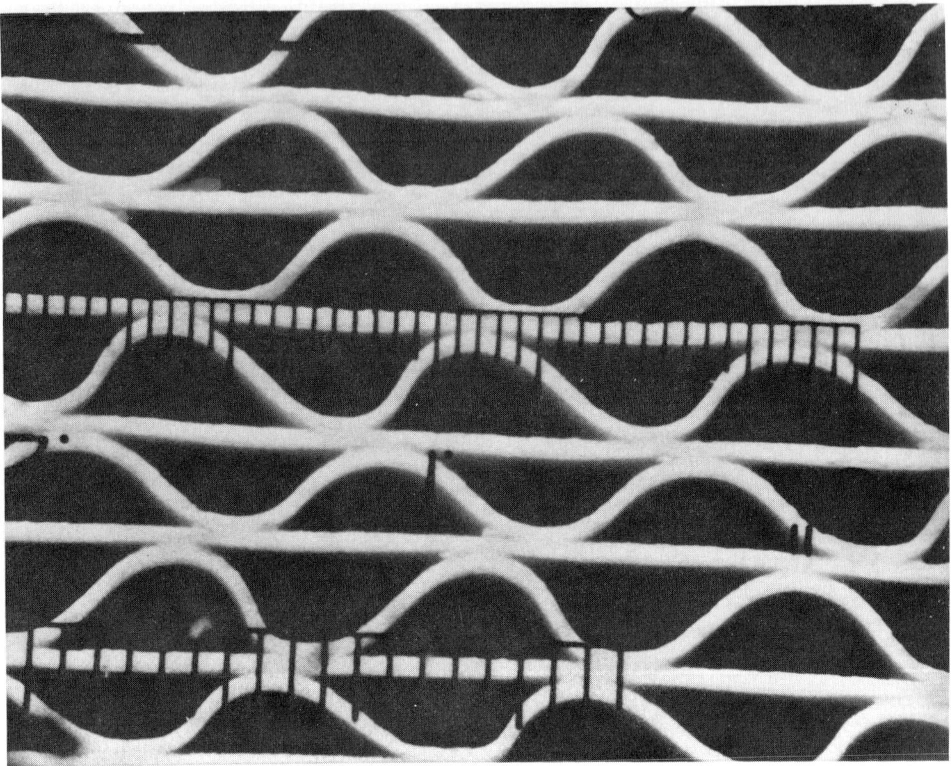

Fig. 2. Rotary regenerator matrix structure. Scale is marked at 0.1 in. increments.

can be over 1000°F. Materials for thermal shock-resistant applications can be selected on either of two bases: (1) resistance to crack initiation or (2) resistance to crack propagation. In cases where the thermal environment is so severe that thermal stress fracture cannot be prevented, materials are selected to avoid crack propagation. The first condition is described by

$$R' = \sigma_f(1 - \nu)k/E\alpha \tag{1}$$

where R' is the maximum allowable heat flux through the body before fracture initiation, σ_f the fracture strength, E the elastic modulus, α the linear coefficient of thermal expansion, ν Poisson's ratio, and k the thermal conductivity. The second condition is described by

$$R'''' = E\gamma_f/\alpha^2(1 - \nu) \tag{2}$$

where R'''' is the minimum degree of crack propagation after crack initiation and γ_f is the fracture surface energy.

Since mechanical stresses are high in the turbine, Eq. (1) must be used in designing regenerator matrix materials. The thermal shock resistance can be increased by increasing σ_f and decreasing E. It has been shown, however, that as σ_f is increased, E increases by a similar amount (Cleveland *et al.*, 1977). Since it is difficult to alter v, the thermal expansion remains as the major factor to be modified in order to increase the thermal shock resistance of regenerator materials. The coefficient of thermal expansion also affects the magnitude of the stress in the core. This results because the center of the core is hot and the material expands in that region. It does not expand at the cooler perimeter of the core and the magnitude of the stress is directly related to the magnitude of the coefficient of thermal expansion.

Thermal expansion is a function of the composition and microstructure of the body. Techniques for reducing thermal expansion through modification of these parameters will be described. Because of the importance of thermal expansion in designing regenerators for turbine applications, emphasis will be concentrated on the use of phase diagrams in the development of compositions for low thermal expansion ceramics.

Tests were run to confirm the effect of these parameters on shock resistance (Smoke, 1969). To determine the effect of thermal conductivity, three bodies were studied. The thermal conductivities varied between 0.04 and 0.27 cal/cm-sec-°C, but the flexural strength and thermal expansions were about the same. All three bodies had the same general thermal shock resistance, all failing between 200 and 230°C quenching cycles. It was concluded that thermal conductivity had little effect on thermal shock resistance of the LAS ceramics tested.

The effect of strength on the thermal shock resistance of two materials having values of 10,000 and 24,000 psi also was relatively small in this system. All these specimens failed between 200 and 260°C quenching cycles.

The strength and thermal conductivity effects were studied together. One body had 90% higher transverse strength and 44 times higher thermal conductivity than the other. Again, there was little difference in thermal shock resistance.

The effect of modulus of elasticity (MOE) was also studied. The MOE varied from 7 to 22×10^6 psi but the thermal shock resistance values were similar.

The thermal expansion was studied and the thermal shock resistance was found to increase significantly as the thermal expansion decreased. The thermal shock resistance did not show much of a decrease until the coefficient of expansion was 4.0×10^{-6}/°C or lower. The most dramatic increase in the thermal shock resistance resulted when the thermal expansion approached zero. This effect is shown in Fig. 3. Bodies with high negative thermal expansions had poor thermal shock resistance.

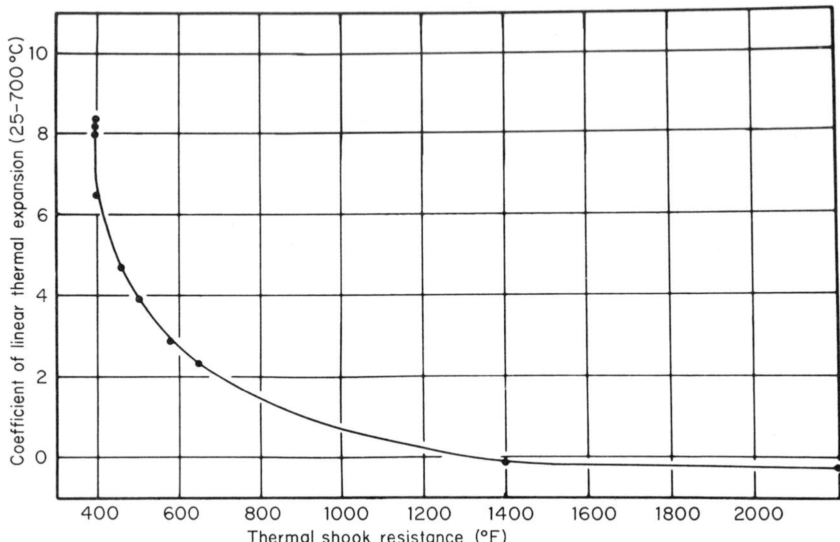

Fig. 3. Thermal shock resistance as a function of coefficient of thermal expansion (after Smoke, 1969).

III. LITHIUM ALUMINUM SILICATES

Lithium aluminum silicates were studied because of their potential for attaining very low, even negative, thermal expansion.

A. Determination of the Phase Diagram

The compounds eucryptite, spodumene, and petalite were investigated before 1914. Kracek (1930), Hatch (1943), and Bowen and Greig (1924) established phase relations along the joins Li_2O–SiO_2, $Li_2O \cdot Al_2O_3$–SiO_2, and Al_2O_3–SiO_2, respectively.

The binary phase diagram of $Li_2O \cdot SiO_2$–SiO_2 [after Kracek (1930)] is shown in Fig. 4 and binary phase diagrams along $Li_2O \cdot Al_2O_3 \cdot 2SiO_2$–$SiO_2$ [after Hatch (1943) and Roy *et al.* (1950)] are shown in Figs. 5 and 6, respectively.

Hatch found that glasses in the β-spodumene and β-eucryptite fields devitrify readily when heated between 1000 and 1100°C for 15 min. This has led to the development of many products with low expansions based on the glass-ceramic technique.

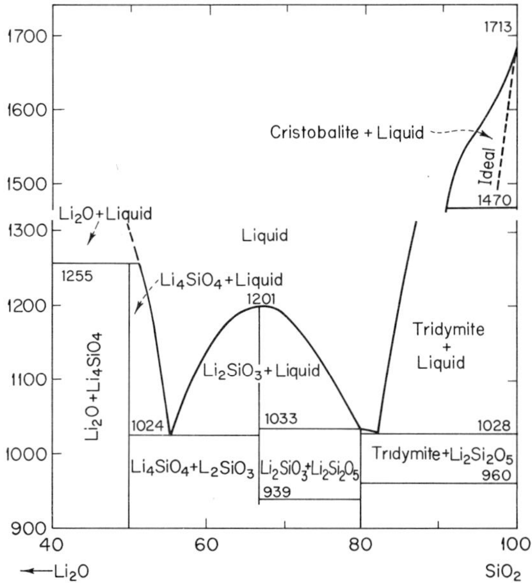

Fig. 4. $Li_2O \cdot SiO_2$–SiO_2 phase diagram (after Kracek, 1930).

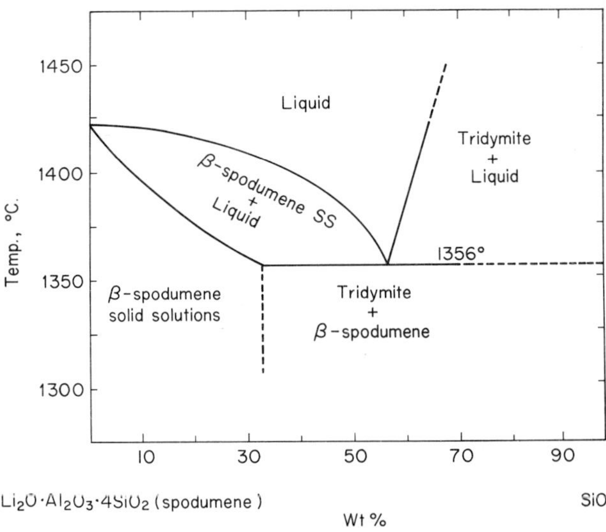

Fig. 5. $Li_2O \cdot Al_2O_3 \cdot 4SiO_2$–$SiO_2$ phase diagram (after Hatch, 1943).

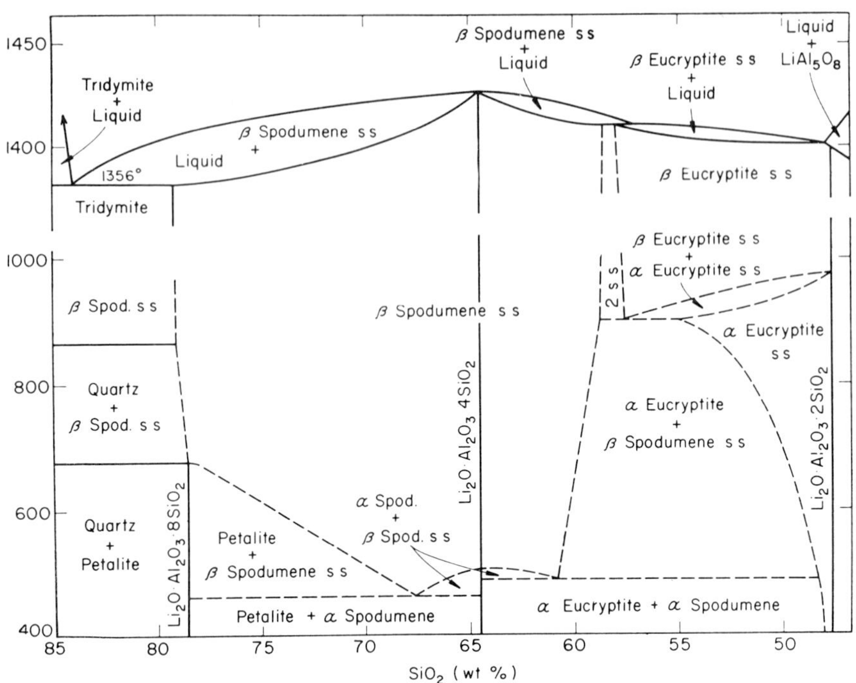

Fig. 6. $Li_2O \cdot Al_2O_3 \cdot 2SiO_2$–$SiO_2$ phase diagram (after Roy *et al.*, 1950).

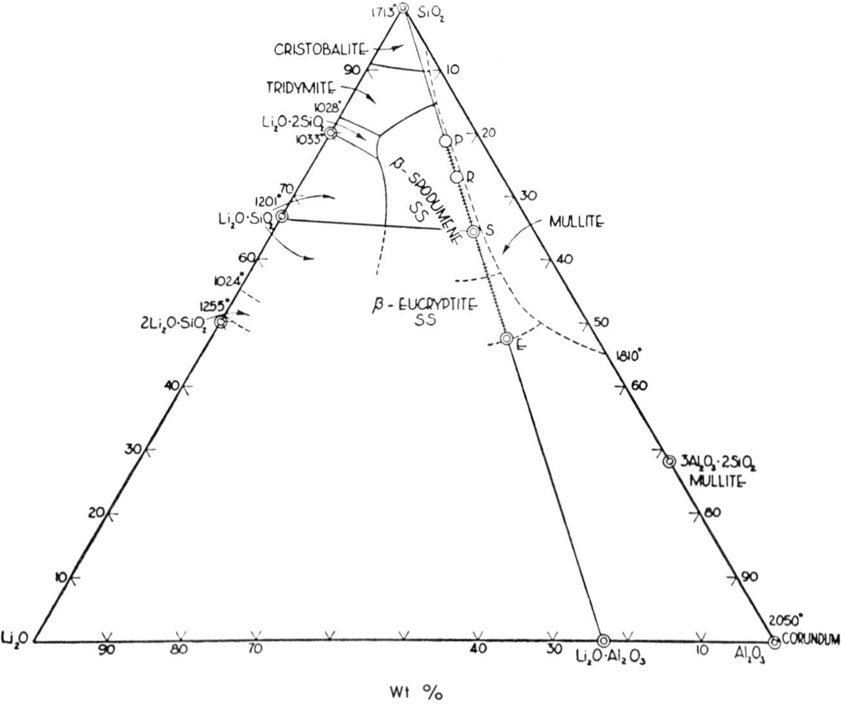

Fig. 7. Li_2O–Al_2O_3–SiO_2 phase diagram (after Levin *et al.*, 1969). P, $Li_2O \cdot Al_2O_3 8SiO_2$ petalite; R, $Li_2O \cdot Al_2O_3 \cdot 6SiO_2$ "lithium orthoclase"; S, $Li_2O \cdot Al_2O_3 \cdot 4SiO_2$ spodumene; E, $Li_2O \cdot Al_2O_3 2SiO_2$ eucryptite.

1. β-SPODUMENE SOLID SOLUTION FIELD

β-spodumene solid solutions near the solidus give x-ray patterns similar to those of devitrified glasses but more high angle lines appear. The β-spodumene solid solution possibly evolves a second crystalline phase. Evidence for β-spodumene solid solution is (1) an increase in the refractive indices of the crystalline products from 64.6 to 60% SiO_2, (2) a small but regular shift in the high angle x-ray powder reflections, and (3) the presence of a β-eucryptitelike phase as a rim around the β-spodumene phase.

Below the inversion temperature two phases react slowly to give a β-spodumene solid solution which contains only a trace of the exsolution product. Above the inversion temperature, the β-spodumene solid solution dissociates completely and rapidly, producing intergrowth with the exsolution product (Hatch, 1943).

The low temperature form of spodumene did not appear in the study by Roy and Osborn (1949) nor has it been possible to synthesize it. Naturally occurring spodumene inverts to β-spodumene, the high-temperature polymorph, rapidly above 900°C but it occurs at temperatures as low as 700°C. The high-temperature form of β-spodumene appears as small anhedral grains.

2. β-EUCRYPTITE SOLID SOLUTION FIELD

Evidence for eucryptite solid solution is that (1) the refractive indices of the solid solution increase slightly with decreasing silica and (2) the high angle x-ray lines show a regular shift (Hatch, 1943).

A single phase exists in this field although a fine-grained fibrous material exsolves with long heating. These products seem to be the same as the fibrous material in the β-spodumene field.

β-eucryptite is unstable and dissociates just below the melting point inverting to a β-eucryptite-like phase that contains γ-Al_2O_3 in the fibrouslike phase. The melting point of natural β-eucryptite is difficult to measure because it always appears to grow as an intergrowth with albite. The exsolution of the fibrous product above and below the solidus indicates the instability of the solid solutions.

The Li_2O–Al_2O_3–SiO_2 ternary system after Kracek, Hatch, Bowen and Greig, and Roy and Osborn is shown in Fig. 7 (Levin *et al.*, 1969a).

The Li_2O–Al_2O_3–SiO_2 ternary is similar to the MgO–Al_2O_3–SiO_2 (MAS) system. The mullite field approaches the join where the ternary compounds exist but does not cross it. The β-spodumene solid solution resembles the cordierite solid solution series in MAS system, and an extensive solid solution exists between cordierite and β-spodumene.

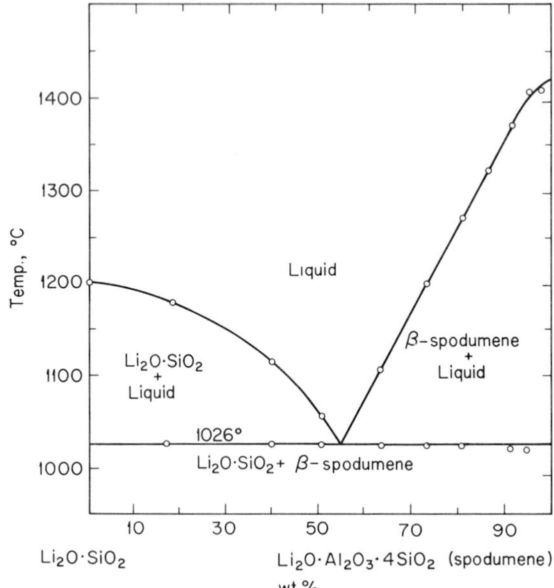

Fig. 8. $Li_2O \cdot SiO_2$–$Li_2O \cdot Al_2O_3 \cdot 4SiO_2$ phase diagram (after Roy and Osborn, 1949).

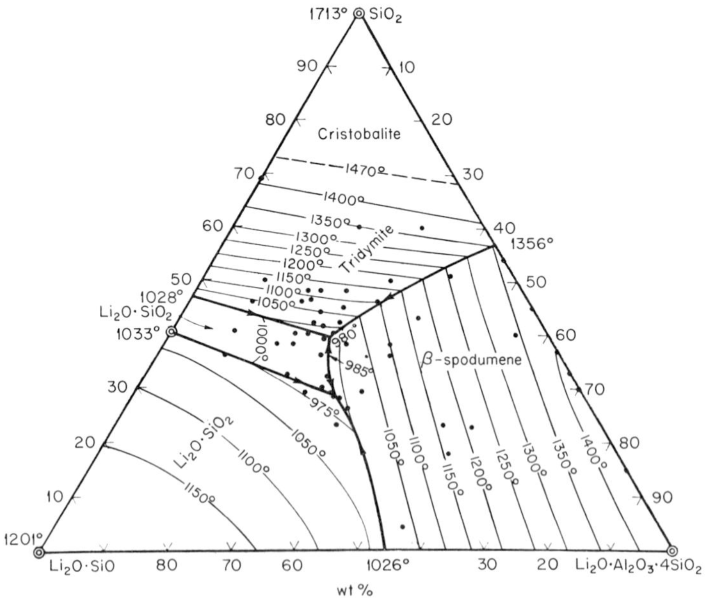

Fig. 9. $Li_2O \cdot SiO_2$–$Li_2O \cdot Al_2O_3 \cdot 4SiO_2$–$SiO_2$ phase diagram (after Roy and Osborn, 1949).

Petalite ($Li_2O \cdot Al_2O_3 \cdot 8SiO_2$) was not observed in the study by Roy *et al.* (1950). Their data on establishing the limit of the solid solution series extending from β-spodumene toward SiO_2 confirmed the data of Hatch. Roy and Osborn (1949) studied the join $Li_2O \cdot SiO_2$–$Li_2O \cdot Al_2O_3 \cdot 4SiO_2$ shown in Fig. 8. The melting point of spodumene is 1423°C which is in agreement with the data of Hatch (1943). Roy and Osborn (1949) also studied the $Li_2O \cdot SiO_2$–$Li_2O \cdot Al_2O_3 \cdot 4SiO_2$–$SiO_2$ area of the Li_2O–Al_2O_3–SiO_2 ternary shown in Fig. 9. The ternary contains two invariant points, both eutectics and a maximum on the boundary joining these points.

At one invariant point, $Li_2O \cdot 2SiO_2$, tridymite and β-spodumene solid solution (33% SiO_2) are in equilibrium at 980°C with a liquid of composition 39.2% SiO_2, 26% spodumene, and 34.8% $Li_2O \cdot SiO_2$.

At the other invariant point at 985°C, $Li_2O \cdot SiO_2$, $Li_2O \cdot 2SiO_2$, and β-spodumene solid solution (24% SiO_2) are in equilibrium with a liquid of composition 36% SiO_2, 37% $Li_2O \cdot SiO_2$, and 27% spodumene.

B. Thermal Expansion of Compounds in the Phase Diagram

The primary reason for interest in the lithium–aluminum–silicate (LAS) ternary system has been for fabricating low positive or negative thermal expansion ceramics. Hummel (1951) reported the thermal expansions for several compounds in the LAS system.

$Li_2O \cdot 5Al_2O_3$	82×10^{-7}/°C	(25–1000°C)
$Li_2O \cdot Al_2O_3$	124×10^{-7}/°C	(25–1000°C)
$Li_2O \cdot Al_2O_3 \cdot 4SiO_2$	9×10^{-7}/°C	(25–1000°C)
$Li_2O \cdot Al_2O_3 \cdot 6SiO_2$	5×10^{-7}/°C	(25–1000°C)
$Li_2O \cdot Al_2O_3 \cdot 8SiO_2$	3×10^{-7}/°C	(25–1000°C)
$Li_2O \cdot Al_2O_3 \cdot 10SiO_2$	5×10^{-7}/°C	(25–1000°C)

The actual curves must be referred to in order to get a complete picture of the thermal expansion behavior of these minerals since these curves are often nonlinear. These curves are shown in Figs. 10 and 11.

$Li_2O \cdot Al_2O_3 \cdot 2SiO_2$ and $Li_2O \cdot Al_2O_3 \cdot 3SiO_2$ show a large negative thermal expansion. The slopes for spodumene, $Li_2O \cdot Al_2O_3 \cdot 6SiO_2$, $Li_2O \cdot Al_2O \cdot 8SiO_2$, at $Li_2O \cdot Al_2O_3 \cdot 10SiO_2$ are slightly positive. The expansion of $Li_2O \cdot Al_2O_3 \cdot 10SiO_2$ is also slightly positive and quartz and cristobalite begin to appear. This indicates the limit of the amount of SiO_2 that can be accommodated in the β-spodumene crystal lattice ($Li_2O \cdot Al_2O_3 \cdot 8SiO_2$).

Petalite ($Li_2O \cdot Al_2O_3 \cdot 8SiO_2$) has a very low negative coefficient of thermal expansion (-4.9×10^{-7}/°C). This body is very difficult to sinter because it melts very close to the required sintering temperatures. Overfiring results in a crazed body formed of a glass with a coefficient of expansion of

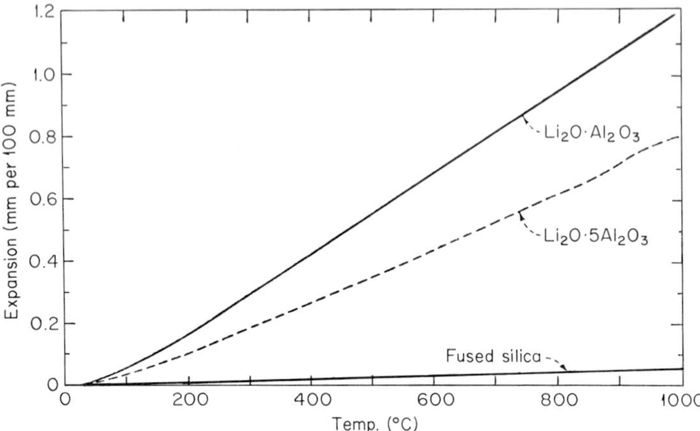

Fig. 10. Thermal expansion curves of $Li_2O \cdot Al_2O_3$, $Li_2O \cdot 5Al_2O_3$, and fused silica (after Hummel, 1951).

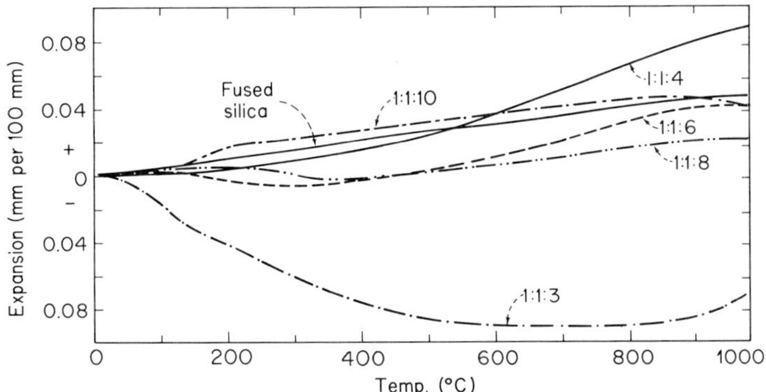

Fig. 11. Thermal expansion curves of $Li_2O \cdot Al_2O_3 \cdot SiO_2$ in the molar ratios of 1:1:3, 1:1:4, 1:1:6, 1:1:8, 1:1:10 and of fused silica (after Hummel, 1951).

45×10^{-7}/°C and β-spodumene crystals with a coefficient of about zero. Underfiring results in a low strength, porous body.

Hummel (1951) made successive runs on the $Li_2O \cdot Al_2O_3 \cdot 12SiO_2$ sample and the expansion curve could not be duplicated (Fig. 12). Apparently some of the cristobalite in the sample transformed at 1300°C to quartz after heating to 1000°C. Alkaline earth oxides will cause amorphous silicic acid to transform to quartz at 1300°C. The thermal expansions of β-eucryptite and its solid solutions are shown in Fig. 13.

The bulk thermal expansion of β-eucryptite is negative. Beta-eucryptite is strongly anisotropic. Its thermal expansion is negative parallel to the

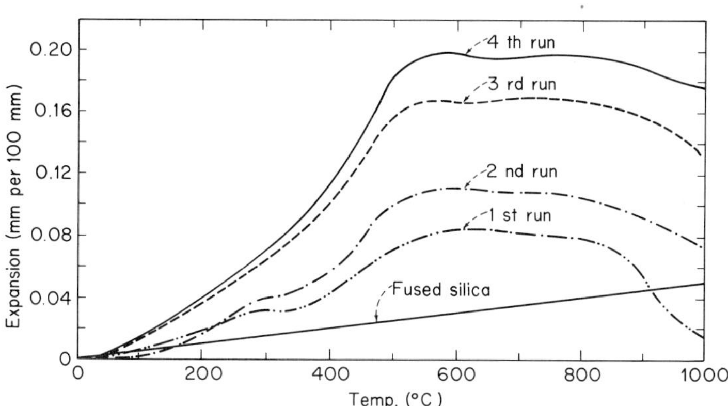

Fig. 12. Successive thermal expansion runs on $Li_2O \cdot Al_2O_3 \cdot 12SiO_2$ (after Hummel, 1951).

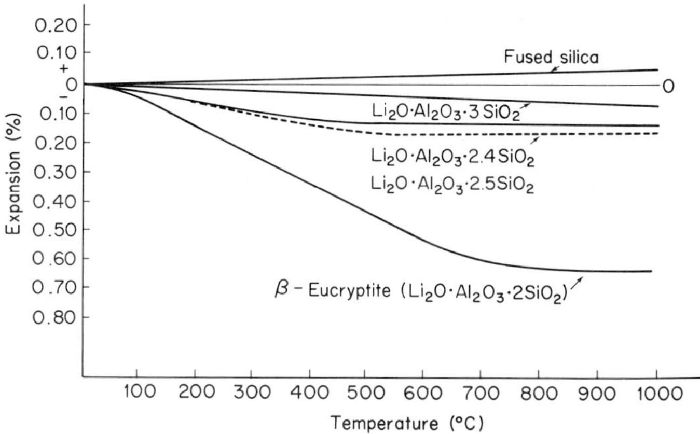

Fig. 13. Thermal expansion curves of $Li_2O \cdot Al_2O_3 \cdot SiO_2$. Molar ratios of 1:1:2, 1:1:2.5, 1:1:3, and fused silica, (after Hummel, 1960), 1:1:2.4 (after Moya *et al.*, 1974).

c axis and positive perpendicular to the *c* axis. Gillery and Bush (1959) attributed this behavior to elastic properties; Schulz (1974) explained it in terms of thermal excitation.

Hummel first reported the negative thermal expansion in β-eucryptite in 1951. The dilatometer measurement on a polycrystalline specimen measures only the linear bulk effect. Using x-ray methods, Gillery and Bush obtained the following coefficients of expansion:

$$\alpha_{11} = -17.6 \times 10^{-6}, \qquad \text{parallel to } c \text{ axis}$$
$$\alpha = 8.21 \times 10^{-6}, \qquad \text{perpendicular to } c \text{ axis}$$

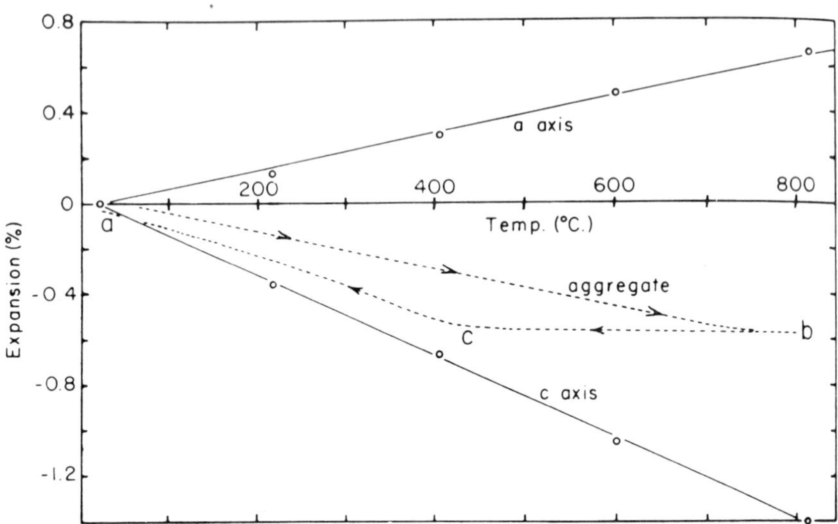

Fig. 14. Thermal expansion of β-eucryptite using x-ray and dilatometric measurement techniques (after Gillery and Bush, 1959).

These results are summarized in Fig. 14. Also shown in this figure are the dilatometer results. The volume expansion coefficient of the hexagonal unit cell is $\alpha_{11} + 2\alpha$. In the cooling part of the dilatometer curve, this coefficient closely approximates the true average linear expansion coefficient of the lattice $(\alpha_{11} + 2\alpha)/3$. This occurs because the crystallites are sintered and behave as a solid mass. Because of the high anisotropy, the stresses reach the same magnitude as the strength of the material. It ruptures, forming voids or microcracks, and the bend in the dilatometric curve results. The expansion in the c direction dominates and the thermal expansion tends to approach the value for the c direction. During the heating part of the curve, the microcracks begin to heal and the expansion in the a direction is approached. Gillery and Bush (1959) point out that the β-eucryptite structure is similar to the high quartz structure. Smyth (1955) explains the negative thermal expansion of vitreous silica in terms of the transverse vibrational frequency of the oxygen atoms. It is possible that the negative thermal expansion phenomenon could be explained by this mechanical analogy of atomic behavior.

The healing effect is confirmed in further work by Bush and Hummel (1959) in which the modulus of rupture was measured as a function of temperature. Their results are shown in Fig. 15. The strength is low at room temperature because of the internal fractures that occur during cooling from firing. The strength increases during heating because of the

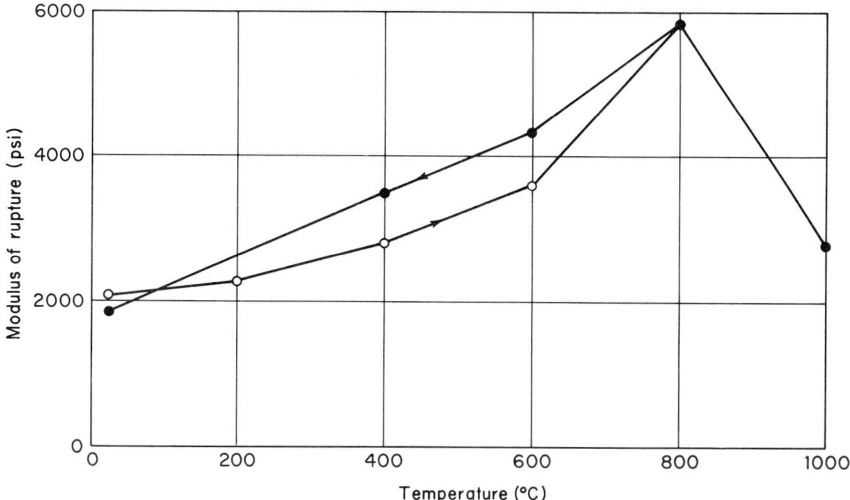

Fig. 15. Modulus of rupture versus temperature of β-eucryptite (after Bush and Hummel, 1959).

healing of the microcracks. A maximum is obtained at 800°C. At higher temperatures the decrease in strength is explained as due to plasticity or additional internal disruptions.

The thermal expansion coefficients measured by Tscherry *et al.* (1972) are in agreement with those obtained by Gillery and Bush (1959) and Tien and Hummel (1964). Schultz (1974) measured the lattice constants for a heat treated and nonheat treated single crystal of β-eucryptite. The values are as follows:

		a	c
(O°C)	Nonheat treated	10.5015	11.185
(650°C)	Heat treated	10.5088	11.147

The resultant thermal expansions were calculated as follows:

	Nonheat treated	Heat treated
a direction	7.8×10^{-6}	6.9×10^{-6}
c direction	17.5×10^{-6}	-15.5×10^{-6}

The thermal expansion model proposed is based on changes in the Li atoms between the sites of fourfold and sixfold oxygen coordination. At higher temperatures the Li atoms jump from one type of site in fourfold oxygen coordination into fourfold and sixfold oxygen coordination sites which were empty at room temperature. This process is reversible.

The *a* axis increases because the oxygen positions change to increase the Li–O distance and the *c* axis decreases. These magnitudes increase with increasing temperature.

Moya *et al.* (1974) measured the dilatometric thermal expansion of β-eucryptite and obtained results similar to those of Hummel (1960). They point out that if the longitudinal compressibility parallel to the axis of symmetry is negative, the negative bulk thermal expansion is an elastic effect as assumed by Gillery and Bush (1959). If the compressibility is positive, the Gruneisen function is negative and the negative bulk thermal expansion has nothing to do with elasticity.

Moya *et al.* (1974) state that the negative bulk thermal expansion in β-eucryptite may be due to a tension effect in the chain structure. The O–Al–O angle is tetrahedral and the equilibrium angle is octahedral. Substituting Al by Si decreases this tension, decreasing the Gruneisen function and the thermal expansion.

Hortal *et al.* (1975) measured the linear compressibility of β-eucryptite and concluded that it seems most likely that the sign of the Gruneisen function is positive. Therefore, they conclude that the negative thermal expansion results mainly from an elastic effect as proposed by Gillery and Bush (1959). Determination of the exact mechanism of this negative thermal expansion effect has been difficult but the foregoing conclusion appears plausible.

C. Application of the Phase Diagram to Development of Low Thermal Expansion Lithium–Aluminum–Silicate Ceramics

The negative thermal expansion area of the LAS phase diagram after Smoke (1969) is shown in Fig. 16; it ranges from 3 to 6 wt % Li_2O, 10 to 52 wt % Al_2O, and 38 to 82 wt % SiO_2. The composition corresponding to each thermal expansion curve is located at the intersection of the ordinate and abscissa of each graph. The silica content had the greatest effect on thermal expansion. The negative thermal expansion was high, between 38 and 62% SiO_2.

The best results for obtaining high thermal shock resistance were obtained on materials with a low negative thermal expansion. High negative thermal expansion was as detrimental as a high positive thermal expansion.

Development of ceramic materials with optimal properties for use in gas turbine engine applications may be accomplished by first investigating compositions in this region of known low expansion compounds in the phase diagram. The phase diagram can be used to decide how to alter the

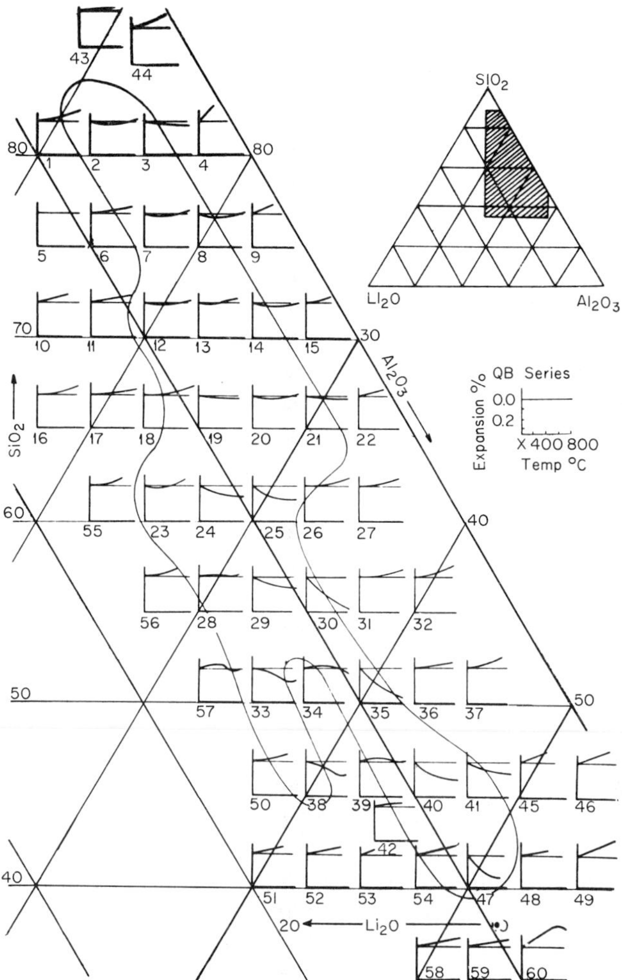

Fig. 16. Negative thermal expansion region of the $Li_2O \cdot Al_2O_3 \cdot SiO_2$ phase diagram (after Smoke, 1969).

starting compositions to attain desired low expansion phases and to eliminate undesirable high expansion second phases. Once a composition with good properties is derived, it is possible to further improve these properties through microstructure variations. It must be kept in mind that impurities present in the raw materials can alter the phase relationships as can the cooling cycle since it is impractical in most cases to cool slowly enough to approach equilibrium conditions.

In practice, most lithium aluminum silicates are formed using the glass-ceramic process. This process involves melting minerals or oxides which form a glass upon quenching. The glass is subsequently milled and then formed into a part and fired. The part is nucleated during the firing cycle and crystallized to form a ceramic. The quenching process results in non-equilibrium conditions. The sluggish nature of silicate reactions also prevents the reactants from following the crystallization paths in the equilibrium diagrams.

Much of the work done on the development of low expansion LAS ceramics is contained in the patent literature. These references give the starting composition, resultant phases, and resultant thermal expansion. Trends observed in the literature will be discussed in this section. Even though heat treatment cycles and resulting microstructure have significant effects on the resultant phases and properties, little is mentioned about these parameters in the patent literature.

Smoke (1969) reported that in the negative thermal expansion region of the LAS phase diagram bodies in the range 40–55% SiO_2 had β-eucryptite as the only phase present, bodies with 55–60% SiO_2 had a β-eucryptite solid solution phase present. At 65% SiO_2, a body with 7.5% Li_2O had only β-spodumene present while compositions at 5 and 10% Li_2O had β-spodumene and an unidentified phase. At 2.5 and 12.5% Li_2O (outside the negative expansion area), β-spodumene was no longer present. It was concluded that negative thermal expansion was dependent on the presence of either β-spodumene or β-eucryptite in the body. At 75% SiO_2, two crystalline phases appeared. One had a much higher index of refraction than β-spodumene and the other had an index close to β-spodumene but did not appear to be β-spodumene. At 80% SiO_2, two crystalline phases also appeared; one had a high index of refraction and the other had a slightly lower index than β-spodumene, possibly indicating a solid solution. At 86%, the high index phase appeared but the index of refraction shifted from 75 to 86, indicating the possibility of another solid solution series. This high index phase was postulated to be a metastable or low-temperature form of β-spodumene.

Smith (1966) combined a thermally devitrifiable lithia–alumino–silicate glass and petalite in the ratio: 25:75 w %. The resulting body had a relatively long firing range, high strength, and low coefficient of expansion (-0.4×10^{-7}/°C). The composition of the mixture is 77.7 wt % SiO_2, 18.1 wt % Al_2O_3, 4.2 wt % Li_2O and is shown in Fig. 17. The solid circles are starting compositions for the references cited. The minor additives were not included in the calculations. The fact that the SiO_2 phase appeared with this starting composition which had 77% SiO_2 and that it did not appear until 80% in the study by Smoke may be due to different firing cycles or raw materials.

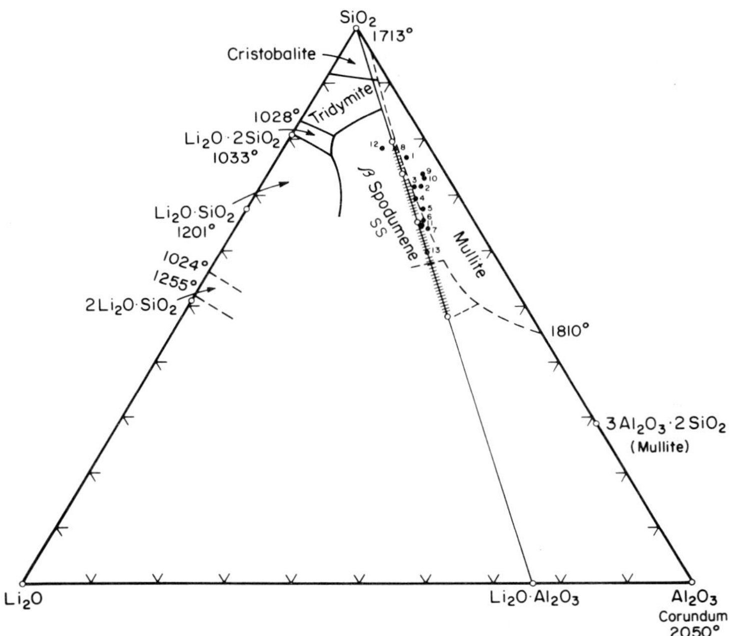

Fig. 17. Selected low thermal expansion compositions in the $Li_2O \cdot Al_2O_3 \cdot SiO_2$ phase diagram. Smith (1966) 1, Smith (1966); 2–8, Beall and Duke (1971); 9, Petticrew (1971); 10, Confer and McTaggart (1973); 11, Planchock *et al.* (1974); 12–13, Fritsch and Cleveland (1973).

Upon heating to 1240°C, the petalite, which is only stable below 700°C, converts to a solid solution of silica and β-spodumene and these phases are retained upon rapid cooling.

Beall and Duke (1971) found that a very narrow range of compositions in the LAS system yields a body with a very low coefficient of thermal expansion (0–15 × 10^{-7}/°C over the range 0–1000°C) and excellent dimensional stability (400 ppm after 1500 hr at 800°C). Crystallization must take place above 1000°C or β-eucryptite solid solution will appear. Several starting compositions are shown in Fig. 16 and the resultant phases are β-spodumene solid solution and mullite. The residual glass that is present is dispersed within the crystal phases and is different in composition from the parent glass, being highly siliceous with a minor amount of Al_2O_3. Good dimensional stability occurs because there is less than 10% glass phase present and it is highly siliceous and has a high viscosity. The mullite that forms is less than 15 wt % and appears to grow along the β-spodumene solid solution grain boundaries. This probably retards secondary grain growth and keeps the grain size less than 10 μm.

Petticrew (1971) describes a composition (shown in Fig. 17) that yielded a body with a coefficient of expansion less than $11 \times 10^{-7}/°C$ over the range (0–300°C). The minor amounts of ZnO, TiO_2, ZnO_2, Na_2O, K_2O, and F added were not included in the calculation of the starting composition in Fig. 17. When the highest temperature of crystallization is 900°C, the coefficient of expansion is less than 11×10^{-7} and the main crystalline phase is β-eucryptitelike crystals. β-spodumenelike crystals result when the crystallization is conducted above 910°C. Petticrew uses the terms β-spodumenelike and β-eucryptitelike to designate phases that have the structures of β-spodumene and β-eucryptite, respectively, but which have a shift in the x-ray diffraction peaks. This is indicative of more or less SiO_2 than the respective stoichiometric compositions (solid solutions). He also notes that it is probable that none of these crystalline products is formed under equilibrium conditions. Petticrew did not report finding any mullite even though his composition was close to those of Beall and Duke where mullite was reported.

Confer and McTaggart (1973) found that by using 45–55 wt % of a material having the composition SiO, 5.0 wt % Li_2O, 23.0 wt % Al_2O_3, and 71.8 wt % SiO_2 and a material with the composition 2.6 wt % Li_2O, 17.9 wt % Al_2O_3, and 69.7 wt % SiO_2 that a body with essentially zero coefficient of expansion over the range 0–35°C results. The resulting composition is shown in Fig. 17. Minor amounts of TiO_2, MgO, ZnO, and others that were added were not included in the calculations. The composition is very close to the composition used by Petticrew. Confer and McTaggart, as with Beall and Duke, found mullite as a minor phase. The major crystalline phase is a β-spodumene solid solution, containing mainly spodumene, alumina, and silica, minor crystalline phases are mullite, cordierite, and rutile. The glass is fired at 1250–1270°C for 1 hr to form the glass ceramic. The coefficient of expansion is approximately $0.2 \times 10^{-7}/°C$ over the range 0–35°C.

Planchock (1974) patented a range of compositions in the LAS phase diagram. One composition, characterized by a low coefficients of expansion ($6.8 \times 10^{-7}/°C$), good dimensional stability, and a MOR of >10,000 psi is very close to the compound spodumene as shown in Fig. 17. In the range 720–870°C, a high quartz solution crystallizes. Between 870 and 1100°C, the high quartz phase undergoes a solid state transformation to the keatite solid solution phase. No other results are presented with respect to the phases present but presumably β-spodumene would be present in addition to the SiO_2.

The previous section illustrated examples of the glass ceramic approach to the development of low expansion bodies in the LAS system. Another approach to the development of low expansion bodies in this system is to utilize conventional ceramic processing not involving glass-ceramic processing. This approach was utilized by Fritsch and Cleveland (1973) in the development of LAS ceramics. Two starting compositions developed by Fritsch and Cleveland are as follows:

composition 12

Li_2O	7.1 wt %
Al_2O_3	13.1 wt %
SiO_2	79.8 wt %

composition 13

Li_2O	9.45 wt %
Al_2O_3	30.40 wt %
SiO_2	60.12 wt %

These compositions resulted in bodies having thermal expansions of 1.06×10^{-6} and $0.9 \times 10^{-6}/°C$ and strengths of 7000 and 8000 psi, respectively. Using the glass-ceramic approach, β-spodumene solid solution and possibly SiO_2 are the expected phases from work reported previously. The resultant phases for composition 12 were β-spodumene, cristobalite, $Li_2O \cdot SiO_2$, and possibly $Li_2O \cdot 5Al_2O_3$. Using composition 13 and the glass-ceramic approach, β-spodumene would be the expected phase. The phases resulting in composition 13, using the conventional sintering process, were β-spodumene, $Li_2O \cdot SiO_2$, and possibly $Li_2O \cdot 5Al_2O_3$ (see Fig. 17).

The conventional sintering processing approach does not have the rapid quench of the glass-ceramic approach. The phases resulting when the sintering process is used are more likely to be determined by the nature of the raw materials, and the soak temperature and time. The phases resulting when the glass-ceramic process is used are more likely to be determined by the crystallization temperature and cooling rates.

In the previous section, results were described of the many experiments to develop negative or very low thermal expansion ceramics made from LAS, using primarily the glass-ceramic processing technique. The use of phase diagrams in developing low expansion bodies by means of composition changes in these bodies was discussed. Other factors such as microstructures and additives also affect thermal expansion.

Rapp (1973) showed the effect of residual glass in devitrified LAS glass ceramics on the thermal expansion. Pure LAS glass has a thermal expansion near $40 \times 10^{-7}/°C$ and becomes more negative as it becomes more crystalline. Rapp concludes that the fractional change in the thermal expansion coefficient observed in some glasses in the LAS system is approximately equal to the fractional change in the percentage of crystallinity of the glass.

Beall (1975) described the microstructure and phase changes that occur in a $Li_2O \cdot Al_2O_3 \cdot 7SiO_2$ glass with 4 mole % TiO_2 to provide nucleation. On heating to 825°C, nuclei of aluminum titanate develop. Heat treatment at 900°C allows crystallization of a metastable β-quartz solid solution on the nuclei ($\approx$0.1-μm diameter). The metastable quartz phase transforms above 950°C to stable β-spodumene solid solution crystals. The grain size is now about 1 μm and rutile precipitates. At 1200°C the grain size is about 5 μm. Severe microstresses develop when the grain size reaches 10 μm and thermal cycling creates microcracks.

D. Modified Lithium Aluminum Silicates

Lithium–aluminum–silicate matrices having approximately 1000 cells/in.^{2} and 0.005-in. thick walls were operated in several gas turbine engine programs. The strength of the regenerators was adequate to survive hub drive and rim drive stresses and the thermal expansion was sufficiently low for the regenerator to survive the thermal shock of start-up/shut-down cycles. Chemical durability then became a major problem. Sodium from the road salt attacks the LAS on the hot face of the core. Sulfur from the fuel forms sulfuric acid vapor during combustion and condenses on the cold face. The hydrogen cations exchanged with the lithium cations, which caused a volume expansion and a decrease in thermal expansion. This resulted in severe radial cracking in the cores.

In order to overcome this problem, a modified LAS has been developed (Grossman and Ritter, 1977). This material is processed by leaching an LAS body with H_2SO_4 or HNO_3 and replacing Li cations with H ions. This results in a hydroxy-alumino-silicate phase. Growth of the structure occurs up to 0.6% as the lithium is removed. Upon subsequent firing up to 1000°C water is removed from the structure and the remaining alumino-silicate phase has a structure similar to keatite, a polymorph of silica. "Microcleavage openings," reported by Grossman and Lanning (1977), that result during the leaching process are consolidated during the firing treatment; however, remnants are visible.

Heating this aluminous keatite over 1000°C results in a phase transformation from the keatite structure to mullite. The thermal expansion of this body can be brought from -2.0×10^{-6}/°C (25–800°C) to essentially zero expansion by controlling the amount of mullite developed by means of the heat treating cycle. Cristobalite is formed at temperatures over 1200°C. Performance of this material in engines has proved to be much superior to the earlier LAS bodies because of its very low thermal expansion and improved chemical durability (Grossman and Lanning, 1977).

IV. MAGNESIUM ALUMINUM SILICATES

A review of materials with potentially higher thermal shock resistance by Hummel (1955) has pointed out most of the logical candidate materials for heat exchanger applications with the exception of the less known family of nitride ceramics. These would, however, have to be classified as materials with intermediate thermal expansion in Hummel's review of low, intermediate, and high expansion materials. Among those belonging to the group of low thermal expansion materials, are some compounds in $Li_2O \cdot Al_2O_3 \cdot SiO_2$ (which were discussed earlier) and $MgO \cdot Al_2O_3 \cdot SiO_2$ systems that emerge as most promising. Since compounds of intermediate thermal expansion are not considered, the number of materials to be discussed is considerably narrowed down, as in the case of $MgO \cdot Al_2O_3 \cdot SiO_2$ ceramics, to one type, namely, cordierite ceramics. Because of their chemical nature, cordierite ceramics exhibit better corrosion resistance than lithium aluminum silicates in a gas turbine engine environment. This fact, together with relatively low thermal expansion coefficient and good thermal shock properties, makes cordierite ceramic a material of particular interest for gas turbine heat exchanger applications.

A. Cordierite

The first exploration of the $MgO \cdot Al_2O_3 \cdot SiO_2$ system was conducted by Rankin and Merwin (1918). Since then, phase relations in the system and hence the equilibrium diagram of the system have undergone considerable refinement. Due to the fact that the primary crystallization field of cordierite occupies the central position in this ternary system and due to the complex physical and chemical behavior of cordierite, this portion of the system was examined many times. The reproduction of the $MgO \cdot Al_2O_3 \cdot SiO_2$ ternary diagram revised by Osborn and Muan as it appeared in "Phase Diagrams for Ceramists" (Levin *et al.*, 1969b) is given in Fig. 18. Although most major features of the phase diagram have been defined, many fine points still need to be clarified.

Rankin and Merwin (1918) in their early work on the $MgO \cdot Al_2O_3 \cdot SiO_2$ system have indicated the possibility of a solid solution of the α-cordierite type to lie in the stoichiometric range from $2MgO \cdot 2Al_2O_3 \cdot 5SiO_2$ (2:2:5) to $2MgO \cdot 2Al_2O_3 \cdot 6SiO_2$ (2:2:6); Hummel and Reid (1951) extended this possibility to 1:1:4 and Iiyama (1933) indicated that it is possible that solid solutions extend nearly to 1:1:6. On the other hand, Schreyer and Schairer (1961) in their extensive study of composition and structural states of anhydrous cordierites concluded that the composition of cordierite is 2:2:5

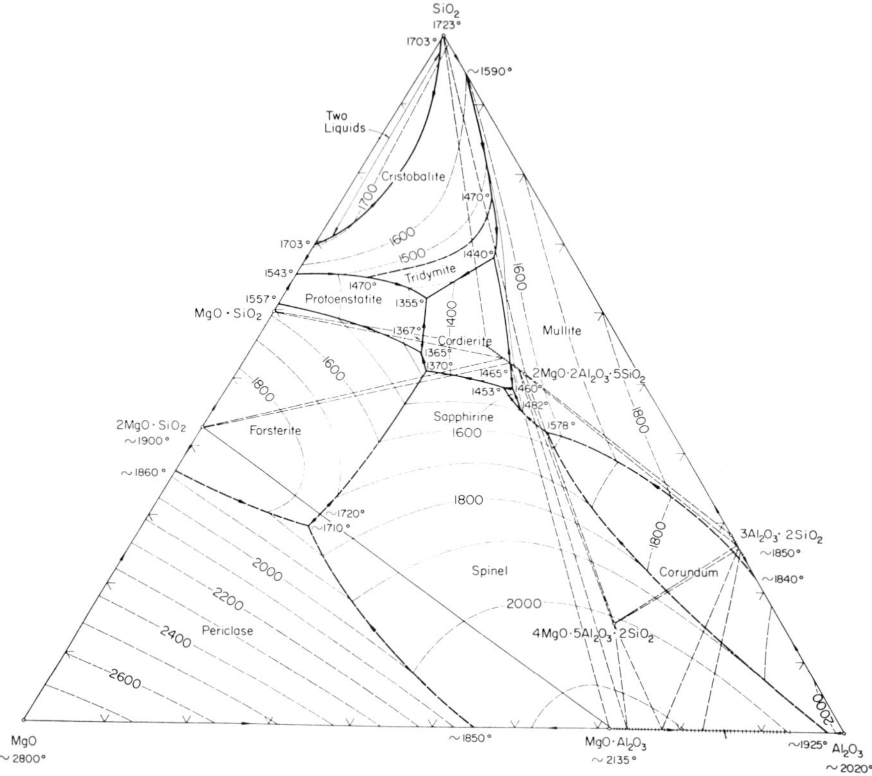

Fig. 18. System MgO–Al_2O–SiO_2 (revised by Osborn and Muan, 1969). Temperatures up to approximately 1500°C are on the Geophysical Laboratory scale; those above 1500°C are on the 1948 International scale.

Crystalline Phases			
Notations	Oxide Formulas	Notations	Oxide Formulas
Cristobalite	SiO_2	Spinel	$MgO \cdot Al_2O_3$
Tridymite		Corundum	Al_2O_3
Protoenstatite	$MgO \cdot SiO_2$	Mullite	$3Al_2O_3 \cdot 2SiO_2$
Forsterite	$2MgO \cdot SiO$	Cordierite	$2MgO \cdot 2Al_2O_35SiO_2$
Periclase	MgO	Sapphirine	$2MgO \cdot 5Al_2O_3 \cdot 2SiO_2$

and that there is no solid solution within the range of experimental error that would imply that all reported solid solutions are indeed metastable phases.

The classification and systematization of cordierite polymorphic forms has proved to be a no less difficult task. Cordierite is known to exist in six, or at least five, polymorphic forms.

Rankin and Merwin (1918) have already distinguished two polymorphic forms of the ternary compound $2MgO \cdot 2Al_2O_3 \cdot 5SiO_2$: a stable α form and an unstable μ cordierite form. Much later Karkhanavala and Hummel (1953) indicated a considerable resemblance between μ-cordierite and β-spodumene and defined β-cordierite as a stable low-temperature form phase similar to one prepared by Yoder (1952) and having a reversible inversion point at 830°C which could be prepared from either μ or α phase or glass of the same composition. In later years Miyashiro and Iiyama (1954) and Miyashiro *et al.* (1955) reported that cordierite of composition 2:2:5 has at least three modifications: indialite, high cordierite, and low cordierite. Both α and β synthetic cordierite show hexagonal symmetry in contrast to natural cordierites which are orthorombic or according to Miyashiro *et al.* possibly monoclinic, with exception of one found in India which was mentioned by Fermor (1924) and described by Venkatesh (1952, 1954). This form, called by Miyashiro and co-workers (1957), indialite proved the fact that continuous gradation exists between hexagonal and orthorhombic cordierites and used this deviation from hexagonal structure and refractive indexes to classify five or possibly six polymorphs. Iiyama (1958) published data indicating that the high–low inversion of cordierite is due to dehydration, thus is nonpolymorphic, a fact which was later confirmed by Schreyer and Yoder (1960). Later Schreyer and Schairer (1961) proposed a modified classification of cordierites according to a distortion index into high cordierite (hexagonal), low cordierites with highest deviation from hexagonal symmetry, and intermediate state cordierites. They use the terms as indicators of change in structure rather than the indicators of water content used by Miyashiro (1957).

B. Thermal Expansion of Cordierite

High cordierite has an anisotropic thermal expansion exhibiting contraction along the *c* axis and expansion along the *a* axis. Fisher *et al.* (1974) have measured thermal expansion of hexagonal cordierite by x-ray diffraction methods and concluded that in the range of temperatures from 20 to 1200°C, $\alpha = 29.4 \times 10^{-7}/°C^{-1}$. Later, Lee and Pentecost (1976) reported results of their measurements on single-crystal cordierite, obtained by

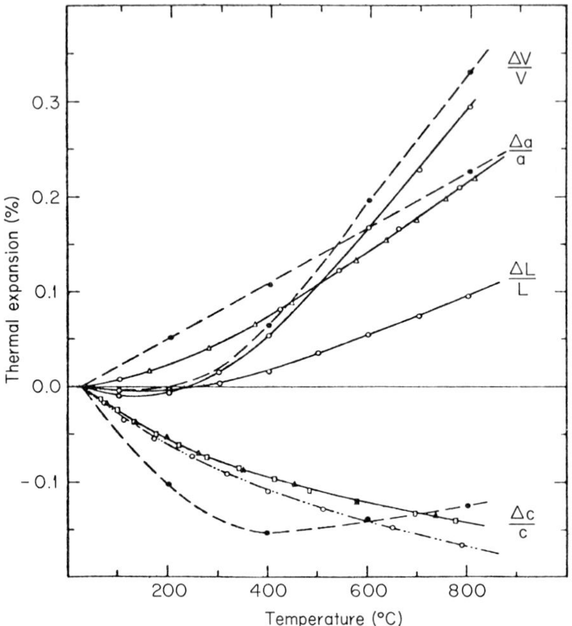

Fig. 19. Axial, volume, and linear thermal expansion of hexagonal and orthorhombic cordierite (after Lee and Pentecost, 1976). ——, hexagonal; – – – –, orthorhombic; –·–·–, Fischer's work (hexagonal).

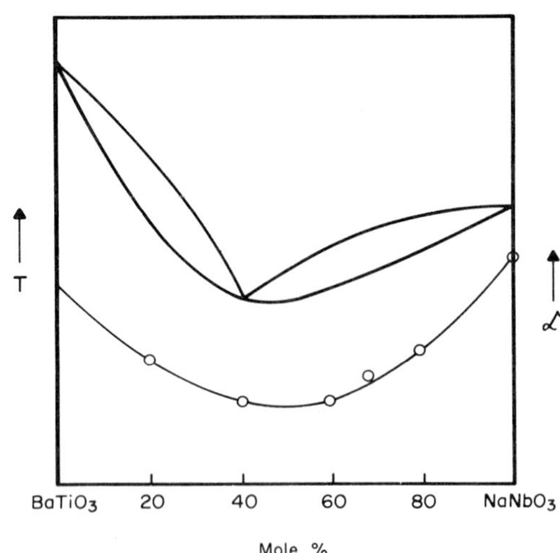

Fig. 20. Thermal expansion of solid solutions in $BaTiO_3$–$NaNbO_3$ system (after Buljan and Kleiner, 1975).

using a dilatation interferometer, which are, in terms of volume expansion, in fair agreement with results obtained by Fisher *et al.* (1974). Their measurements, however, indicate continuous thermal contraction of the *c* axis in contrast to the observations of Fisher *et al.* which indicate the existence of an inversion point. The results of both studies are shown in Fig. 19. As can be seen from the results shown, the temperature dependence of the thermal expansion in cordierite is an anything but simple relationship. Work done on the mechanism of order–disorder transition and the solid solution effect on thermal expansion: observations by Schulz (1974) on thermal expansion of ordered and disordered β-eucryptite and the effect of solid solutions on the thermal expansion coefficient, such as one shown in Fig. 20 for the $NbNSO_3$–$BaTiO_3$ system (Buljan and Kleiner, 1975), are just a couple of examples of lattice thermal expansion modifications. It stands to reason then that such mechanisms, including the $MgO \cdot Al_2O_3 \cdot SiO_2$ cordierite solid solution, should be considered as a tool for cordierite lattice expansion modification. This, however, may prove to be a very difficult task and will require a better definition of the cordierite solid solubility range and polymorphism.

C. Cordierite Ceramics

The first synthetic cordierite was prepared by L. Burgois in 1883 but it was not until 1917 that the first industrial cordierite-based ceramic bodies of low thermal expansion appeared. The thermal expansion of commercial cordierite ceramics produced in the period before 1940 ranges anywhere from 1.3 to 2.8×10^{-6}/°C and was as low as 0.1×10^{-6}/°C (0–200°C) as reported by Singer (1946). However, the temperature dependence of the thermal expansion coefficient in cordierite ceramics is not a simple function and any conclusions drawn by comparing the thermal expansion coefficients of two cordierite ceramics could be quite misleading if those coefficients were not evaluated over the same temperature range.

Cordierite ceramics have been prepared by sintering or using the glass-ceramic approach and a variety of raw materials. Bulk ceramics, depending on the fabrication method, could contain one or more metastable phases or metastable extensions of equilibrium phases (such as a metastable extension of terminal or intermediate solid solutions), structures with various degrees of order, and short-range order phases. Their properties depend on composition, raw materials, impurities, and additives, as well as on fabrication methods and conditions. It is understandable then that a review of the properties of some commercial ceramics, and for that matter a review of the thermal expansion characteristics and other properties of cordierite

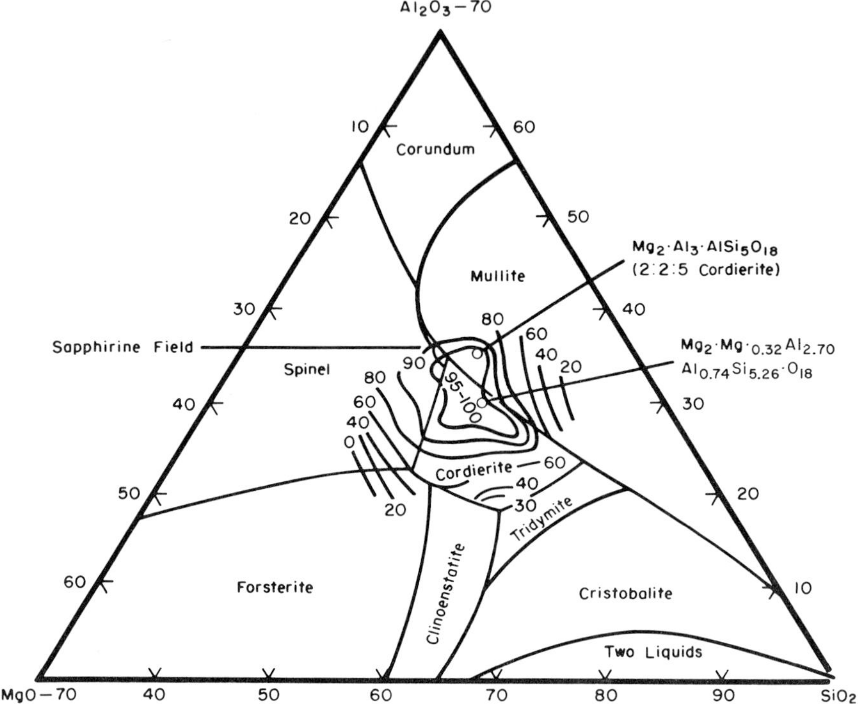

Fig. 21 Cordierite content in MgO–Al_2O_3–SiO_2 glasses devitrified at 1110°C (after Tirrell *et al.*, 1961).

ceramics, shows a vast variation of properties depending on those conditions. It is not always possible to distinguish the individual effects of such a variety of factors, which makes direct correlation and systematization of the literature data extremely difficult because of the lack of definition of all the necessary parameters.

Figure 21 is the $MgO \cdot Al_2O_3 \cdot SiO_2$ ternary phase diagrams, showing cordierite bearing compositions obtained by devitrification of glasses (Tirrel *et al.*, 1961), and Fig. 22 illustrates the correlation of thermal expansion and composition at 50% SiO_2 cross section of the $MgO \cdot Al_2O_3 \cdot SiO_2$ system (Buljan and Kleiner 1975). From the results obtained and reported to date it appears that materials of lowest thermal expansion in the $MgO \cdot Al_2O_3 \cdot SiO_2$ system are most likely to be found near the $2MgO \cdot 2Al_2O_3 \cdot 5SiO_2$ (2:2:5) composition representing stoichiometric cordierite.

Figure 22 also shows the correlation between the coefficient of thermal expansion and the liquidus temperature, where compositions with lower melting temperatures exhibit lower thermal expansions. Compounds with

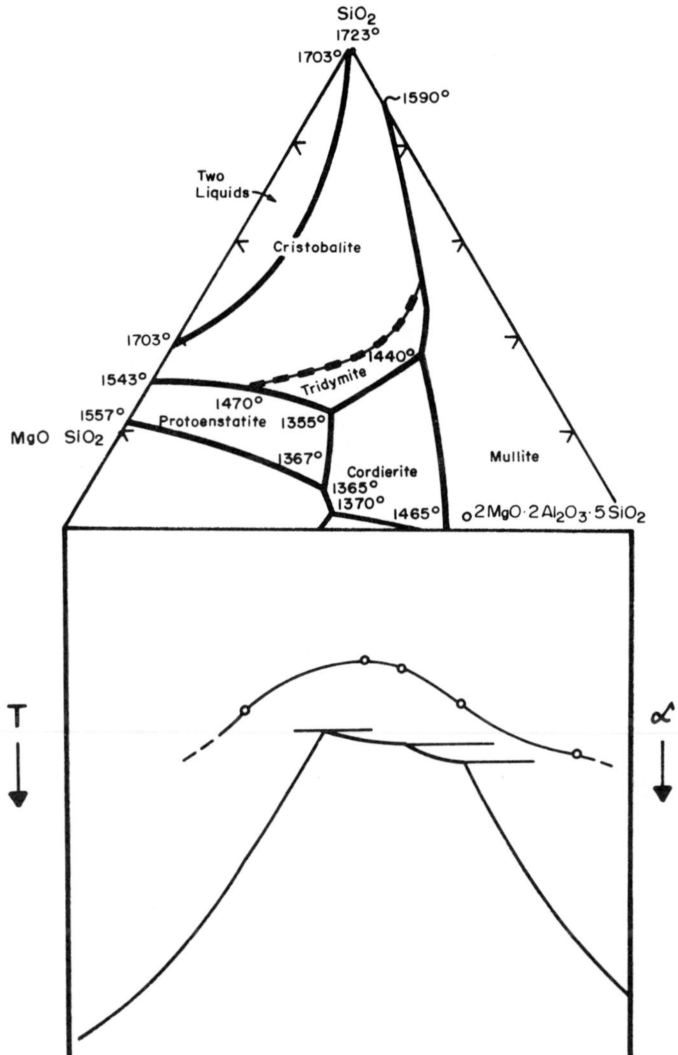

Fig. 22. Thermal expansion and liquidus temperature at 50% SiO_2 cross section of MgO–Al_2O_3–SiO_2 system (after Buljan and Kleiner, 1975).

lower melting points often have lower thermal expansions (Fig. 23). Exceptions to this rule, however, are not uncommon (e.g., alkali halides).

It has been stated before that the properties of bulk ceramics could vary strongly, depending on the raw materials, impurity content, and processing conditions. These factors interact strongly and it is not always possible to

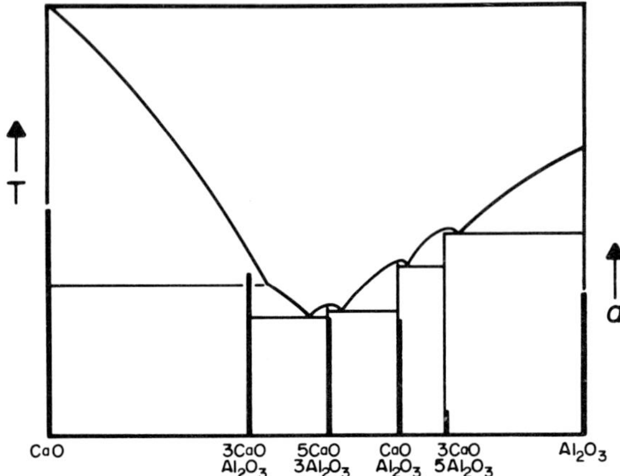

Fig. 23. Thermal expansion of compounds in the $CaO-Al_2O_3$ system (after Austin, 1952).

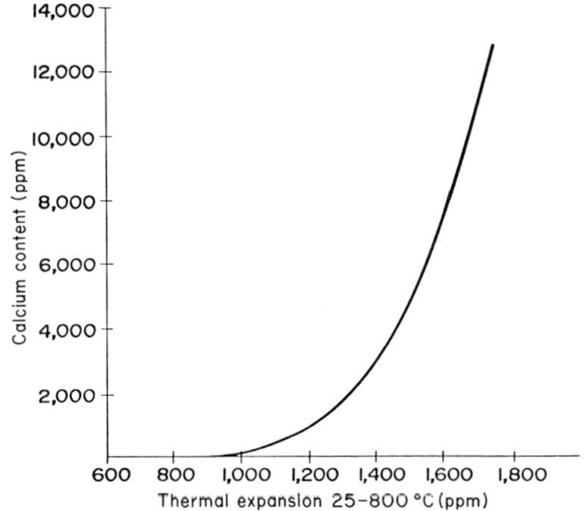

Fig. 24. Influence of presence of calcium oxide on thermal expansion of 2:2:5 cordierite ceramic sintered at 1375°C for 10 hr (after Fritsch and Buljan, 1975).

determine their individual effects. A few examples that follow are an illustration of the extent of the effects of impurities in raw materials and of processing conditions on the bulk thermal expansion of 2:2:5 cordierite ceramics. The presence of Ca, for example, in amounts as low as 200 ppm raises the bulk thermal expansion of cordierite ceramics considerably. The correlation between calcium content and thermal expansion of cordierite bodies prepared from talc, clay, and alumina mixtures (2:2:5 composition) is given in Fig. 24 for cordierite bodies fired at 1375°C for 10 hrs. Results obtained at 1350°C are essentially the same.

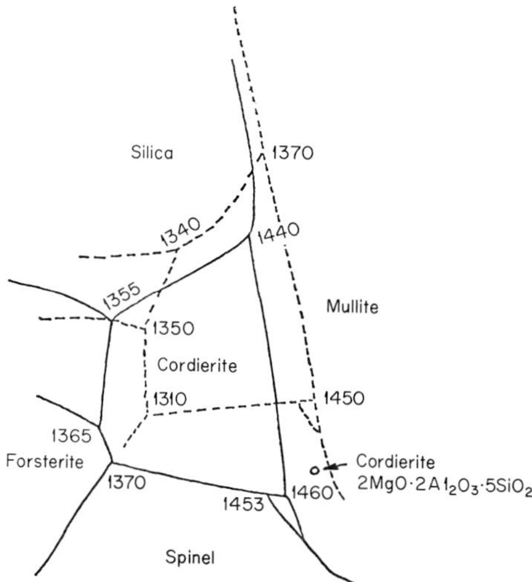

Fig. 25. Effect of 5% CaO addition in the MgO–Al_2O_3–SiO_2 system (after Osborn *et al.*, 1954).

Calcium oxide in amounts as high as 3–5% reduces liquidus temperatures by over 50°C and results in a stable glass in the fired cordierite body (Fig. 25). The solid lines of Fig. 25 show a primary phase region surrounding cordierite in the $MgO \cdot Al_2O_3 \cdot SiO_2$ ternary system. Assuming a constant alumina content and extrapolating the position and temperature of the primary phase fields (dashed lines), the effect of 5% CaO on equilibrium in the $MgO \cdot Al_2O_3 \cdot SiO_2$ system may be estimated. The appearance of the liquid, about 50°C lower in temperature, and the increased stability of the glass phase with calcium oxide tend to decrease the percentage of cordierite in the fired body.

It is not surprising then that materials of very similar composition, prepared under virtually the same conditions, often show different dilational properties. This holds true in particular with cordierite ceramics prepared from clay–talc mixtures, the most common precursor materials of commercial cordierite ceramics, which, being naturally occurring substances, could contain up to 3% impurities.

However, even if material is prepared from relatively pure precursor materials, variations in thermal processing could result in a material having substantially different dilational properties. The simplest example of the variation of thermal expansion, resulting from variation in phase assembly due to heat treatment, is a simple two-phase system glass cordierite (2:2:5) that was studied in a series of experiments by Buljan and Kleiner (1975).

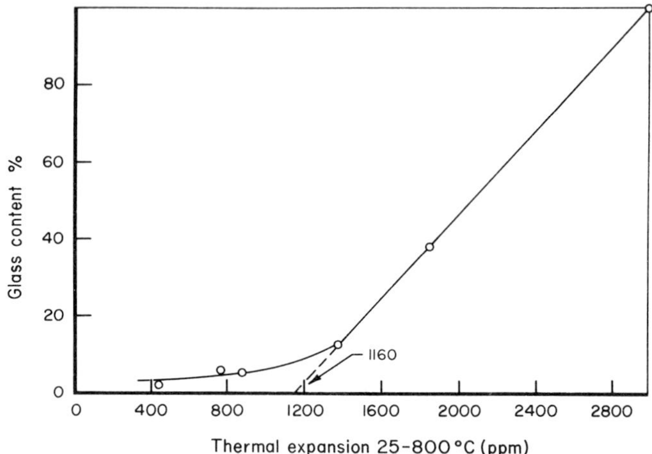

Fig. 26. Thermal expansion and glass content in the (2:2:5) cordierite-glass system (after Buljan and Kleiner, 1975).

The glass content was determined, using scanning electron microscopy, on polished and etched samples and by x-ray diffraction techniques.

Glass of cordierite composition has a relatively high thermal expansion coefficient of $3.6 \times 10^{-6}/°C$ or 3000 ppm between 25 and 800°C. The thermal expansion is expected to be equal to the sum of the products of the thermal expansion of the individual phases and their respective volume fractions. By varying the glass-cordierite ratio, using the appropriate heat treatment, it is expected that the thermal expansion of the obtained material would change as a linear function of the glass content. The experimental results are shown in Fig. 26, in which thermal expansion is plotted as a function of glass content. It can be seen that the linear dependence is, in fact, observed on the high-glass content portion of the curve. The thermal expansion value for 100% cordierite, obtained by extrapolation, is approximately 1160 ppm between 25 and 800°C. The $\frac{1}{3}(\Delta V/V)$ value calculated, using lattice thermal expansion data obtained by Fisher *et al.* (1974), is in good agreement with the extrapolated value from the diagram in Fig. 26. Deviation from linearity in the low-glass content portion of the curve, corresponding to values of thermal expansion down to 400 ppm, far below the value for 100% cordierite, can be attributed to factors other than phase assembly, namely, microstructural parameters governing thermal expansion of polycrystalline ceramics. Microstructural features other than phase assembly, such as presence of microcracks and preferential orientation of crystallites, have a strong influence on the dilational properties of cordierite ceramics and are highly dependent on processing and precursor materials. Their

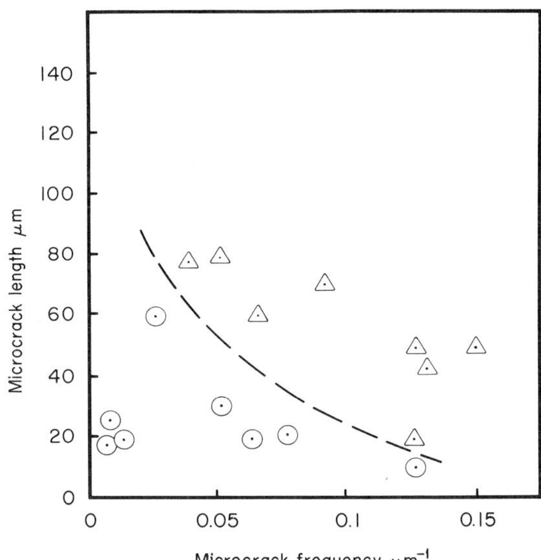

Fig. 27. Thermal expansion as a function of average microcrack length and frequency. △ Bodies with TE lower than 1000 ppm; ⊙, bodies with TE higher than 1000 ppm.

influence on bulk thermal expansion is very strong. Figure 27 gives the rough correlation between thermal expansion and microcrack size, and density for 2:2:5 cordierite ceramics. Microcrack size and density were determined by analysis of SEM photomicrographs of polished and etched samples. The dashed line indicates bodies with a thermal expansion of 1000 ppm between 25 and 800°C.

The mechanism of microcrack formation in cordierite ceramics is still not well defined. However, results obtained to date indicate that the mechanism can be attributed to intercrystalline stresses resulting from the thermal expansion anisotropy of crystalline cordierite.

This effect could be exploited in order to lower the bulk thermal expansion of cordierite ceramics. Cordierite ceramics, having thermal expansions lower than that calculated from the lattice thermal expansion, and as low as 200 ppm between 25 and 800°C, and extraordinary thermal shock properties, have been prepared by controlled introduction of microcracks (Fritsch and Buljan, 1976). Such bodies exhibit an isotropic thermal expansion.

A choice of raw materials and processing methods could also result in an anisotropic polycrystalline aggregate because of the preferential orientation of cordierite crystallites. This feature was used by Lachman and Lewis (1975) to prepare an anisotropic cordierite monolith from delaminated kaolin and talc mixtures by extrusion. The resulting bodies exhibit a low expansion in one plane and a higher thermal expansion in a direction perpendicular to this plane.

These few examples clearly show the extent of effects of impurities and microstructural parameters resulting from processing conditions, and illustrate the necessity and benefits of better understanding and control of factors influencing properties of bulk ceramics. Further improvement of the thermal expansion properties of cordierite ceramics could come through a better understanding and a better definition of cordierite solid solution and polymorphism, and the possibility of modifying the lattice thermal expansion of cordierite, which, combined with an understanding of parameters governing thermal expansion of polycrystalline ceramics, could bring about substantial improvements of thermal shock properties necessary for heat exchanger applications.

The physical–chemical behavior of silicates is one of the most complicated problems in inorganic chemical research. The abundant phenomena, characterized by slow speed of reaction, can only be investigated systematically with any prospect of success if the relationships in the ideal case of actual equilibria are known. The degree of deviation from these in each given case could then be judged. Well-defined phase relations and their thermal stability are a necessary prerequisite for enlightened material design, allowing intelligent selection of materials and manipulation of the composition-phase assembly-controlled properties.

REFERENCES

Anderson, D. H., Fucinari, C. A., Rahnke, C. J., and Rossi, L. R. (1975a). U.S. Energy Res. Dev. Admin. Contract No. E(11-1)2630.

Anderson, D. H., Fucinari, C. A., Rahnke, C. J., and Rossi, L. R. (1975b). U.S. Energy Res. Dev. Admin. Contract No. 68-03-2150.

Austin, J. B. (1952). *J. Am. Ceram. Soc.*, **35**, 243.

Beall, G. H. (1975). *In* "Microstructure of Ceramics." *Br. Ceram. Soc.*

Beall, G. H., and Duke, D. A. (1971). U.S. Patent No. 3,600,204.

Bowen, M. L., and Greig J. W. (1924). *J. Am. Ceram. Soc.* **7**, 238.

Buljan, S. T. and Kleiner, R. N. (1975). *ASME Publ.* 75-*GT*-66

Bush, E. A. and Hummel, F. A., (1959). *J. Am. Ceram. Soc.* **42**, 388.

Cleveland, J. J., Fritsch, C. W., and Kleiner, R. N., (1977). *ASME Publ.* 77-GT-98.

Confer, J. O., and McTaggart, G. D. (1973). U.S. Patent No. 3,715,220.

Fermor, L. L. (1924). *Q. J. Geol. Soc. London* **80**, 70–71.

Fisher, G. R., Evans, D. C., and Geiger, J. E. (1974). Abstract B-18, *Am. Cryst. Assoc. Prog. Abstr. Ser.* 2, **2**, 214.

Fritsch, C. W., and Buljan, S. T. (1976). U.S. Patent No. 3,979,216.

Fritsch, C. W., and Cleveland, J. J. (1973). Private communication.

Gilley, F. H., and Bush, E. A. (1959). *J. Am. Ceram. Soc.* **42**, 175.

Grossman, D. G., and Lanning, J. G. (1977). *J. Am. Ceram. Soc.* **56**, 474.
Grossman, D. G., and Rittler, H. L. (1974). U.S. Patent No. 3,834,981.
Hasselman, D. P. H. (1970). *J. Am. Ceram. Soc.* **49**, 1033.
Hatch, R. A. (1943). *Am. Minerol.* **28**, (9, 10), 471.
Hortal, M., Villar, R., Vieira, S., and Moya, J. S. (1975). *J. Am. Ceram. Soc.* **58**, 262.
Hummel, F. A. (1951), *J. Am. Ceram. Soc.* **34**, 235.
Hummel, F. A. (1955). *Ceram. Ind.*, **65** (6), 84–86.
Hummel, F. A. (1960). U.S. Patent, Re. 24,795.
Hummel, F. A., and Reid, H. W. (1951). *J. Am. Ceram. Soc.* **34**, 319–321.
Iiyama, T. (1955). *Proc. Imp. Acad. Tokyo* **31**, 166–168.
Iiyama, T. (1958). *C. R. Acad. Sci.* **246**, 795–798.
Karkhanavala, M. D., and Hummel, F. A. (1953). *J. Am. Ceram. Soc.* **36**, 389–392.
Kracek, F. C. (1930). *J. Phys. Chem.* **34**, (Pt. II), 2645.
Lachman, I. M., and Lewis, R. M. (1975). U.S. Patent No. 3,885,977.
Lee, J. D., and Pentecost, T. L. (1976). *J. Am. Ceram. Soc.* **59**, 183.
Levin, E. M., Robbins, C. R., and McMurdie, H. F. (1969a). "Phase Diagrams for Ceramists," 2nd ed., Fig. 449. Am. Ceram. Soc., Columbus, Ohio.
Levin, E. M., Robbins, C. R., and McMurdie, H. F. (1969b). "Phase Diagrams for Ceramists," 2nd ed., Fig. 712. Am. Ceram. Soc., Columbus, Ohio.
Miyashiro, A. (1957). *Am. J. Sci.* **225**, 43–62.
Miyashiro, A., and Iiyama, T. (1954). *Proc. Imp. Acad. Tokyo* **30**, 746–751.
Miyashiro, A., Iiyama, T., Yamasaki, M., and Miyashiro, T. (1955). *Am. J. Sci.* **253**, 185–208.
Moya, J. S., Verduch, A. G., and Hortal, M. (1974). *Trans. Br. Ceram. Soc.* **73**, 177.
Osborn, E. F., Devries, R. C., Gee, K. H., and Kraner, H. M. (1954). *Trans. AIME* **200**, 38–39.
Osborn, E. E., and Muan, A. (1969). "Phase Diagrams for Ceramists" (edited by E. M. Levin, C. R. Robbins, and H. F. McMurdie) Am. Ceram. Soc., Columbus, Ohio.
Petticrew, R. W. (1971). U.S. Patent No. 3,625,718.
Planchock, J. L., Stewart, D. R., and Brock, T. W. (1974). U.S. Patent No. 3,841,950.
Rankin, G. A., and Merwin, H. E. (1918). *Am. J. Sci. Ser.* 4, **45**, 301–325.
Rapp, J. E. (1973). *Bull. Am. Ceram. Soc.* **52**, 499.
Roy, R., and Osborn, E. F. (1949). *J. Am. Chem. Soc.* **71**, 2086.
Roy, R., Roy, D. H., and Osborn, E. F. (1950). *J. Am. Ceram. Soc.* **33**, 152.
Schreyer, W., and Schairer, J. F. (1961). *J. Petrol.* **2**, 324–406.
Schreyer, W., and Yoder, H. S. (1960). *Carnegie Inst. Washington Yearb.* **59**, 91–94.
Schulz, H. (1974). *J. Am. Ceram. Soc.* **57**, 313.
Singer, F. (1946). *Trans. Can. Ceram. Soc.* **15**, Commun. No. 124.
Smith, G. P. (1966). U.S. Patent No. 3,246,972.
Smoke, E. J. (1969). U.S. Army Electron. Command Contract DAAB 07-67-C-0232.
Smyth, H. T. (1955). *J. Am. Ceram. Soc.* **38**, 140.
Tien, T. Y., and Hummel, F. A. (1964). *J. Am. Ceram. Soc.* **47**, 582.
Tirrell, M. E., Gibbs, G. V., and Shell, H. R. (1961). *U.S. Bur. Min. Bull.* 594.
Tscherry, V., Schulz, H., and Czank, M. (1972). *Ber. Deut. Keram. Ges.* **49**, 153.
Venkatesh, V. (1952). *Am. Min.* **37**, 831–848.
Venkatesh, V. (1954). *Am. Min.* **39**, 636–646.
Yoder, H. S., Jr. (1952). *Am. J. Sci.* **Bowen Vol.**, 569–627.

Index

C

D

E

M

N

O

P

Q

R

A
B
C 8
D 9
E 0
F 1
G 2
H 3
I 4
J 5